D1553252

VOLUME FIVE HUNDRED AND TWENTY NINE

METHODS IN ENZYMOLOGY

Laboratory Methods in Enzymology: DNA

METHODS IN ENZYMOLOGY

Editors-in-Chief

JOHN N. ABELSON and MELVIN I. SIMON
Division of Biology
California Institute of Technology
Pasadena, California

Founding Editors

SIDNEY P. COLOWICK and NATHAN O. KAPLAN

VOLUME FIVE HUNDRED AND TWENTY NINE

METHODS IN ENZYMOLOGY

Laboratory Methods in Enzymology: DNA

Edited by

JON LORSCH

Johns Hopkins University School of Medicine
Baltimore, MD, USA

AMSTERDAM • BOSTON • HEIDELBERG • LONDON
NEW YORK • OXFORD • PARIS • SAN DIEGO
SAN FRANCISCO • SINGAPORE • SYDNEY • TOKYO

Academic Press is an imprint of Elsevier

Academic Press is an imprint of Elsevier
225 Wyman Street, Waltham, MA 02451, USA
525 B Street, Suite 1800, San Diego, CA 92101-4495, USA
Radarweg 29, PO Box 211, 1000 AE Amsterdam, The Netherlands
The Boulevard, Langford Lane, Kidlington, Oxford, OX5 1GB, UK
32 Jamestown Road, London NW1 7BY, UK

First edition 2013

For information on all Academic Press publications
visit our website at store.elsevier.com

ISBN: 978-0-12-418687-3
ISSN: 0076-6879

Printed and bound in United States of America
13 14 15 16 11 10 9 8 7 6 5 4 3 2 1

CONTENTS

Contributors *xv*
Preface *xix*
Volumes in Series *xxi*

1. Explanatory Chapter: PCR Primer Design **1**
Rubén Álvarez-Fernández
1. Theory 2
Acknowledgments 19
References 19

2. Explanatory Chapter: How Plasmid Preparation Kits Work **23**
Laura Koontz
1. Theory 23
2. Equipment 24
3. Materials 24
4. Protocol 24
5. Step 1 Pellet Bacteria and Resuspend in Resuspension Buffer 24
6. Step 2 Lyse Bacteria 25
7. Step 3 Neutralize the Solution 26
8. Step 4 Apply Clarified Lysate to Column 26
9. Step 5 Pass Binding Buffer Through Column 27
10. Step 6 Wash the Column 27
11. Step 7 Elute the DNA 27
References 28

3. Explanatory Chapter: Introducing Exogenous DNA into Cells **29**
Laura Koontz
1. Theory 29
2. Protocol 30
Source Reference 34

4. Agarose Gel Electrophoresis **35**
Laura Koontz
1. Theory 36
2. Equipment 36

3. Materials 36
4. Protocol 39
5. Step 1 Casting an Agarose Gel 40
6. Step 2 Loading and Running an Agarose Gel 42
7. Step 3 Visualization of Samples on an Agarose Gel 44

5. Analysis of DNA by Southern Blotting **47**

Gary Glenn and Lefkothea-Vasiliki Andreou

1. Theory 48
2. Equipment 49
3. Materials 49
4. Protocol 53
5. Step 1 Restriction Digestion of Genomic DNA and Agarose Gel Electrophoresis 54
6. Step 2 Denature DNA and Transfer It to a Nylon Membrane 54
7. Step 3A Label the Probe with [α-^{32}P]-dCTP Using a Random Primed Labeling Kit 56
8. Step 3B Label the Probe with Digoxigenin-11-dUTP for Nonradioactive Detection 59
9. Step 4 Hybridize the Labeled Probe to the Membrane 59
10. Step 5 Detect the Location of the Hybridized Probe 61

References 63

6. Purification of DNA Oligos by Denaturing Polyacrylamide Gel Electrophoresis (PAGE) **65**

Sara Lopez-Gomollon and Francisco Esteban Nicolas

1. Theory 66
2. Equipment 69
3. Materials 70
4. Protocol 72
5. Step 1 Butanol Precipitation of Crude Oligonucleotide 72
6. Step 2 Preparative Polyacrylamide Gel Electrophoresis 75
7. Step 3 Isolation of Oligonucleotides from Polyacrylamide Gels 76
8. Step 4A Elution of Oligonucleotide from Polyacrylamide Gel by Diffusion 78
9. Step 4B Elution of Oligonucleotide from Polyacrylamide Gel by Electroelution 80
10. Step 5 Ethanol Precipitation 82

References 83

7. Molecular Cloning **85**

Juliane C. Lessard

1. Theory 86
2. Equipment 88
3. Materials 88
4. Protocol 90
5. Step 1 Restriction Digests of Vector and Insert 91
6. Step 2 Ligation 93
7. Step 3 Transformation into Chemically Competent *E. coli* 94
8. Step 4 Identify Successful Ligation Events 96

References 98

8. Rapid Creation of Stable Mammalian Cell Lines for Regulated Expression of Proteins Using the Gateway® Recombination Cloning Technology and Flp-In T-REx® Lines **99**

Jessica Spitzer, Markus Landthaler, and Thomas Tuschl

1. Theory 100
2. Equipment 104
3. Materials 105
4. Protocol 110
5. Step 1 Molecular Cloning of the Gene of Interest into Gateway Expression Vectors 112
6. Step 2 Establishing Stable Cell Lines Expressing the Protein of Interest 118

References 123

Source References 123

9. Restrictionless Cloning **125**

Mikkel A. Algire

1. Theory 126
2. Equipment 126
3. Materials 126
4. Protocol 127
5. Step 1 Cloning Vector Preparation 128
6. Step 2 PCR Primer Design and PCR 130
7. Step 3 Insert Denaturation and Annealing 132
8. Step 4 Ligation and Transformation 133

References 134

10. Isolation of Plasmid DNA from Bacteria **135**

Lefkothea-Vasiliki Andreou

1. Theory 136
2. Equipment 136

3. Materials 137
4. Protocol 138
5. Step 1 Harvesting the Bacteria 139
6. Step 2 Cell Lysis and Isolation of Plasmid DNA 139
References 142

11. Preparation of Genomic DNA from Bacteria **143**
Lefkothea-Vasiliki Andreou

1. Theory 144
2. Equipment 144
3. Materials 144
4. Protocol 145
5. Step 1 Cell Lysis 146
6. Step 2 Organic Extraction and Ethanol Precipitation of DNA 147
7. Step 3 DNA Quantity and Quality Assessment 149
References 151

12. Preparation of Genomic DNA from *Saccharomyces cerevisiae* **153**
Jessica S. Dymond

1. Theory 154
2. Equipment 154
3. Materials 154
4. Protocol 155
5. Step 1 Harvesting Cells from the Overnight Culture 155
6. Step 2 Initial DNA Extraction 156
7. Step 3 Purification of the Crude DNA Preparation 157
References 160
Source References 160

13. Isolation of Genomic DNA from Mammalian Cells **161**
Cheryl M. Koh

1. Theory 162
2. Equipment 162
3. Materials 162
4. Protocol 164
5. Step 1 Collection of Cells 165
6. Step 2 Cell Lysis 166
7. Step 3 Organic Extraction 167
8. Step 4 Ethanol Precipitation 167
References 169

14. Sanger Dideoxy Sequencing of DNA 171

Sarah E. Walker and Jon Lorsch

1. Theory 172
2. Equipment 172
3. Materials 173
4. Protocol 175
5. Step 1 Anneal Primer to DNA 176
6. Step 2 Labeling 177
7. Step 3 Extension/Termination 178
8. Step 4 Resolution of Labeled Products by Urea-PAGE 179

References 184

15. Preparation of Fragment Libraries for Next-Generation Sequencing on the Applied Biosystems SOLiD Platform 185

Srinivasan Yegnasubramanian

1. Theory 186
2. Equipment 187
3. Materials 188
4. Protocol 190
5. Step 1 Shear the DNA to Generate Random Fragments 191
6. Step 2 End-Repair the Fragmented DNA 193
7. Step 3 Adaptor Ligation 194
8. Step 4 Size Selection of Library 196
9. Step 5 Amplification of the Library 197

References 200

16. Explanatory Chapter: Next Generation Sequencing 201

Srinivasan Yegnasubramanian

1. Theory 202

References 208

17. Generating Mammalian Stable Cell Lines by Electroporation 209

Patti A. Longo, Jennifer M. Kavran, Min-Sung Kim, and Daniel J. Leahy

1. Theory 210
2. Equipment 211
3. Materials 212
4. Protocol 214
5. Step 1 Dilute Plasmid DNA 215
6. Step 2 Prepare Cells for Electroporation 216

7. Step 3 Electroporate the Cells 219
8. Step 4 Plating Electroporated Cells 220
9. Step 5 Picking Single Colonies of Cells 221
10. Step 6 Methotrexate Amplification 223
References 225

18. Transient Mammalian Cell Transfection with Polyethylenimine (PEI) 227

Patti A. Longo, Jennifer M. Kavran, Min-Sung Kim, and Daniel J. Leahy

1. Theory 228
2. Equipment 229
3. Materials 229
4. Protocol 231
5. Step 1 Small-Scale Transient Transfection 232
6. Step 1.1 Seed Adherent Cells for Transfection 232
7. Step 1.2 Transiently Transfect Cells 234
8. Step 1.3 Harvest Cells and Analyze Protein Expression 235
9. Step 2 Large-Scale Transient Transfection of Suspension Cells 236
10. Step 2.1 Prepare the Cells To Be Transfected 236
11. Step 2.2 Transfect Cells 237
12. Step 2.3 Harvest Cells and Process Protein as Needed 238
References 240

19. Site-Directed Mutagenesis 241

Julia Bachman

1. Theory 242
2. Equipment 242
3. Materials 242
4. Protocol 243
5. Step 1 Setting up and Run the PCR 243
6. Step 2 Digestion of Template DNA 246
7. Step 3 Transformation into Chemically Competent *E. coli* 246
8. Step 4 Colony Screening 248
References 248

20. PCR-Based Random Mutagenesis 249

Jessica S. Dymond

1. Theory 250
2. Equipment 250

3. Materials 250
4. Protocol 251
5. Step 1 PCR Setup 252
6. Step 2 Generation of Additional Template for Mutagenesis 253
7. Step 3 Mutagenic PCR 254
8. Step 4 Subclone and Sequence the PCR Products 255
References 257

21. Megaprimer Method for Mutagenesis of DNA **259**

Craig W. Vander Kooi

1. Theory 260
2. Equipment 260
3. Materials 261
4. Protocol 261
5. Step 1 Design Mutagenic Primer 262
6. Step 2 First Round of PCR 263
7. Step 3 Gel Purification of the Megaprimer 265
8. Step 4 Second Round of PCR 265
9. Step 5 Gel Purification of the Mutagenized Gene 267
10. Step 6 Subclone the Mutagenized DNA and Verify the Mutation by DNA Sequencing 268
References 269

22. Explanatory Chapter: Troubleshooting PCR **271**

Kirstie Canene-Adams

1. Theory 271
2. Equipment 272
3. Materials 272
4. Protocol 272
References 278

23. Explanatory Chapter: Quantitative PCR **279**

Jessica S. Dymond

1. Theory 280
2. Terminology 281
3. Equipment 281
4. Materials 282
5. Protocol 282
6. Step 1 Primer Design 282

7. Step 2 Control Gene 283
8. Step 3 Amplification Efficiency 283
9. Step 4 Probe Choice 284
10. Step 5 Data Analysis 287
References 288

24. General PCR 291

Kirstie Canene-Adams

1. Theory 292
2. Equipment 293
3. Materials 293
4. Protocol 294
5. Step 1 Polymerase Chain Reaction 295
6. Step 2 Analyze PCR Products 297
References 298

25. Colony PCR 299

Megan Bergkessel and Christine Guthrie

1. Theory 300
2. Equipment 300
3. Materials 301
4. Protocol 303
5. Step 1 Extraction of DNA from a Colony 304
6. Step 2 PCR 307
7. Step 3 Visualization of Product on an Agarose Gel 308
References 309

26. Chemical Transformation of Yeast 311

Megan Bergkessel and Christine Guthrie

1. Theory 312
2. Equipment 312
3. Materials 313
4. Protocol 314
5. Step 1 Preparation of Competent Cells 315
6. Step 2 Transformation of Yeast Cells 317
References 320

27. Transformation of *E. coli* Via Electroporation **321**

Juliane C. Lessard

1. Theory 321
2. Equipment 322
3. Materials 322
4. Protocol 322
5. Step 1 Create Electro-Competent *E. coli* 324
6. Step 2 Electroporation of *E. coli* 324

References 327

28. Transformation of Chemically Competent *E. coli* **329**

Rachel Green and Elizabeth J. Rogers

1. Theory 330
2. Equipment 330
3. Materials 330
4. Protocol 332
5. Step 1 Prepare Competent Cells 333
6. Step 2 Transform Competent Cells 334

References 336

Author Index *337*

Subject Index *341*

CONTRIBUTORS

Mikkel A. Algire
J. Craig Venter Institute, Synthetic Biology Group, Rockville, MD, USA

Rubén Álvarez-Fernández
Department of Plant Sciences, University of Cambridge, Cambridge, United Kingdom

Lefkothea-Vasiliki Andreou
Ear Institute, University College London, London, United Kingdom

Julia Bachman
Department of Neuroscience, Johns Hopkins University School of Medicine, Baltimore, MD, USA

Megan Bergkessel
Department of Biochemistry and Biophysics, University of California, San Francisco, CA, USA

Kirstie Canene-Adams
Department of Pathology, Johns Hopkins University School of Medicine, Baltimore, MD, USA

Jessica S. Dymond
The High Throughput Biology Center and Department of Molecular Biology and Genetics, Johns Hopkins University School of Medicine, Baltimore, MD, USA

Gary Glenn
Ear Institute, University College London, London, United Kingdom

Rachel Green
Johns Hopkins School of Medicine/HHMI, Molecular Biology and Genetics, Baltimore, MD, USA

Christine Guthrie
Department of Biochemistry and Biophysics, University of California, San Francisco, CA, USA

Jennifer M. Kavran
Johns Hopkins University School of Medicine, Baltimore, MD, USA

Min-Sung Kim
Johns Hopkins University School of Medicine, Baltimore, MD, USA

Cheryl M. Koh
Department of Pathology, The Johns Hopkins University School of Medicine, Baltimore, MD, USA

Laura Koontz
Department of Molecular Biology and Genetics, Johns Hopkins University School of Medicine, Baltimore, MD, USA

Markus Landthaler
Berlin Institute for Medical Systems Biology, Max-Delbruck-Center for Molecular Medicine, Berlin, Germany

Daniel J. Leahy
Johns Hopkins University School of Medicine, Baltimore, MD, USA

Juliane C. Lessard
Department of Biochemistry and Molecular Biology, Johns Hopkins School of Public Health, Baltimore, MD, USA

Patti A. Longo
Johns Hopkins University School of Medicine, Baltimore, MD, USA

Sara Lopez-Gomollon
University of East Anglia, School of Biological Sciences, Norwich, United Kingdom

Jon Lorsch
Department of Biophysics and Biophysical Chemistry, Johns Hopkins University School of Medicine, Baltimore, MD, USA

Francisco Esteban Nicolas
University of East Anglia, School of Biological Sciences, Norwich, United Kingdom

Elizabeth J. Rogers
Johns Hopkins School of Medicine/HHMI, Molecular Biology and Genetics, Baltimore, MD, USA

Jessica Spitzer
Howard Hughes Medical Institute, Laboratory of RNA Molecular Biology, The Rockefeller University, New York, NY, USA

Thomas Tuschl
Howard Hughes Medical Institute, Laboratory of RNA Molecular Biology, The Rockefeller University, New York, NY, USA

Craig W. Vander Kooi
Department of Molecular and Cellular Biochemistry and Center for Structural Biology, University of Kentucky, Lexington, KY, USA

Sarah E. Walker
Department of Biophysics and Biophysical Chemistry, Johns Hopkins University School of Medicine, Baltimore, MD, USA

Srinivasan Yegnasubramanian
Sidney Kimmel Comprehensive Cancer Center, Johns Hopkins University School of Medicine, Baltimore, MD, USA

Methods in Enzymology volumes provide an indispensable tool for the researcher. Each volume is carefully written and edited by experts to contain state-of-the-art reviews and step-by-step protocols.

In this volume we have brought together a number of core protocols concentrating on DNA, complimenting the traditional content which is found in past, present and future Methods in Enzymology volumes.

PREFACE

These volumes of *Methods in Enzymology* contain the protocols that made up the on-line *Methods Navigator*. Our philosophy when we selected the protocols to include in the *Navigator* was that they should be for techniques useful in any biomedical laboratory, regardless of the system the lab studies. Each protocol was written by researchers who use the technique routinely, and in many cases by the people who actually developed the procedure in the first place. The protocols are very detailed and contain recipes for the necessary buffers and reagents, as well as flow-charts outlining the steps involved. Many of the chapters have accompanying videos demonstrating key parts of the procedures. In a few cases, detailed protocols for certain important approaches could not be generated either because they are instrument-specific (e.g., next-generation sequencing) or because they are proprietary (e.g., column-based nucleic acid purifications). In these cases we have included "explanatory chapters" that outline the theoretical basis for each technique without giving a detailed protocol. The volumes are broken into distinct areas: DNA methods; Cell-based methods; lipid, carbohydrate and miscellaneous methods; RNA methods; protein methods. Our goal is that these protocols will be useful for everyone in the lab, from undergraduates and rotation students to seasoned post-doctoral fellows. We hope that these volumes will become dog-eared and well-worn in your laboratory, either physically or electronically.

Professor Jon Lorsch
Johns Hopkins University
School of Medicine

METHODS IN ENZYMOLOGY

VOLUME I. Preparation and Assay of Enzymes
Edited by SIDNEY P. COLOWICK AND NATHAN O. KAPLAN

VOLUME II. Preparation and Assay of Enzymes
Edited by SIDNEY P. COLOWICK AND NATHAN O. KAPLAN

VOLUME III. Preparation and Assay of Substrates
Edited by SIDNEY P. COLOWICK AND NATHAN O. KAPLAN

VOLUME IV. Special Techniques for the Enzymologist
Edited by SIDNEY P. COLOWICK AND NATHAN O. KAPLAN

VOLUME V. Preparation and Assay of Enzymes
Edited by SIDNEY P. COLOWICK AND NATHAN O. KAPLAN

VOLUME VI. Preparation and Assay of Enzymes (*Continued*)
Preparation and Assay of Substrates
Special Techniques
Edited by SIDNEY P. COLOWICK AND NATHAN O. KAPLAN

VOLUME VII. Cumulative Subject Index
Edited by SIDNEY P. COLOWICK AND NATHAN O. KAPLAN

VOLUME VIII. Complex Carbohydrates
Edited by ELIZABETH F. NEUFELD AND VICTOR GINSBURG

VOLUME IX. Carbohydrate Metabolism
Edited by WILLIS A. WOOD

VOLUME X. Oxidation and Phosphorylation
Edited by RONALD W. ESTABROOK AND MAYNARD E. PULLMAN

VOLUME XI. Enzyme Structure
Edited by C. H. W. HIRS

VOLUME XII. Nucleic Acids (Parts A and B)
Edited by LAWRENCE GROSSMAN AND KIVIE MOLDAVE

VOLUME XIII. Citric Acid Cycle
Edited by J. M. LOWENSTEIN

VOLUME XIV. Lipids
Edited by J. M. LOWENSTEIN

Volume XXXII. Biomembranes (Part B)
Edited by Sidney Fleischer and Lester Packer

Volume XXXIII. Cumulative Subject Index Volumes I-XXX
Edited by Martha G. Dennis and Edward A. Dennis

Volume XXXIV. Affinity Techniques (Enzyme Purification: Part B)
Edited by William B. Jakoby and Meir Wilchek

Volume XXXV. Lipids (Part B)
Edited by John M. Lowenstein

Volume XXXVI. Hormone Action (Part A: Steroid Hormones)
Edited by Bert W. O'Malley and Joel G. Hardman

Volume XXXVII. Hormone Action (Part B: Peptide Hormones)
Edited by Bert W. O'Malley and Joel G. Hardman

Volume XXXVIII. Hormone Action (Part C: Cyclic Nucleotides)
Edited by Joel G. Hardman and Bert W. O'Malley

Volume XXXIX. Hormone Action (Part D: Isolated Cells, Tissues, and Organ Systems)
Edited by Joel G. Hardman and Bert W. O'Malley

Volume XL. Hormone Action (Part E: Nuclear Structure and Function)
Edited by Bert W. O'Malley and Joel G. Hardman

Volume XLI. Carbohydrate Metabolism (Part B)
Edited by W. A. Wood

Volume XLII. Carbohydrate Metabolism (Part C)
Edited by W. A. Wood

Volume XLIII. Antibiotics
Edited by John H. Hash

Volume XLIV. Immobilized Enzymes
Edited by Klaus Mosbach

Volume XLV. Proteolytic Enzymes (Part B)
Edited by Laszlo Lorand

Volume XLVI. Affinity Labeling
Edited by William B. Jakoby and Meir Wilchek

Volume XLVII. Enzyme Structure (Part E)
Edited by C. H. W. Hirs and Serge N. Timasheff

Volume XLVIII. Enzyme Structure (Part F)
Edited by C. H. W. Hirs and Serge N. Timasheff

Volume XLIX. Enzyme Structure (Part G)
Edited by C. H. W. Hirs and Serge N. Timasheff

Volume L. Complex Carbohydrates (Part C)
Edited by Victor Ginsburg

Volume LI. Purine and Pyrimidine Nucleotide Metabolism
Edited by Patricia A. Hoffee and Mary Ellen Jones

Volume LII. Biomembranes (Part C: Biological Oxidations)
Edited by Sidney Fleischer and Lester Packer

Volume LIII. Biomembranes (Part D: Biological Oxidations)
Edited by Sidney Fleischer and Lester Packer

Volume LIV. Biomembranes (Part E: Biological Oxidations)
Edited by Sidney Fleischer and Lester Packer

Volume LV. Biomembranes (Part F: Bioenergetics)
Edited by Sidney Fleischer and Lester Packer

Volume LVI. Biomembranes (Part G: Bioenergetics)
Edited by Sidney Fleischer and Lester Packer

Volume LVII. Bioluminescence and Chemiluminescence
Edited by Marlene A. DeLuca

Volume LVIII. Cell Culture
Edited by William B. Jakoby and Ira Pastan

Volume LIX. Nucleic Acids and Protein Synthesis (Part G)
Edited by Kivie Moldave and Lawrence Grossman

Volume LX. Nucleic Acids and Protein Synthesis (Part H)
Edited by Kivie Moldave and Lawrence Grossman

Volume 61. Enzyme Structure (Part H)
Edited by C. H. W. Hirs and Serge N. Timasheff

Volume 62. Vitamins and Coenzymes (Part D)
Edited by Donald B. McCormick and Lemuel D. Wright

Volume 63. Enzyme Kinetics and Mechanism (Part A: Initial Rate and Inhibitor Methods)
Edited by Daniel L. Purich

Volume 79. Interferons (Part B)
Edited by Sidney Pestka

Volume 80. Proteolytic Enzymes (Part C)
Edited by Laszlo Lorand

Volume 81. Biomembranes (Part H: Visual Pigments and Purple Membranes, I)
Edited by Lester Packer

Volume 82. Structural and Contractile Proteins (Part A: Extracellular Matrix)
Edited by Leon W. Cunningham and Dixie W. Frederiksen

Volume 83. Complex Carbohydrates (Part D)
Edited by Victor Ginsburg

Volume 84. Immunochemical Techniques (Part D: Selected Immunoassays)
Edited by John J. Langone and Helen Van Vunakis

Volume 85. Structural and Contractile Proteins (Part B: The Contractile Apparatus and the Cytoskeleton)
Edited by Dixie W. Frederiksen and Leon W. Cunningham

Volume 86. Prostaglandins and Arachidonate Metabolites
Edited by William E. M. Lands and William L. Smith

Volume 87. Enzyme Kinetics and Mechanism (Part C: Intermediates, Stereo-chemistry, and Rate Studies)
Edited by Daniel L. Purich

Volume 88. Biomembranes (Part I: Visual Pigments and Purple Membranes, II)
Edited by Lester Packer

Volume 89. Carbohydrate Metabolism (Part D)
Edited by Willis A. Wood

Volume 90. Carbohydrate Metabolism (Part E)
Edited by Willis A. Wood

Volume 91. Enzyme Structure (Part I)
Edited by C. H. W. Hirs and Serge N. Timasheff

Volume 92. Immunochemical Techniques (Part E: Monoclonal Antibodies and General Immunoassay Methods)
Edited by John J. Langone and Helen Van Vunakis

Volume 107. Posttranslational Modifications (Part B)
Edited by Finn Wold and Kivie Moldave

Volume 108. Immunochemical Techniques (Part G: Separation and Characterization of Lymphoid Cells)
Edited by Giovanni Di Sabato, John J. Langone, and Helen Van Vunakis

Volume 109. Hormone Action (Part I: Peptide Hormones)
Edited by Lutz Birnbaumer and Bert W. O'Malley

Volume 110. Steroids and Isoprenoids (Part A)
Edited by John H. Law and Hans C. Rilling

Volume 111. Steroids and Isoprenoids (Part B)
Edited by John H. Law and Hans C. Rilling

Volume 112. Drug and Enzyme Targeting (Part A)
Edited by Kenneth J. Widder and Ralph Green

Volume 113. Glutamate, Glutamine, Glutathione, and Related Compounds
Edited by Alton Meister

Volume 114. Diffraction Methods for Biological Macromolecules (Part A)
Edited by Harold W. Wyckoff, C. H. W. Hirs, and Serge N. Timasheff

Volume 115. Diffraction Methods for Biological Macromolecules (Part B)
Edited by Harold W. Wyckoff, C. H. W. Hirs, and Serge N. Timasheff

Volume 116. Immunochemical Techniques
(Part H: Effectors and Mediators of Lymphoid Cell Functions)
Edited by Giovanni Di Sabato, John J. Langone, and Helen Van Vunakis

Volume 117. Enzyme Structure (Part J)
Edited by C. H. W. Hirs and Serge N. Timasheff

Volume 118. Plant Molecular Biology
Edited by Arthur Weissbach and Herbert Weissbach

Volume 119. Interferons (Part C)
Edited by Sidney Pestka

Volume 120. Cumulative Subject Index Volumes 81–94, 96–101

Volume 121. Immunochemical Techniques (Part I: Hybridoma Technology and Monoclonal Antibodies)
Edited by John J. Langone and Helen Van Vunakis

VOLUME 122. Vitamins and Coenzymes (Part G)
Edited by FRANK CHYTIL AND DONALD B. MCCORMICK

VOLUME 123. Vitamins and Coenzymes (Part H)
Edited by FRANK CHYTIL AND DONALD B. MCCORMICK

VOLUME 124. Hormone Action (Part J: Neuroendocrine Peptides)
Edited by P. MICHAEL CONN

VOLUME 125. Biomembranes (Part M: Transport in Bacteria, Mitochondria, and Chloroplasts: General Approaches and Transport Systems)
Edited by SIDNEY FLEISCHER AND BECCA FLEISCHER

VOLUME 126. Biomembranes (Part N: Transport in Bacteria, Mitochondria, and Chloroplasts: Protonmotive Force)
Edited by SIDNEY FLEISCHER AND BECCA FLEISCHER

VOLUME 127. Biomembranes (Part O: Protons and Water: Structure and Translocation)
Edited by LESTER PACKER

VOLUME 128. Plasma Lipoproteins (Part A: Preparation, Structure, and Molecular Biology)
Edited by JERE P. SEGREST AND JOHN J. ALBERS

VOLUME 129. Plasma Lipoproteins (Part B: Characterization, Cell Biology, and Metabolism)
Edited by JOHN J. ALBERS AND JERE P. SEGREST

VOLUME 130. Enzyme Structure (Part K)
Edited by C. H. W. HIRS AND SERGE N. TIMASHEFF

VOLUME 131. Enzyme Structure (Part L)
Edited by C. H. W. HIRS AND SERGE N. TIMASHEFF

VOLUME 132. Immunochemical Techniques (Part J: Phagocytosis and Cell-Mediated Cytotoxicity)
Edited by GIOVANNI DI SABATO AND JOHANNES EVERSE

VOLUME 133. Bioluminescence and Chemiluminescence (Part B)
Edited by MARLENE DELUCA AND WILLIAM D. MCELROY

VOLUME 134. Structural and Contractile Proteins (Part C: The Contractile Apparatus and the Cytoskeleton)
Edited by RICHARD B. VALLEE

VOLUME 135. Immobilized Enzymes and Cells (Part B)
Edited by KLAUS MOSBACH

Volume 136. Immobilized Enzymes and Cells (Part C)
Edited by Klaus Mosbach

Volume 137. Immobilized Enzymes and Cells (Part D)
Edited by Klaus Mosbach

Volume 138. Complex Carbohydrates (Part E)
Edited by Victor Ginsburg

Volume 139. Cellular Regulators (Part A: Calcium- and Calmodulin-Binding Proteins)
Edited by Anthony R. Means and P. Michael Conn

Volume 140. Cumulative Subject Index Volumes 102–119, 121–134

Volume 141. Cellular Regulators (Part B: Calcium and Lipids)
Edited by P. Michael Conn and Anthony R. Means

Volume 142. Metabolism of Aromatic Amino Acids and Amines
Edited by Seymour Kaufman

Volume 143. Sulfur and Sulfur Amino Acids
Edited by William B. Jakoby and Owen Griffith

Volume 144. Structural and Contractile Proteins (Part D: Extracellular Matrix)
Edited by Leon W. Cunningham

Volume 145. Structural and Contractile Proteins (Part E: Extracellular Matrix)
Edited by Leon W. Cunningham

Volume 146. Peptide Growth Factors (Part A)
Edited by David Barnes and David A. Sirbasku

Volume 147. Peptide Growth Factors (Part B)
Edited by David Barnes and David A. Sirbasku

Volume 148. Plant Cell Membranes
Edited by Lester Packer and Roland Douce

Volume 149. Drug and Enzyme Targeting (Part B)
Edited by Ralph Green and Kenneth J. Widder

Volume 150. Immunochemical Techniques (Part K: *In Vitro* Models of B and T Cell Functions and Lymphoid Cell Receptors)
Edited by Giovanni Di Sabato

VOLUME 151. Molecular Genetics of Mammalian Cells
Edited by MICHAEL M. GOTTESMAN

VOLUME 152. Guide to Molecular Cloning Techniques
Edited by SHELBY L. BERGER AND ALAN R. KIMMEL

VOLUME 153. Recombinant DNA (Part D)
Edited by RAY WU AND LAWRENCE GROSSMAN

VOLUME 154. Recombinant DNA (Part E)
Edited by RAY WU AND LAWRENCE GROSSMAN

VOLUME 155. Recombinant DNA (Part F)
Edited by RAY WU

VOLUME 156. Biomembranes (Part P: ATP-Driven Pumps and Related Transport: The Na, K-Pump)
Edited by SIDNEY FLEISCHER AND BECCA FLEISCHER

VOLUME 157. Biomembranes (Part Q: ATP-Driven Pumps and Related Transport: Calcium, Proton, and Potassium Pumps)
Edited by SIDNEY FLEISCHER AND BECCA FLEISCHER

VOLUME 158. Metalloproteins (Part A)
Edited by JAMES F. RIORDAN AND BERT L. VALLEE

VOLUME 159. Initiation and Termination of Cyclic Nucleotide Action
Edited by JACKIE D. CORBIN AND ROGER A. JOHNSON

VOLUME 160. Biomass (Part A: Cellulose and Hemicellulose)
Edited by WILLIS A. WOOD AND SCOTT T. KELLOGG

VOLUME 161. Biomass (Part B: Lignin, Pectin, and Chitin)
Edited by WILLIS A. WOOD AND SCOTT T. KELLOGG

VOLUME 162. Immunochemical Techniques (Part L: Chemotaxis and Inflammation)
Edited by GIOVANNI DI SABATO

VOLUME 163. Immunochemical Techniques (Part M: Chemotaxis and Inflammation)
Edited by GIOVANNI DI SABATO

VOLUME 164. Ribosomes
Edited by HARRY F. NOLLER, JR., AND KIVIE MOLDAVE

VOLUME 165. Microbial Toxins: Tools for Enzymology
Edited by SIDNEY HARSHMAN

VOLUME 166. Branched-Chain Amino Acids
Edited by ROBERT HARRIS AND JOHN R. SOKATCH

VOLUME 167. Cyanobacteria
Edited by LESTER PACKER AND ALEXANDER N. GLAZER

VOLUME 168. Hormone Action (Part K: Neuroendocrine Peptides)
Edited by P. MICHAEL CONN

VOLUME 169. Platelets: Receptors, Adhesion, Secretion (Part A)
Edited by JACEK HAWIGER

VOLUME 170. Nucleosomes
Edited by PAUL M. WASSARMAN AND ROGER D. KORNBERG

VOLUME 171. Biomembranes (Part R: Transport Theory: Cells and Model Membranes)
Edited by SIDNEY FLEISCHER AND BECCA FLEISCHER

VOLUME 172. Biomembranes (Part S: Transport: Membrane Isolation and Characterization)
Edited by SIDNEY FLEISCHER AND BECCA FLEISCHER

VOLUME 173. Biomembranes [Part T: Cellular and Subcellular Transport: Eukaryotic (Nonepithelial) Cells]
Edited by SIDNEY FLEISCHER AND BECCA FLEISCHER

VOLUME 174. Biomembranes [Part U: Cellular and Subcellular Transport: Eukaryotic (Nonepithelial) Cells]
Edited by SIDNEY FLEISCHER AND BECCA FLEISCHER

VOLUME 175. Cumulative Subject Index Volumes 135–139, 141–167

VOLUME 176. Nuclear Magnetic Resonance (Part A: Spectral Techniques and Dynamics)
Edited by NORMAN J. OPPENHEIMER AND THOMAS L. JAMES

VOLUME 177. Nuclear Magnetic Resonance (Part B: Structure and Mechanism)
Edited by NORMAN J. OPPENHEIMER AND THOMAS L. JAMES

VOLUME 178. Antibodies, Antigens, and Molecular Mimicry
Edited by JOHN J. LANGONE

VOLUME 179. Complex Carbohydrates (Part F)
Edited by VICTOR GINSBURG

VOLUME 180. RNA Processing (Part A: General Methods)
Edited by JAMES E. DAHLBERG AND JOHN N. ABELSON

VOLUME 181. RNA Processing (Part B: Specific Methods)
Edited by JAMES E. DAHLBERG AND JOHN N. ABELSON

VOLUME 182. Guide to Protein Purification
Edited by MURRAY P. DEUTSCHER

VOLUME 183. Molecular Evolution: Computer Analysis of Protein and Nucleic Acid Sequences
Edited by RUSSELL F. DOOLITTLE

VOLUME 184. Avidin-Biotin Technology
Edited by MEIR WILCHEK AND EDWARD A. BAYER

VOLUME 185. Gene Expression Technology
Edited by DAVID V. GOEDDEL

VOLUME 186. Oxygen Radicals in Biological Systems (Part B: Oxygen Radicals and Antioxidants)
Edited by LESTER PACKER AND ALEXANDER N. GLAZER

VOLUME 187. Arachidonate Related Lipid Mediators
Edited by ROBERT C. MURPHY AND FRANK A. FITZPATRICK

VOLUME 188. Hydrocarbons and Methylotrophy
Edited by MARY E. LIDSTROM

VOLUME 189. Retinoids (Part A: Molecular and Metabolic Aspects)
Edited by LESTER PACKER

VOLUME 190. Retinoids (Part B: Cell Differentiation and Clinical Applications)
Edited by LESTER PACKER

VOLUME 191. Biomembranes (Part V: Cellular and Subcellular Transport: Epithelial Cells)
Edited by SIDNEY FLEISCHER AND BECCA FLEISCHER

VOLUME 192. Biomembranes (Part W: Cellular and Subcellular Transport: Epithelial Cells)
Edited by SIDNEY FLEISCHER AND BECCA FLEISCHER

VOLUME 193. Mass Spectrometry
Edited by JAMES A. MCCLOSKEY

Volume 194. Guide to Yeast Genetics and Molecular Biology
Edited by Christine Guthrie and Gerald R. Fink

Volume 195. Adenylyl Cyclase, G Proteins, and Guanylyl Cyclase
Edited by Roger A. Johnson and Jackie D. Corbin

Volume 196. Molecular Motors and the Cytoskeleton
Edited by Richard B. Vallee

Volume 197. Phospholipases
Edited by Edward A. Dennis

Volume 198. Peptide Growth Factors (Part C)
Edited by David Barnes, J. P. Mather, and Gordon H. Sato

Volume 199. Cumulative Subject Index Volumes 168–174, 176–194

Volume 200. Protein Phosphorylation (Part A: Protein Kinases: Assays, Purification, Antibodies, Functional Analysis, Cloning, and Expression)
Edited by Tony Hunter and Bartholomew M. Sefton

Volume 201. Protein Phosphorylation (Part B: Analysis of Protein Phosphorylation, Protein Kinase Inhibitors, and Protein Phosphatases)
Edited by Tony Hunter and Bartholomew M. Sefton

Volume 202. Molecular Design and Modeling: Concepts and Applications (Part A: Proteins, Peptides, and Enzymes)
Edited by John J. Langone

Volume 203. Molecular Design and Modeling: Concepts and Applications (Part B: Antibodies and Antigens, Nucleic Acids, Polysaccharides, and Drugs)
Edited by John J. Langone

Volume 204. Bacterial Genetic Systems
Edited by Jeffrey H. Miller

Volume 205. Metallobiochemistry (Part B: Metallothionein and Related Molecules)
Edited by James F. Riordan and Bert L. Vallee

Volume 206. Cytochrome P450
Edited by Michael R. Waterman and Eric F. Johnson

Volume 207. Ion Channels
Edited by Bernardo Rudy and Linda E. Iverson

VOLUME 208. Protein–DNA Interactions
Edited by ROBERT T. SAUER

VOLUME 209. Phospholipid Biosynthesis
Edited by EDWARD A. DENNIS AND DENNIS E. VANCE

VOLUME 210. Numerical Computer Methods
Edited by LUDWIG BRAND AND MICHAEL L. JOHNSON

VOLUME 211. DNA Structures (Part A: Synthesis and Physical Analysis of DNA)
Edited by DAVID M. J. LILLEY AND JAMES E. DAHLBERG

VOLUME 212. DNA Structures (Part B: Chemical and Electrophoretic Analysis of DNA)
Edited by DAVID M. J. LILLEY AND JAMES E. DAHLBERG

VOLUME 213. Carotenoids (Part A: Chemistry, Separation, Quantitation, and Antioxidation)
Edited by LESTER PACKER

VOLUME 214. Carotenoids (Part B: Metabolism, Genetics, and Biosynthesis)
Edited by LESTER PACKER

VOLUME 215. Platelets: Receptors, Adhesion, Secretion (Part B)
Edited by JACEK J. HAWIGER

VOLUME 216. Recombinant DNA (Part G)
Edited by RAY WU

VOLUME 217. Recombinant DNA (Part H)
Edited by RAY WU

VOLUME 218. Recombinant DNA (Part I)
Edited by RAY WU

VOLUME 219. Reconstitution of Intracellular Transport
Edited by JAMES E. ROTHMAN

VOLUME 220. Membrane Fusion Techniques (Part A)
Edited by NEJAT DÜZGÜNEŞ

VOLUME 221. Membrane Fusion Techniques (Part B)
Edited by NEJAT DÜZGÜNEŞ

VOLUME 222. Proteolytic Enzymes in Coagulation, Fibrinolysis, and Complement Activation (Part A: Mammalian Blood Coagulation

Factors and Inhibitors)
Edited by Laszlo Lorand and Kenneth G. Mann

Volume 223. Proteolytic Enzymes in Coagulation, Fibrinolysis, and Complement Activation (Part B: Complement Activation, Fibrinolysis, and Nonmammalian Blood Coagulation Factors)
Edited by Laszlo Lorand and Kenneth G. Mann

Volume 224. Molecular Evolution: Producing the Biochemical Data
Edited by Elizabeth Anne Zimmer, Thomas J. White, Rebecca L. Cann, and Allan C. Wilson

Volume 225. Guide to Techniques in Mouse Development
Edited by Paul M. Wassarman and Melvin L. DePamphilis

Volume 226. Metallobiochemistry (Part C: Spectroscopic and Physical Methods for Probing Metal Ion Environments in Metalloenzymes and Metalloproteins)
Edited by James F. Riordan and Bert L. Vallee

Volume 227. Metallobiochemistry (Part D: Physical and Spectroscopic Methods for Probing Metal Ion Environments in Metalloproteins)
Edited by James F. Riordan and Bert L. Vallee

Volume 228. Aqueous Two-Phase Systems
Edited by Harry Walter and Göte Johansson

Volume 229. Cumulative Subject Index Volumes 195–198, 200–227

Volume 230. Guide to Techniques in Glycobiology
Edited by William J. Lennarz and Gerald W. Hart

Volume 231. Hemoglobins (Part B: Biochemical and Analytical Methods)
Edited by Johannes Everse, Kim D. Vandegriff, and Robert M. Winslow

Volume 232. Hemoglobins (Part C: Biophysical Methods)
Edited by Johannes Everse, Kim D. Vandegriff, and Robert M. Winslow

Volume 233. Oxygen Radicals in Biological Systems (Part C)
Edited by Lester Packer

Volume 234. Oxygen Radicals in Biological Systems (Part D)
Edited by Lester Packer

Volume 235. Bacterial Pathogenesis (Part A: Identification and Regulation of Virulence Factors)
Edited by Virginia L. Clark and Patrik M. Bavoil

Volume 236. Bacterial Pathogenesis (Part B: Integration of Pathogenic Bacteria with Host Cells)
Edited by Virginia L. Clark and Patrik M. Bavoil

Volume 237. Heterotrimeric G Proteins
Edited by Ravi Iyengar

Volume 238. Heterotrimeric G-Protein Effectors
Edited by Ravi Iyengar

Volume 239. Nuclear Magnetic Resonance (Part C)
Edited by Thomas L. James and Norman J. Oppenheimer

Volume 240. Numerical Computer Methods (Part B)
Edited by Michael L. Johnson and Ludwig Brand

Volume 241. Retroviral Proteases
Edited by Lawrence C. Kuo and Jules A. Shafer

Volume 242. Neoglycoconjugates (Part A)
Edited by Y. C. Lee and Reiko T. Lee

Volume 243. Inorganic Microbial Sulfur Metabolism
Edited by Harry D. Peck, Jr., and Jean LeGall

Volume 244. Proteolytic Enzymes: Serine and Cysteine Peptidases
Edited by Alan J. Barrett

Volume 245. Extracellular Matrix Components
Edited by E. Ruoslahti and E. Engvall

Volume 246. Biochemical Spectroscopy
Edited by Kenneth Sauer

Volume 247. Neoglycoconjugates (Part B: Biomedical Applications)
Edited by Y. C. Lee and Reiko T. Lee

Volume 248. Proteolytic Enzymes: Aspartic and Metallo Peptidases
Edited by Alan J. Barrett

Volume 249. Enzyme Kinetics and Mechanism (Part D: Developments in Enzyme Dynamics)
Edited by Daniel L. Purich

Volume 250. Lipid Modifications of Proteins
Edited by Patrick J. Casey and Janice E. Buss

VOLUME 251. Biothiols (Part A: Monothiols and Dithiols, Protein Thiols, and Thiyl Radicals)
Edited by LESTER PACKER

VOLUME 252. Biothiols (Part B: Glutathione and Thioredoxin; Thiols in Signal Transduction and Gene Regulation)
Edited by LESTER PACKER

VOLUME 253. Adhesion of Microbial Pathogens
Edited by RON J. DOYLE AND ITZHAK OFEK

VOLUME 254. Oncogene Techniques
Edited by PETER K. VOGT AND INDER M. VERMA

VOLUME 255. Small GTPases and Their Regulators (Part A: Ras Family)
Edited by W. E. BALCH, CHANNING J. DER, AND ALAN HALL

VOLUME 256. Small GTPases and Their Regulators (Part B: Rho Family)
Edited by W. E. BALCH, CHANNING J. DER, AND ALAN HALL

VOLUME 257. Small GTPases and Their Regulators (Part C: Proteins Involved in Transport)
Edited by W. E. BALCH, CHANNING J. DER, AND ALAN HALL

VOLUME 258. Redox-Active Amino Acids in Biology
Edited by JUDITH P. KLINMAN

VOLUME 259. Energetics of Biological Macromolecules
Edited by MICHAEL L. JOHNSON AND GARY K. ACKERS

VOLUME 260. Mitochondrial Biogenesis and Genetics (Part A)
Edited by GIUSEPPE M. ATTARDI AND ANNE CHOMYN

VOLUME 261. Nuclear Magnetic Resonance and Nucleic Acids
Edited by THOMAS L. JAMES

VOLUME 262. DNA Replication
Edited by JUDITH L. CAMPBELL

VOLUME 263. Plasma Lipoproteins (Part C: Quantitation)
Edited by WILLIAM A. BRADLEY, SANDRA H. GIANTURCO, AND JERE P. SEGREST

VOLUME 264. Mitochondrial Biogenesis and Genetics (Part B)
Edited by GIUSEPPE M. ATTARDI AND ANNE CHOMYN

VOLUME 265. Cumulative Subject Index Volumes 228, 230–262

Volume 266. Computer Methods for Macromolecular Sequence Analysis
Edited by Russell F. Doolittle

Volume 267. Combinatorial Chemistry
Edited by John N. Abelson

Volume 268. Nitric Oxide (Part A: Sources and Detection of NO; NO Synthase)
Edited by Lester Packer

Volume 269. Nitric Oxide (Part B: Physiological and Pathological Processes)
Edited by Lester Packer

Volume 270. High Resolution Separation and Analysis of Biological Macromolecules (Part A: Fundamentals)
Edited by Barry L. Karger and William S. Hancock

Volume 271. High Resolution Separation and Analysis of Biological Macromolecules (Part B: Applications)
Edited by Barry L. Karger and William S. Hancock

Volume 272. Cytochrome P450 (Part B)
Edited by Eric F. Johnson and Michael R. Waterman

Volume 273. RNA Polymerase and Associated Factors (Part A)
Edited by Sankar Adhya

Volume 274. RNA Polymerase and Associated Factors (Part B)
Edited by Sankar Adhya

Volume 275. Viral Polymerases and Related Proteins
Edited by Lawrence C. Kuo, David B. Olsen, and Steven S. Carroll

Volume 276. Macromolecular Crystallography (Part A)
Edited by Charles W. Carter, Jr., and Robert M. Sweet

Volume 277. Macromolecular Crystallography (Part B)
Edited by Charles W. Carter, Jr., and Robert M. Sweet

Volume 278. Fluorescence Spectroscopy
Edited by Ludwig Brand and Michael L. Johnson

Volume 279. Vitamins and Coenzymes (Part I)
Edited by Donald B. McCormick, John W. Suttie, and Conrad Wagner

Volume 280. Vitamins and Coenzymes (Part J)
Edited by Donald B. McCormick, John W. Suttie, and Conrad Wagner

VOLUME 297. Photosynthesis: Molecular Biology of Energy Capture
Edited by LEE MCINTOSH

VOLUME 298. Molecular Motors and the Cytoskeleton (Part B)
Edited by RICHARD B. VALLEE

VOLUME 299. Oxidants and Antioxidants (Part A)
Edited by LESTER PACKER

VOLUME 300. Oxidants and Antioxidants (Part B)
Edited by LESTER PACKER

VOLUME 301. Nitric Oxide: Biological and Antioxidant Activities (Part C)
Edited by LESTER PACKER

VOLUME 302. Green Fluorescent Protein
Edited by P. MICHAEL CONN

VOLUME 303. cDNA Preparation and Display
Edited by SHERMAN M. WEISSMAN

VOLUME 304. Chromatin
Edited by PAUL M. WASSARMAN AND ALAN P. WOLFFE

VOLUME 305. Bioluminescence and Chemiluminescence (Part C)
Edited by THOMAS O. BALDWIN AND MIRIAM M. ZIEGLER

VOLUME 306. Expression of Recombinant Genes in Eukaryotic Systems
Edited by JOSEPH C. GLORIOSO AND MARTIN C. SCHMIDT

VOLUME 307. Confocal Microscopy
Edited by P. MICHAEL CONN

VOLUME 308. Enzyme Kinetics and Mechanism (Part E: Energetics of Enzyme Catalysis)
Edited by DANIEL L. PURICH AND VERN L. SCHRAMM

VOLUME 309. Amyloid, Prions, and Other Protein Aggregates
Edited by RONALD WETZEL

VOLUME 310. Biofilms
Edited by RON J. DOYLE

VOLUME 311. Sphingolipid Metabolism and Cell Signaling (Part A)
Edited by ALFRED H. MERRILL, JR., AND YUSUF A. HANNUN

VOLUME 312. Sphingolipid Metabolism and Cell Signaling (Part B)
Edited by ALFRED H. MERRILL, JR., AND YUSUF A. HANNUN

Volume 313. Antisense Technology
(Part A: General Methods, Methods of Delivery, and RNA Studies)
Edited by M. Ian Phillips

Volume 314. Antisense Technology (Part B: Applications)
Edited by M. Ian Phillips

Volume 315. Vertebrate Phototransduction and the Visual Cycle
(Part A)
Edited by Krzysztof Palczewski

Volume 316. Vertebrate Phototransduction and the Visual Cycle (Part B)
Edited by Krzysztof Palczewski

Volume 317. RNA–Ligand Interactions (Part A: Structural Biology Methods)
Edited by Daniel W. Celander and John N. Abelson

Volume 318. RNA–Ligand Interactions (Part B: Molecular Biology Methods)
Edited by Daniel W. Celander and John N. Abelson

Volume 319. Singlet Oxygen, UV-A, and Ozone
Edited by Lester Packer and Helmut Sies

Volume 320. Cumulative Subject Index Volumes 290–319

Volume 321. Numerical Computer Methods (Part C)
Edited by Michael L. Johnson and Ludwig Brand

Volume 322. Apoptosis
Edited by John C. Reed

Volume 323. Energetics of Biological Macromolecules (Part C)
Edited by Michael L. Johnson and Gary K. Ackers

Volume 324. Branched-Chain Amino Acids (Part B)
Edited by Robert A. Harris and John R. Sokatch

Volume 325. Regulators and Effectors of Small GTPases
(Part D: Rho Family)
Edited by W. E. Balch, Channing J. Der, and Alan Hall

Volume 326. Applications of Chimeric Genes and Hybrid Proteins
(Part A: Gene Expression and Protein Purification)
Edited by Jeremy Thorner, Scott D. Emr, and John N. Abelson

Volume 327. Applications of Chimeric Genes and Hybrid Proteins (Part B: Cell Biology and Physiology)
Edited by Jeremy Thorner, Scott D. Emr, and John N. Abelson

Volume 328. Applications of Chimeric Genes and Hybrid Proteins (Part C: Protein–Protein Interactions and Genomics)
Edited by Jeremy Thorner, Scott D. Emr, and John N. Abelson

Volume 329. Regulators and Effectors of Small GTPases (Part E: GTPases Involved in Vesicular Traffic)
Edited by W. E. Balch, Channing J. Der, and Alan Hall

Volume 330. Hyperthermophilic Enzymes (Part A)
Edited by Michael W. W. Adams and Robert M. Kelly

Volume 331. Hyperthermophilic Enzymes (Part B)
Edited by Michael W. W. Adams and Robert M. Kelly

Volume 332. Regulators and Effectors of Small GTPases (Part F: Ras Family I)
Edited by W. E. Balch, Channing J. Der, and Alan Hall

Volume 333. Regulators and Effectors of Small GTPases (Part G: Ras Family II)
Edited by W. E. Balch, Channing J. Der, and Alan Hall

Volume 334. Hyperthermophilic Enzymes (Part C)
Edited by Michael W. W. Adams and Robert M. Kelly

Volume 335. Flavonoids and Other Polyphenols
Edited by Lester Packer

Volume 336. Microbial Growth in Biofilms (Part A: Developmental and Molecular Biological Aspects)
Edited by Ron J. Doyle

Volume 337. Microbial Growth in Biofilms (Part B: Special Environments and Physicochemical Aspects)
Edited by Ron J. Doyle

Volume 338. Nuclear Magnetic Resonance of Biological Macromolecules (Part A)
Edited by Thomas L. James, Volker Dötsch, and Uli Schmitz

Volume 339. Nuclear Magnetic Resonance of Biological Macromolecules (Part B)
Edited by Thomas L. James, Volker Dötsch, and Uli Schmitz

Volume 340. Drug–Nucleic Acid Interactions
Edited by Jonathan B. Chaires and Michael J. Waring

Volume 341. Ribonucleases (Part A)
Edited by Allen W. Nicholson

Volume 342. Ribonucleases (Part B)
Edited by Allen W. Nicholson

Volume 343. G Protein Pathways (Part A: Receptors)
Edited by Ravi Iyengar and John D. Hildebrandt

Volume 344. G Protein Pathways (Part B: G Proteins and Their Regulators)
Edited by Ravi Iyengar and John D. Hildebrandt

Volume 345. G Protein Pathways (Part C: Effector Mechanisms)
Edited by Ravi Iyengar and John D. Hildebrandt

Volume 346. Gene Therapy Methods
Edited by M. Ian Phillips

Volume 347. Protein Sensors and Reactive Oxygen Species (Part A: Selenoproteins and Thioredoxin)
Edited by Helmut Sies and Lester Packer

Volume 348. Protein Sensors and Reactive Oxygen Species (Part B: Thiol Enzymes and Proteins)
Edited by Helmut Sies and Lester Packer

Volume 349. Superoxide Dismutase
Edited by Lester Packer

Volume 350. Guide to Yeast Genetics and Molecular and Cell Biology (Part B)
Edited by Christine Guthrie and Gerald R. Fink

Volume 351. Guide to Yeast Genetics and Molecular and Cell Biology (Part C)
Edited by Christine Guthrie and Gerald R. Fink

Volume 352. Redox Cell Biology and Genetics (Part A)
Edited by Chandan K. Sen and Lester Packer

Volume 353. Redox Cell Biology and Genetics (Part B)
Edited by Chandan K. Sen and Lester Packer

VOLUME 354. Enzyme Kinetics and Mechanisms (Part F: Detection and Characterization of Enzyme Reaction Intermediates)
Edited by DANIEL L. PURICH

VOLUME 355. Cumulative Subject Index Volumes 321–354

VOLUME 356. Laser Capture Microscopy and Microdissection
Edited by P. MICHAEL CONN

VOLUME 357. Cytochrome P450, Part C
Edited by ERIC F. JOHNSON AND MICHAEL R. WATERMAN

VOLUME 358. Bacterial Pathogenesis (Part C: Identification, Regulation, and Function of Virulence Factors)
Edited by VIRGINIA L. CLARK AND PATRIK M. BAVOIL

VOLUME 359. Nitric Oxide (Part D)
Edited by ENRIQUE CADENAS AND LESTER PACKER

VOLUME 360. Biophotonics (Part A)
Edited by GERARD MARRIOTT AND IAN PARKER

VOLUME 361. Biophotonics (Part B)
Edited by GERARD MARRIOTT AND IAN PARKER

VOLUME 362. Recognition of Carbohydrates in Biological Systems (Part A)
Edited by YUAN C. LEE AND REIKO T. LEE

VOLUME 363. Recognition of Carbohydrates in Biological Systems (Part B)
Edited by YUAN C. LEE AND REIKO T. LEE

VOLUME 364. Nuclear Receptors
Edited by DAVID W. RUSSELL AND DAVID J. MANGELSDORF

VOLUME 365. Differentiation of Embryonic Stem Cells
Edited by PAUL M. WASSAUMAN AND GORDON M. KELLER

VOLUME 366. Protein Phosphatases
Edited by SUSANNE KLUMPP AND JOSEF KRIEGLSTEIN

VOLUME 367. Liposomes (Part A)
Edited by NEJAT DÜZGÜNEŞ

VOLUME 368. Macromolecular Crystallography (Part C)
Edited by CHARLES W. CARTER, JR., AND ROBERT M. SWEET

VOLUME 369. Combinational Chemistry (Part B)
Edited by GUILLERMO A. MORALES AND BARRY A. BUNIN

Volume 370. RNA Polymerases and Associated Factors (Part C)
Edited by Sankar L. Adhya and Susan Garges

Volume 371. RNA Polymerases and Associated Factors (Part D)
Edited by Sankar L. Adhya and Susan Garges

Volume 372. Liposomes (Part B)
Edited by Nejat Düzgüneş

Volume 373. Liposomes (Part C)
Edited by Nejat Düzgüneş

Volume 374. Macromolecular Crystallography (Part D)
Edited by Charles W. Carter, Jr., and Robert W. Sweet

Volume 375. Chromatin and Chromatin Remodeling Enzymes (Part A)
Edited by C. David Allis and Carl Wu

Volume 376. Chromatin and Chromatin Remodeling Enzymes (Part B)
Edited by C. David Allis and Carl Wu

Volume 377. Chromatin and Chromatin Remodeling Enzymes (Part C)
Edited by C. David Allis and Carl Wu

Volume 378. Quinones and Quinone Enzymes (Part A)
Edited by Helmut Sies and Lester Packer

Volume 379. Energetics of Biological Macromolecules (Part D)
Edited by Jo M. Holt, Michael L. Johnson, and Gary K. Ackers

Volume 380. Energetics of Biological Macromolecules (Part E)
Edited by Jo M. Holt, Michael L. Johnson, and Gary K. Ackers

Volume 381. Oxygen Sensing
Edited by Chandan K. Sen and Gregg L. Semenza

Volume 382. Quinones and Quinone Enzymes (Part B)
Edited by Helmut Sies and Lester Packer

Volume 383. Numerical Computer Methods (Part D)
Edited by Ludwig Brand and Michael L. Johnson

Volume 384. Numerical Computer Methods (Part E)
Edited by Ludwig Brand and Michael L. Johnson

Volume 385. Imaging in Biological Research (Part A)
Edited by P. Michael Conn

Volume 386. Imaging in Biological Research (Part B)
Edited by P. Michael Conn

VOLUME 403. GTPases Regulating Membrane Targeting and Fusion
Edited by WILLIAM E. BALCH, CHANNING J. DER, AND ALAN HALL

VOLUME 404. GTPases Regulating Membrane Dynamics
Edited by WILLIAM E. BALCH, CHANNING J. DER, AND ALAN HALL

VOLUME 405. Mass Spectrometry: Modified Proteins and Glycoconjugates
Edited by A. L. BURLINGAME

VOLUME 406. Regulators and Effectors of Small GTPases: Rho Family
Edited by WILLIAM E. BALCH, CHANNING J. DER, AND ALAN HALL

VOLUME 407. Regulators and Effectors of Small GTPases: Ras Family
Edited by WILLIAM E. BALCH, CHANNING J. DER, AND ALAN HALL

VOLUME 408. DNA Repair (Part A)
Edited by JUDITH L. CAMPBELL AND PAUL MODRICH

VOLUME 409. DNA Repair (Part B)
Edited by JUDITH L. CAMPBELL AND PAUL MODRICH

VOLUME 410. DNA Microarrays (Part A: Array Platforms and Web-Bench Protocols)
Edited by ALAN KIMMEL AND BRIAN OLIVER

VOLUME 411. DNA Microarrays (Part B: Databases and Statistics)
Edited by ALAN KIMMEL AND BRIAN OLIVER

VOLUME 412. Amyloid, Prions, and Other Protein Aggregates (Part B)
Edited by INDU KHETERPAL AND RONALD WETZEL

VOLUME 413. Amyloid, Prions, and Other Protein Aggregates (Part C)
Edited by INDU KHETERPAL AND RONALD WETZEL

VOLUME 414. Measuring Biological Responses with Automated Microscopy
Edited by JAMES INGLESE

VOLUME 415. Glycobiology
Edited by MINORU FUKUDA

VOLUME 416. Glycomics
Edited by MINORU FUKUDA

VOLUME 417. Functional Glycomics
Edited by MINORU FUKUDA

VOLUME 418. Embryonic Stem Cells
Edited by IRINA KLIMANSKAYA AND ROBERT LANZA

VOLUME 419. Adult Stem Cells
Edited by IRINA KLIMANSKAYA AND ROBERT LANZA

VOLUME 420. Stem Cell Tools and Other Experimental Protocols
Edited by IRINA KLIMANSKAYA AND ROBERT LANZA

VOLUME 421. Advanced Bacterial Genetics: Use of Transposons and Phage for Genomic Engineering
Edited by KELLY T. HUGHES

VOLUME 422. Two-Component Signaling Systems, Part A
Edited by MELVIN I. SIMON, BRIAN R. CRANE, AND ALEXANDRINE CRANE

VOLUME 423. Two-Component Signaling Systems, Part B
Edited by MELVIN I. SIMON, BRIAN R. CRANE, AND ALEXANDRINE CRANE

VOLUME 424. RNA Editing
Edited by JONATHA M. GOTT

VOLUME 425. RNA Modification
Edited by JONATHA M. GOTT

VOLUME 426. Integrins
Edited by DAVID CHERESH

VOLUME 427. MicroRNA Methods
Edited by JOHN J. ROSSI

VOLUME 428. Osmosensing and Osmosignaling
Edited by HELMUT SIES AND DIETER HAUSSINGER

VOLUME 429. Translation Initiation: Extract Systems and Molecular Genetics
Edited by JON LORSCH

VOLUME 430. Translation Initiation: Reconstituted Systems and Biophysical Methods
Edited by JON LORSCH

VOLUME 431. Translation Initiation: Cell Biology, High-Throughput and Chemical-Based Approaches
Edited by JON LORSCH

VOLUME 432. Lipidomics and Bioactive Lipids: Mass-Spectrometry–Based Lipid Analysis
Edited by H. ALEX BROWN

Volume 433. Lipidomics and Bioactive Lipids: Specialized Analytical Methods and Lipids in Disease
Edited by H. Alex Brown

Volume 434. Lipidomics and Bioactive Lipids: Lipids and Cell Signaling
Edited by H. Alex Brown

Volume 435. Oxygen Biology and Hypoxia
Edited by Helmut Sies and Bernhard Brüne

Volume 436. Globins and Other Nitric Oxide-Reactive Protiens (Part A)
Edited by Robert K. Poole

Volume 437. Globins and Other Nitric Oxide-Reactive Protiens (Part B)
Edited by Robert K. Poole

Volume 438. Small GTPases in Disease (Part A)
Edited by William E. Balch, Channing J. Der, and Alan Hall

Volume 439. Small GTPases in Disease (Part B)
Edited by William E. Balch, Channing J. Der, and Alan Hall

Volume 440. Nitric Oxide, Part F Oxidative and Nitrosative Stress in Redox Regulation of Cell Signaling
Edited by Enrique Cadenas and Lester Packer

Volume 441. Nitric Oxide, Part G Oxidative and Nitrosative Stress in Redox Regulation of Cell Signaling
Edited by Enrique Cadenas and Lester Packer

Volume 442. Programmed Cell Death, General Principles for Studying Cell Death (Part A)
Edited by Roya Khosravi-Far, Zahra Zakeri, Richard A. Lockshin, and Mauro Piacentini

Volume 443. Angiogenesis: *In Vitro* Systems
Edited by David A. Cheresh

Volume 444. Angiogenesis: *In Vivo* Systems (Part A)
Edited by David A. Cheresh

Volume 445. Angiogenesis: *In Vivo* Systems (Part B)
Edited by David A. Cheresh

Volume 446. Programmed Cell Death, The Biology and Therapeutic Implications of Cell Death (Part B)
Edited by Roya Khosravi-Far, Zahra Zakeri, Richard A. Lockshin, and Mauro Piacentini

VOLUME 447. RNA Turnover in Bacteria, Archaea and Organelles
Edited by LYNNE E. MAQUAT AND CECILIA M. ARRAIANO

VOLUME 448. RNA Turnover in Eukaryotes: Nucleases, Pathways and Analysis of mRNA Decay
Edited by LYNNE E. MAQUAT AND MEGERDITCH KILEDJIAN

VOLUME 449. RNA Turnover in Eukaryotes: Analysis of Specialized and Quality Control RNA Decay Pathways
Edited by LYNNE E. MAQUAT AND MEGERDITCH KILEDJIAN

VOLUME 450. Fluorescence Spectroscopy
Edited by LUDWIG BRAND AND MICHAEL L. JOHNSON

VOLUME 451. Autophagy: Lower Eukaryotes and Non-Mammalian Systems (Part A)
Edited by DANIEL J. KLIONSKY

VOLUME 452. Autophagy in Mammalian Systems (Part B)
Edited by DANIEL J. KLIONSKY

VOLUME 453. Autophagy in Disease and Clinical Applications (Part C)
Edited by DANIEL J. KLIONSKY

VOLUME 454. Computer Methods (Part A)
Edited by MICHAEL L. JOHNSON AND LUDWIG BRAND

VOLUME 455. Biothermodynamics (Part A)
Edited by MICHAEL L. JOHNSON, JO M. HOLT, AND GARY K. ACKERS (RETIRED)

VOLUME 456. Mitochondrial Function, Part A: Mitochondrial Electron Transport Complexes and Reactive Oxygen Species
Edited by WILLIAM S. ALLISON AND IMMO E. SCHEFFLER

VOLUME 457. Mitochondrial Function, Part B: Mitochondrial Protein Kinases, Protein Phosphatases and Mitochondrial Diseases
Edited by WILLIAM S. ALLISON AND ANNE N. MURPHY

VOLUME 458. Complex Enzymes in Microbial Natural Product Biosynthesis, Part A: Overview Articles and Peptides
Edited by DAVID A. HOPWOOD

VOLUME 459. Complex Enzymes in Microbial Natural Product Biosynthesis, Part B: Polyketides, Aminocoumarins and Carbohydrates
Edited by DAVID A. HOPWOOD

Volume 474. Thiol Redox Transitions in Cell Signaling, Part B Cellular Localization and Signaling
Edited by Enrique Cadenas and Lester Packer

Volume 475. Single Molecule Tools, Part B: Super-Resolution, Particle Tracking, Multiparameter, and Force Based Methods
Edited by Nils G. Walter

Volume 476. Guide to Techniques in Mouse Development, Part A Mice, Embryos, and Cells, 2nd Edition
Edited by Paul M. Wassarman and Philippe M. Soriano

Volume 477. Guide to Techniques in Mouse Development, Part B Mouse Molecular Genetics, 2nd Edition
Edited by Paul M. Wassarman and Philippe M. Soriano

Volume 478. Glycomics
Edited by Minoru Fukuda

Volume 479. Functional Glycomics
Edited by Minoru Fukuda

Volume 480. Glycobiology
Edited by Minoru Fukuda

Volume 481. Cryo-EM, Part A: Sample Preparation and Data Collection
Edited by Grant J. Jensen

Volume 482. Cryo-EM, Part B: 3-D Reconstruction
Edited by Grant J. Jensen

Volume 483. Cryo-EM, Part C: Analyses, Interpretation, and Case Studies
Edited by Grant J. Jensen

Volume 484. Constitutive Activity in Receptors and Other Proteins, Part A
Edited by P. Michael Conn

Volume 485. Constitutive Activity in Receptors and Other Proteins, Part B
Edited by P. Michael Conn

Volume 486. Research on Nitrification and Related Processes, Part A
Edited by Martin G. Klotz

Volume 487. Computer Methods, Part C
Edited by Michael L. Johnson and Ludwig Brand

Volume 488. Biothermodynamics, Part C
Edited by Michael L. Johnson, Jo M. Holt, and Gary K. Ackers

Volume 503. Protein Engineering for Therapeutics, Part B
Edited by K. Dane Wittrup and Gregory L. Verdine

Volume 504. Imaging and Spectroscopic Analysis of Living Cells
Optical and Spectroscopic Techniques
Edited by P. Michael Conn

Volume 505. Imaging and Spectroscopic Analysis of Living Cells
Live Cell Imaging of Cellular Elements and Functions
Edited by P. Michael Conn

Volume 506. Imaging and Spectroscopic Analysis of Living Cells
Imaging Live Cells in Health and Disease
Edited by P. Michael Conn

Volume 507. Gene Transfer Vectors for Clinical Application
Edited by Theodore Friedmann

Volume 508. Nanomedicine
Cancer, Diabetes, and Cardiovascular, Central Nervous System, Pulmonary and Inflammatory Diseases
Edited by Nejat Düzgüneş

Volume 509. Nanomedicine
Infectious Diseases, Immunotherapy, Diagnostics, Antifibrotics, Toxicology and Gene Medicine
Edited by Nejat Düzgüneş

Volume 510. Cellulases
Edited by Harry J. Gilbert

Volume 511. RNA Helicases
Edited by Eckhard Jankowsky

Volume 512. Nucleosomes, Histones & Chromatin, Part A
Edited by Carl Wu and C. David Allis

Volume 513. Nucleosomes, Histones & Chromatin, Part B
Edited by Carl Wu and C. David Allis

Volume 514. Ghrelin
Edited by Masayasu Kojima and Kenji Kangawa

Volume 515. Natural Product Biosynthesis by Microorganisms and Plants, Part A
Edited by David A. Hopwood

Volume 516. Natural Product Biosynthesis by Microorganisms and Plants, Part B
Edited by David A. Hopwood

Volume 517. Natural Product Biosynthesis by Microorganisms and Plants, Part C
Edited by David A. Hopwood

Volume 518. Fluorescence Fluctuation Spectroscopy (FFS), Part A
Edited by Sergey Y. Tetin

Volume 519. Fluorescence Fluctuation Spectroscopy (FFS), Part B
Edited by Sergey Y. Tetin

Volume 520. G Protein Couple Receptors
Structure
Edited by P. Michael Conn

Volume 521. G Protein Couple Receptors
Trafficking and Oligomerization
Edited by P. Michael Conn

Volume 522. G Protein Couple Receptors
Modeling, Activation, Interactions and Virtual Screening
Edited by P. Michael Conn

Volume 523. Methods in Protein Design
Edited by Amy E. Keating

Volume 524. Cilia, Part A
Edited by Wallace F. Marshall

Volume 525. Cilia, Part B
Edited by Wallace F. Marshall

Volume 526. Hydrogen Peroxide and Cell Signaling, Part A
Edited by Enrique Cadenas and Lester Packer

Volume 527. Hydrogen Peroxide and Cell Signaling, Part B
Edited by Enrique Cadenas and Lester Packer

Volume 528. Hydrogen Peroxide and Cell Signaling, Part C
Edited by Enrique Cadenas and Lester Packer

Volume 529. Laboratory Methods in Enzymology: DNA
Edited by Jon Lorsch

CHAPTER ONE

Explanatory Chapter: PCR Primer Design

Rubén Álvarez-Fernández[1]
Department of Plant Sciences, University of Cambridge, Cambridge, United Kingdom
[1]Corresponding author: e-mail address: rubenaf.uo@gmail.com

Contents

1. Theory 2
   1.1 Thermodynamics 2
   1.2 General guidelines for primer design 3
   1.3 Melting temperature (T_m) and GC content 3
   1.4 Primer length 4
   1.5 Primer sequence and secondary structures 5
   1.6 PCR product length and placement within the target sequence 6
   1.7 Step-by-step guide for primer design 7
   1.8 Troubleshooting and general rule 8
   1.9 Specific applications 8
   1.10 Target cloning 8
   1.11 Reverse-transcription PCR (RT-PCR) 9
   1.12 Real-time PCR (quantitative PCR, qPCR) 10
   1.13 Multiplex PCR 12
   1.14 Sequencing 12
   1.15 Degenerate primers 13
   1.16 Modified primers 16
   1.17 Recombineering 16
   1.18 Primers for homologous recombination 17
   1.19 Software tools list 18

Acknowledgments 19

References 19

Abstract

This chapter is intended as a guide on polymerase chain reaction (PCR) primer design (for information on PCR, see General PCR and Explanatory Chapter: Troubleshooting PCR). In the next section, general guidelines will be provided, followed by a discussion on primer design for specific applications. A list of recommended software tools is shown at the end.

Methods in Enzymology, Volume 529
ISSN 0076-6879
http://dx.doi.org/10.1016/B978-0-12-418687-3.00001-X

1. THEORY

In the spring of 1983, Kary Mullis had the idea to use a pair of synthetic oligonucleotides (primers) to potentially copy *ad infinitum* a target DNA sequence using DNA polymerase; an idea worthy of a Nobel Prize. However, thermal cycling was the essence of the process, and posed a critical drawback: the double-stranded DNA was denatured by heat at the beginning of each cycle, destroying the polymerase in the process, and it had to be replaced in each round. In 1986, Mullis solved the problem by using a thermo-resistant DNA polymerase from *Thermophilus aquaticus* (Taq). With the drawback gone, the PCR became dramatically more affordable and was subject to automation.

Now, PCR is a core technique in molecular biology and an extremely large number of applications have been developed. These methods directly depend on the efficiency of the reaction, which relies on the fine-tuning of its components: DNA template, dNTPs, reaction buffer, and a set of primers that flank the target sequence and are extended by the DNA polymerase.

This chapter focuses on primer design, a critical factor to both the efficiency and the specificity of the PCR.

1.1. Thermodynamics

The changes in two thermodynamic properties of a system (ΔS – change of disorder, and ΔH – change of heat) can be used to calculate ΔG, the Gibb's free energy change. This value shows whether a process is spontaneous, and allows one to determine the equilibrium constant K, with which the concentrations of all the species that are involved in the equilibrium can be worked out. Therefore, the thermodynamic changes that happen in going from random coil to duplex can be known and used to determine the primer quality. A full thermodynamic explanation is given in SantaLucia (2007); Mann et al. (2009).

To quickly sum up, the ΔG is directly related to the equilibrium constant (K) of the primer/template hybridization at a given temperature (T).

$$\Delta G = \Delta H - T\Delta S$$

$$K = [\text{primer plus template duplex}]/([\text{primer}][\text{template}])$$

$$\Delta G = -^{**}RT \times \ln(K)$$

A key parameter that affects the efficiency of a PCR is the amount of primer bound to the template. If the concentration of duplex (primer bound to template) is different for each of the primers, the efficiency will be reduced. From the formulae above, it can be deduced that designing primers with the same ΔG (at a fixed annealing temperature) will render more efficient primers pairs, and also explains why matching T_m's is a less accurate approach than matching ΔG's.

The more negative the ΔG the higher the K, and then the more stable the duplex. Primer software sets thresholds that are used to discard bad primers, as secondary and tertiary structures (primer-dimers, hairpins, etc.) with very low ΔG (very stable) are undesirable.

1.2. General guidelines for primer design

Primer design has two goals: specificity and efficiency of amplification. Specificity is the frequency of proper priming events – mispriming leads to amplifications off-target, though mindfully used it can serve to hunt for related sequences (e.g., degenerate primers, see below). Efficiency is the increase of product in each PCR cycle; the theoretical optimum is a two-fold increase. The application determines the balance between these parameters. For instance in diagnostics, efficiency is sacrificed for a higher specificity, minimizing the false positives at a cost of less PCR product. These two parameters are controlled by factors such as temperature, extension time, template concentration and thermodynamic properties, length of the amplicon, divalent cations (Mg^{2+}, Mn^{2+}), and detergents.

The main parameters affecting primer design and therefore the PCR reaction are discussed below.

1.3. Melting temperature (T_m) and GC content

T_m is the temperature at which half the primer strands are bound to the target. It depends on base composition and can be roughly estimated as

$$T_m = 4(G + C) + 2(A + T)$$

However, this equation neglects the dependence of T_m on strand concentration, salt concentration, and base sequence. Typical errors for this method compared with determining the T_m experimentally can be $>15\ ^\circ C$, thus this equation is not recommended (SantaLucia, 2007).

The best current approach to calculate T_m is the nearest-neighbor (NN) method (SantaLucia, 1998). The previous 1986 NN model is unreliable, but still present in most common primer design tools such as **Primer3**, **OLIGO**, and **Vector NTI®**. When possible, then, the 1998 model should be selected (SantaLucia, 2007).

T_m depends on base content. Two primers with a different length can have the same T_m then, but the longer one will anneal less efficiently than its partner, thereby affecting the efficiency of the PCR. A different T_m between the pair will affect specificity (at lower annealing temperatures, the primer with higher T_m will be more prone to misprime) and efficiency (at higher temperatures, the primer with lower T_m will anneal less frequently or not at all). ΔH° of binding is different for each primer, resulting in different slopes of binding at the T_m, and then in different amplification efficiencies and artifacts. Therefore, the key point is not to match T_m, but equal amounts of primers bound to the target: to do so, design primers with matching ΔG° at the desired annealing temperature (SantaLucia, 2007).

Typically, the annealing temperature to start with is 4–10 °C below the T_m. However, the optimal annealing temperature can be more accurately calculated as

$$T_a = 0.3T_{m,primer} + 0.7T_{m,product} - 14.9$$

with $T_{m,primer}$ being the T_m of the less stable primer-template pair (Rychlik, 1990. Note that T_m calculations are not made with the latest NN model – SantaLucia, 1998).

In summary, for regular PCR, a match of ΔG° (otherwise no more than 2 °C of difference in T_m) and GC content (40–60%) is advised. However, the GC content can be higher and work fine, so the more general rule is that it should be not be lower than that of the amplicon (Rychlik, 1993).

1.4. Primer length

A major factor controlling specificity is the annealing temperature, which is determined by the length and base composition of the primers.

The shorter the primer, the quicker the annealing with the template and the lower the specificity, therefore, primers <18 nucleotides in length are likely to perform badly with complex templates (e.g., genomic DNA). Primers 16–18 nucleotides in length will perform well only for <500 bp amplicons; for longer amplicons, longer primers are recommended, for example, 24 nt primers will do for 5 kb products (Rychlik, 1993).

The longer the primer, the lower the efficiency (a small inefficiency will propagate with the number of cycles). However, a longer primer leads to a higher specificity because of the primer length (4 times per added base) and its melting temperature. Thus, 28–35 nt primers can be used to amplify templates expected to have a high degree of heterogeneity, or for cloning closely related genes. It is important to note here that longer primers are less pure due to inefficiencies in chemical synthesis, so purification better than standard desalting might be required for certain applications (see Purification of DNA Oligos by Denaturing Polyacrylamide Gel Electrophoresis (PAGE)).

In summary, the utilization of primers of a minimal length that ensures $T_m > 54\ ^\circ C$ is recommended to balance specificity and efficiency in general conditions (Dieffenbach et al., 1993).

1.5. Primer sequence and secondary structures

The primer's sequence determines its specificity for the target but can also facilitate interaction with itself and other primers, affecting its performance.

Taq polymerase extends only the 3′-end of the primer, so its sequence is critical. Perfect base pairing for at least the last 4 bases is optimal, and minimal mismatch should exist within the last 5–6 ones (lowering the annealing temperature does not compensate well for mismatches near the 3′-end of the template, but rather it promotes mispriming). Conversely, the 5′-end sequence is not that critical, so 5′ tails (e.g., restriction sites, fluorophores, etc.) can be added to the primers without significantly affecting annealing efficiency.

Primers with low 3′-end stability (AT-rich, $\Delta G > -9\ \text{kcal mol}^{-1}$) but a stable 5′-end perform best in sequencing and PCR, since this structure reduces mispriming (Rychlik, 1993). Therefore, structures with high 3′-end stability such as *GC clamps* (more than 3 G or C bases within the last five 3′-end bases) should be avoided because the stronger bonding raises the likelihood of partial priming.

Avoid also mononucleotide *runs* longer than 4 nucleotides (e.g., GGGG) and *repeats* of more than 4 dinucleotides (e.g., ATATATAT), because they can cause mispriming and polymerase slippage (see Viguera et al., 2001 for more information on slippage).

In general, discard primers that form intramolecular or primer/primer duplexes with negative ΔG, especially if these structures are formed at the 3′-end. Although $T_{m,\text{duplexes}} < T_{m,\text{primer/template}}$ dimerization will occur to some extent, lowering the PCR efficiency by sequestering the polymerase and the primers themselves. For instance, *hairpins* can cause internal primer extension if they form at the 3′-end, eliminating that primer from the reaction (though on the 5′-end do not significantly affect the PCR). Primer-dimers are extended by the polymerase and can be detected on an agarose gel as a low molecular weight band, or on a melting curve (usually performed in qPCR experiments) as extra peaks. 'Fake' primer-dimers can also appear if the primers bind off-target close to one another, which explains why sometimes the band is longer than expected, and why sequencing shows a few extra unexpected bases between the actual primers.

Prevention of dimerization is crucial in qPCR experiments and multiplex PCR. Patches like the hot start method can mitigate it but, if possible, it is preferable to avoid it from the beginning.

An approximation that meets these problems is the proprietary **ThermoBLAST™**, by DNA software, which can be implemented into their **Visual OMP™** primer design package. Although a BLAST search will give information on mispriming, it just considers base pairing and not thermodynamic parameters like duplex stability, so its help is limited. ThermoBLAST™ retains the computational efficiency of BLAST but uses thermodynamic scoring for base pairs, dangling ends, single mismatches, bulges, tandem mismatches, and other motifs, thereby rendering more balanced primer pairs (SantaLucia, 2007). **Pythia** is another software approach in which state-of-the art DNA binding affinity computations are directly integrated into the primer design process. It seems it is comparable to other available software, although it outperforms them on the design for primers on difficult regions (e.g., repeats).

In summary, select those primers less likely to form secondary structures and with low 3′-end stability.

1.6. PCR product length and placement within the target sequence

The product length affects the efficiency of the PCR. Generally, 150–1000 bp products are amplified for target detection, and some techniques like qPCR (see below, *real-time PCR*) have narrower windows for the length of the amplicon. For detection of gene expression, it is preferable to design the primers on contiguous exons (see below, *RT-PCR*).

If possible, avoid regions with long runs, repeats, or secondary structures, because they will make good primer design and matching more complicated. Those structures can be predicted with software such as **UNAFold** (also called Mfold^{++}). Some primer design software, such as **Phytia,** takes into consideration binding and folding energies, outperforming other software for designing primers for repeat regions. Proprietary software like **Visual OMP™** also allows for prediction of secondary structures and can solve the multistate coupled equilibrium of these cases (SantaLucia, 2007).

1.7. Step-by-step guide for primer design

These are some guidelines for RT-PCR primer design, but can mostly be applied to design primers for other purposes.

Conventions:

- Sequences are always written from 5′ to 3′, no matter whether they are forward or reverse primers (therefore, *do not ever* write a primer sequence reversed because it will lead to confusion).
- Polymerase extends from the 3′-end.

Procedure:

1. Get the cDNA sequence. If possible, use RefSeqs.

   Refseq, or **Ref**erence **Seq**uence database from NCBI, is a collection of annotated DNA, RNA, and protein sequences. Each RefSeq is reviewed and represents a single molecule from one organism (http://www.ncbi.nlm.nih.gov/RefSeq/). Sequences other than RefSeqs do not have that assured quality and there is a chance they are truncated, untrimmed, or incorrect. Make sure you are using the latest version of a RefSeq entry (i.e., the accession with the highest version number), or simply just use the accession without the version number, as you will automatically get the latest version.
2. Primers can be designed from the NCBI itself by clicking on 'pick primers,' next to the sequence. The tool combines Primer3 and BLAST and is called **Primer-BLAST**.

   Primer-BLAST allows one to design primers bridging an exon–exon junction or spanning an intron (see below for explanation, on *RT-PCR*), thereby avoiding false positives due to DNA amplification. The BLAST addition allows for the selection of organism-specific primers and selective amplification of splice variants.
3. Select the parameters that suit your specificity goals (see guidelines above) and the software will provide the best primer pairs (Note: Primer-BLAST uses the 1998 model for T_m calculation by default).
4. Primer quality can be double-checked using software such as **NetPrimer**.
5. Alternatively, design the primers using software like **Primer3**, **Primer3Plus**, **Pythia**, **Visual OMP™**, **OLIGO**, or **Vector NTI®** (just a few examples in no particular order. The first 3 are freely available and the last 3 are proprietary).
6. Finally, for RT-PCR, it is recommended that primers be chosen close to the mRNA 3′-end. The reverse transcription reaction starts at this end, and premature termination of the reaction can produce shorter

molecules that lack the 5′-end. Specially if poly-(T) are used, designing the primers close to the 3′-end is suggested to avoid the underestimation of the target concentration.

1.8. Troubleshooting and general rule

If no specific product is achieved in the first attempt, alternative approaches can be used which are, however, beyond the scope of this chapter (see Explanatory Chapter: Troubleshooting PCR). A gradient PCR is very useful to determine empirically the best annealing temperature for the primer set. A touchdown PCR will increase specificity at a cost of lower efficiency. A nested PCR will enrich the sample in specific product. The use of a different polymerase can dramatically affect the results as well (e.g., switching from regular *Taq* to a high-fidelity polymerase). Amplification of some DNA templates works better at extension temperatures lower than 72 °C (especially if they are <500 bp or highly AT-rich, and therefore less stable at high temperatures). The use of two-step PCRs (denaturation plus annealing-extension, by designing primers with a T_m of 68–70 °C) or tweaking the PCR mix (changing the Mg^{2+} concentration, addition of DMSO, etc.) can also help. See Mülhardt (2007) for further information on PCR troubleshooting.

However, the general rule can be: begin with well-designed primers, and if a few troubleshooting attempts fail, consider designing a new set of primers instead of trying to tackle all of the modifications possible.

1.9. Specific applications

Ultimately, primer design strategy is determined by the goal of the PCR method. Several designs are discussed below; it is important to note that PCR depends not only on primer design, but also on primer quality and purity. Chemical synthesis of primers ends up with a salty mixture of oligonucleotides of different lengths, or a mixture of labeled and unlabeled ones. They have to be purified, and the purification method should be chosen in terms of the final application. Therefore, a faulty PCR can happen after a perfect primer design, simply because in the reaction we are adding interfering faulty oligonucleotides. See Purification of DNA Oligos by Denaturing Polyacrylamide Gel Electrophoresis (PAGE) for further explanation.

1.10. Target cloning

There are several approaches for cloning (see Molecular Cloning and Restrictionless cloning). Primers can be designed at the ends of a target and the

PCR product cloned into an A/T vector (even directionally cloned by doing some modifications, see Zhou and Gomez-Sanchez, 2000). This can be done in **Primer3Plus** by choosing the option 'cloning.'

Alternatively, restriction sites can be designed as 5′-tails in the primers, where they do not substantially affect annealing (if necessary, lower the annealing temperature for the first few cycles and then raise it to the annealing temperature considering the whole primer including the tail). Restriction enzymes will not cut efficiently if just the target restriction site is added; adding 4–6 bases at the 5′-end are usually required for best performance. New England Biolabs offers some information on this subject:

http://www.neb.com/nebecomm/tech_reference/restriction_enzymes/cleavage_linearized_vector.asp.

http://www.neb.com/nebecomm/tech_reference/restriction_enzymes/cleavage_olignucleotides.asp (bear in mind that these examples are oligonucleotides. Therefore, only a few bases are needed at the 5′-end of the restriction site since the 3′-end is linked to the primer).

Check that the target sequence does not contain the desired restriction sites, design the primers, and add the 5′-tails containing the restriction sites. If the sequence contains the restriction site(s), you can still use different restriction enzymes for insert and vector that generate compatible ends (e.g., *Sal* I and *Xho* I; see a complete list at http://www.neb.com/nebecomm/tech_reference/restriction_enzymes/compatible_cohesive_overhangs.asp).

For further reference on restriction enzymes, check the NEB website at http://www.neb.com/nebecomm/tech_reference/restriction_enzymes/default.asp?

1.11. Reverse-transcription PCR (RT-PCR)

RT-PCR (see Reverse-transcription PCR (RT-PCR)) is sensitive enough to enable detection and quantitation of RNA from even a single cell. In contrast, two other commonly used techniques for quantifying mRNA levels – Northern blot analysis (see Northern blotting) and nuclease protection assays (see Explanatory Chapter: Nuclease Protection Assays) – require larger of amounts of RNA.

A potential problem with RT-PCR is contamination of the RNA with genomic DNA, which can result in false positives. Consequently, primers should be designed to span introns or bridge an exon–exon junction (Fig. 1.1). When they span one or more introns, the amplified product from genomic DNA will be bigger than expected, or there will be no product at all if the distance between the primers exceeds the DNA polymerase

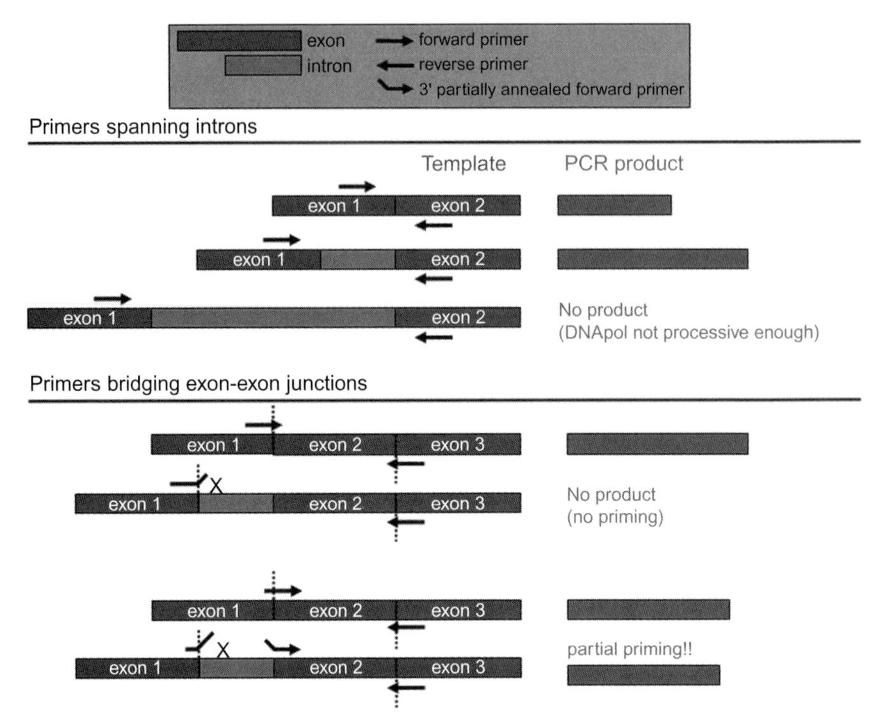

Figure 1.1 *Considerations on RT-PCR primer design.* DNA contamination in RNA can result in false positives. Consequently, primers should be designed to span introns or bridge an exon–exon junction, so the cDNA-derived PCR product is distinguishable from the genomic DNA-derived one. When bridging exon–exon junctions, take care not to design a primer with too long a 3′-end or GC clamps which allow for extension, since you will not be able to distinguish between cDNA- and genomic DNA-derived amplifications (example at the bottom).

processivity. When they bridge an exon–exon junction, they will not amplify genomic DNA template because the intron will hinder the annealing. Make sure you avoid GC clamps and that only a few nucleotides from the 3′-portion of the primer can base pair with the 3′-exon, because a substantial pairing can still prime the polymerase. See above *Step-by-step guide for primer design*, where a RT-PCR-oriented designing protocol is provided.

1.12. Real-time PCR (quantitative PCR, qPCR)

(Adapted from http://www.eurogentec.com).

Guidelines for primers:

- Length: 18–30 nucleotides
- %GC: 30–80% (ideally, 40–60%)

- T_m: 63–67 °C (ideally, 64 °C), so the annealing temperature would be 58–60 °C (Note: as discussed above, matching ΔG is more accurate than matching T_m. Otherwise, avoid T_m differences larger than 4 °C).
- Avoid mismatches, runs, repeats, and self- and primer-primer complementarity.
- Avoid a *T* residue at the 3′-end (allows mismatching).
- Make sure the primers are specific and target a single amplicon by BLAST.
- The considerations given in the section on RT-PCR should also be observed if a RT-qPCR is to be performed.
- For qPCR, it is preferable to design primers bridging exon–exon junctions (see Fig. 1.1), since these will amplify specifically the mRNA. If there is genomic DNA contamination, primers spanning introns will amplify the contaminant template and the resulting product will interfere with the accurate quantification of the target. It is then recommended that a separate control be set up to detect genomic DNA contamination.

Guidelines for probes:

- Length: 18–30 nucleotides (optimal length is 20 nt. If >30 nt are required, use an internal quencher on dT around the 20th nucleotide rather than at the 3′-end).
- %GC: 30–80%.
- T_m: 8–10 °C higher than the T_m of the primers (8 °C for genotyping, 10 °C for expression profiling).
- Select the strand that gives the probe more Cs than Gs.
- Place the probe the close to the primers without overlapping them.
- Avoid mismatches, runs, repeats, and self- and probe-primer complementarity.
- Avoid a G residue at the 5′-end because it quenches the fluorophore.
- Make sure the probe is specific and targets a single amplicon (the same as the primers) by BLAST.
- If genotyping by multiplexing, position the polymorphism in the center of the probe and match the probe's ΔG if possible, otherwise match the T_m.

Guidelines for amplicons:

- Length for SYBR® Green I assays: 80–150 nucleotides (shorter amplicons will increase the PCR efficiency, and longer ones will give a higher ΔRn as more dye will be incorporated).
- Length for 5′ exonuclease probe assays: 80–120 nucleotides (shorter amplicons will increase the efficiencies of the PCR and 5′ nuclease reactions).

- %GC: 30–80% (ideally, 40–60%).
- Avoid secondary structures in the amplicon (they can be checked with **UNAFold**).

It is *critical* that the primers are well designed. No secondary structures and very close ΔG's, no annealing to parts of the template that can result in unbalanced binding, so that the efficiency of amplification is close to 100% and no off-target amplification occurs.

See Explanatory Chapter: Quantitative PCR and Wong and Medrano (2005) for further information regarding real-time PCR.

1.13. Multiplex PCR

The goal of multiplex PCR is to amplify several targets in the same tube. Since first described by Chamberlain et al. (1988), it has had many applications and is typically used for genotyping, where simultaneous analysis of multiple markers is required, for detection of pathogens or genetically modified organisms, or for microsatellite analyses. Multiplex assays are difficult to establish, so they are worthwhile only for analyses that will be repeated many times, since at the end, they save space, money, and time.

Two or more primer pairs are used at a time. They could be designed and work separately, but when put all together, primers from one pair can interact with primers from another one. This makes the optimization of multiplex PCR quite difficult, as each primer pair could have different requirements. To match the T_m and ΔG is just not enough because of the potential interactions, and an effort made in primer design is worthwhile.

Sets of primer pairs can be checked for cross-dimerization using tools like **AutoDimer**.

If starting from scratch, there is software for multiplex PCR primer design: **MIPS** is specially made for designing multiplex PCR experiments for SNP genotyping, **PrimerStation** is specific for the human genome, and the commercially available **Visual OMP™** can deal with multistate equilibrium reactions and multiplex design as well (SantaLucia, 2007).

1.14. Sequencing

As stated above, sequences are written from 5′ to 3′ and the polymerase extends from the 3′-end. Sequencing is done on just one strand, so you have to make sure you design the primer for the right one.

Additionally consider that whilst PCR is exponential, sequencing is not. Therefore, bad primers that work for PCR after tweaking the reaction might not be suited for automated sequencing: mismatched primers will create templates that are perfect matches in a PCR, allowing amplification, but that will not happen with sequencing since only one strand is replicated.

Desired conditions:

- Use accurate sequence data, as mismatching will reduce the specificity and efficiency of the sequencing reaction, leading to bad readings or overlapped sequences. Verify the site-specificity of the primer. The last 8–10 bases at the 3′-end should be unique. If you repeatedly get overlapped sequences, primers might be partially annealing to the vector or on a different place on the insert, and you should redesign them.
- Design the primers no <50 bases from the sequence you are interested in (because the first reads are usually inaccurate), but not farther than 300 bases from it (usable sequence reads reach up to 700–1000 bases). Sequence data is often most accurate about 80–150 nucleotides away from the primer.
- Optimal primer length is 20–25 nucleotides (acceptable range is 18–30 nt).
- %GC content: 40–60%.
- T_m within the range of 55–65 °C.
- Discard candidate primers that self-anneal
- Try to choose primers with <3 Cs or Gs at the last 5 bases at the 3′-end, and that do not end with C or G. This will reduce the likelihood of partial unspecific priming.

Some software, such as **Primer3Plus,** has a specific option for designing sequencing primers.

Additionally, software has been developed specifically for directed sequencing, such as **JCVI Primer Designer**. This particular software is designed for pipelining high-throughput PCR primer design, and high-quality directed sequencing results have been obtained using it (as a note, this software is based on Primer3 but does not use the latest NN T_m calculation method of SantaLucia (1998)).

Anyway, always check with your sequencing facility as they might have specific requirements for primer design, template concentration, etc.

1.15. Degenerate primers

A PCR primer sequence is called 'degenerate' if some of its positions have several possible bases. The 'degeneracy' of the primer is the number of unique sequence combinations it contains. Hence, these primers are supplied as mixtures of similar, but not identical, oligonucleotides. They have various applications such as amplification or cloning of related genes from the same or different organisms, cloning of genes based on protein sequence, or genome walking with techniques such as TAIL-PCR (thermal asymmetric interlaced PCR, Liu and Whittier, 1995).

Sequence comparison of multiple members of protein families has revealed protein motifs and domains that play important roles in protein function. Closely related sequences can be cloned using degenerate primers (a pool of primers containing most, or all, possible nucleotide sequences encoding a conserved amino acid motif) or consensus primers (a single primer containing the most common nucleotide at each codon position within the motif).

The goal is then to design primers that match as many sequences of homologous genes as possible. A naïve solution would be to align the sequences without gaps, count the number of different nucleotides in each position along the alignment, and seek a primer-length window (20–30 bp) where the product of the count is low. Such a solution is insufficient because of gaps, the inappropriate objective function of the alignment, and most notably, the exceedingly high degeneracy: when degeneracy is too high, unrelated sequences may be amplified as well, losing specificity (Linhart and Shamir, 2002; 2005). The degeneracy has to be high to maximize the coverage, but has to be bound to decrease the probability of amplifying unrelated sequences. This problem and the approaches to solve it are addressed in Najafabadi et al. (2008).

Below, some software approaches are suggested for the design of degenerate primers:

PAMPS (Pairwise Alignment for Multiple Primer Selection). In contrast to previous algorithms, this one does not restrict the output to the exact primer length that was given, which allows selecting an appropriate primer in terms of annealing temperature. PAMPS can be used to design degenerate primers for amplifying genes with uncertain sequences, such as new members of gene families or libraries of antibody variable fragments. PAMPS has been shown to outperform previous algorithms (HYDEN and MIPS) in this task. This software is freely available.

Other tactics have been developed based on COnsensus-DEgenerate Hybrid Oligonucleotide Primers (CODEHOPs). A **CODEHOP** is a hybrid primer consisting of a 15–20 nt consensus 5′-*clamp* and a 9–12 nt degenerate 3′-*core* region. The *core* gives a broader specificity for distantly related target gene templates, while the *clamp* allows for a robust amplification during the later cycles of the PCR. CODEHOP works well for small sets of proteins, taking into account the codon usage of the target genome and the desired annealing temperature. However, it is inappropriate for constructing primers with very high degeneracy on large sets of long genomic sequences, and software like PAMPS should be used for that task. An interactive version of the software, iCODEHOP, is freely available.

MAD-DPD (Minimum Accumulative Degeneracy Degenerate Primer Design). MAD-DPD can be considered as a first approach in primer design when a high degeneracy is desired and the number of degenerate primers to be constructed is known or has to be kept low. A drawback of this software is that it does not consider parameters like melting temperature, self-annealing, or secondary structures. This software is freely available.

MIPS (Multiple, Iterative Primer Selector). The software MIPS outperforms HYDEN on the task of designing multiplex PCR experiments for SNP genotyping. This software is freely available.

HYDEN (HighlY DEgeNerate primers) is a software useful for constructing primers and primer pairs with high degeneracy and yet high specificity. This software is freely available.

Use of degenerate primers can greatly reduce the specificity of the PCR amplification, a problem that can be partly solved by using touchdown PCR.

Genetic code (IUPAC)

Code	Amino acid	Stands for
A	A	**A**denine
C	C	**C**ytosine
G	G	**G**uanine
T	T	**T**hymine
U	U	**U**racil
R	A or G	pu**R**ine
Y	C or T (U)	p**Y**rimidine
M	A or C	a**M**ino
K	G or T (U)	**K**eto
S	C or G	**S**trong (triple hydrogen bonds)
W	A or T (U)	**W**eak (double hydrogen bonds)
B	C or G or T (U)	not A

Continued

D	A or G or T (U)	not C
H	A or C or T (U)	not G
V	A or C or G	not T (U)
N	A or C or G or T (U)	a**N**y nucleotide

1.16. Modified primers

Modifications can be positioned at the 3′- or 5′-end of the oligonucleotide, or internally within the sugar-phosphate backbone or at the nucleobases. These modified primers are used in applications such as *in situ* hybridization, sequencing, fragment analysis, real-time PCR, fluorescent assays, chemiluminescent assays, microscopy, antisense experiments, and gene cloning and discovery (e.g., degenerate primers). The modifications can also allow for the oligonucleotides to be immobilized onto a surface, enabling spot-detection (e.g., in microarrays, chemiluminiscent, and fluorescent assays).

These modifications include

- functionalization with amino, phosphate, or thiol groups
- fluorescent dyes
- hapten or enzyme coupling
- unnatural bases, for example, Br-dU, I-dU
- phosphothioates
- multiply-labeled oligonucleotides, for example, doubly labeled probes
- Locked Nucleic Acid (LNA™*)

The applications of each modification fall beyond the scope of this chapter, but they can be found at major oligonucleotide supplier's Web sites (e.g., Invitrogen, Sigma-Aldrich, IDT DNA, GeneLink, Exiqon, Thermo Electron, etc., in no particular order).

1.17. Recombineering

Recombineering stands for *in vivo* homologous **recombi**nation-mediated genetic engi**neering**. This technique relies on bacteriophage-based recombination systems such as λ Red, or RedET, which mediate recombination between the host's genome and a construct. Restriction-ligation techniques

* LNA nucleosides are a class of nucleic acid analogues in which the ribose ring is 'locked' by a methylene bridge connecting the 2′-O atom and the 4′-C atom. This modification enhances the base pairing, increasing the T_m and then the specificity, making them very useful in a variety of applications such as Northern blotting, *in situ* hybridization, and real-time PCR, microarrays (see http://www.exiqon.com/ for further information).

are not used, thereby allowing for manipulation of longer constructs and more versatility, including marker-free approaches (Karsten Tischer et al., 2006). Recombineering with ssDNA oligonucleotides has been shown to be very efficient (Costantino and Court, 2003); however, it can also be performed with dsDNAs.

Primer design is key for efficient recombineering (Sawitzke et al., 2007. See Recombineering: Highly Efficient *in vivo* genetic engineering using single-strand oligos):

- The primer should correspond in sequence to the DNA strand that is replicated discontinuously (the lagging-strand). Twenty-fold reductions in efficiency are expected if the leading strand is chosen.
- It should be 70 bases long, and the base changes no closer than 10 bases from an end. The longer the oligo, the more likely it is to have errors introduced during synthesis. However, further purification is not typically helpful.
- It should be designed to avoid the DNA-mismatch repair (MMR) system (otherwise, the recombineering should be performed in a MMR-deficient strain), following these guidelines:
  - it should create a C/C mispair at the target base or 6 bases away from it (Costantino and Court, 2003),
  - or change 5 bases in a row,
  - or change 4–5 wobble positions in a row, in addition to the designed change.

For troubleshooting and the most up-to-date information on recombineering, see http://redrecombineering.ncifcrf.gov/.

1.18. Primers for homologous recombination

Gene replacement can be performed in yeast by homologous recombination using PCR products. Primer design is critical for this purpose, and some guidelines are given below and those should be taken into account apart from those at the beginning of this chapter (see the full protocol in Yeast - Gene Replacement Using PCR Products).

- Homology to the gene to be replaced or interrupted: 50 nucleotides are recommended and up to 100 nucleotide-long homology regions could be needed. In some cases, as little as 40 nucleotides might work. If the gene is to be replaced, the homology regions have to be upstream and downstream of the coding sequence.
- Homology to the selectable cassette: 20 nucleotides are recommended.

- Since the primers are >70 nucleotides long, additional purification rather than simply desalting might be needed (see Purification of DNA Oligos by Denaturing Polyacrylamide Gel Electrophoresis (PAGE)).

1.19. Software tools list

Please note that no commercial interests have influenced the software choice below, where an attempt has been made to balance the use of freely available and commercial software. This review is not intended as an in-depth comparison and marking of software tools, hence although these ones will perform well for your primer design, better ones might be available in the market for more specific applications.

AutoDimer (Vallone and Butler, 2004). http://www.cstl.nist.gov/strbase/NIJ/AutoDimer.htm

CODEHOP (Rose et al., 2003; Boyce et al., 2009). https://icodehop.cphi.washington.edu/i-codehop-context/Welcome

HYDEN (Linhart and Shamir, 2002; 2005). http://acgt.cs.tau.ac.il/hyden/HYDEN.htm

JCVI Primer Designer (Li et al., 2008). http://sourceforge.net/projects/primerdesigner/

MAD-DPD (Najafabadi et al., 2008). http://bioinf.cs.ipm.ir/download/MAD_DPD08172007.zip

MIPS (Souvenir et al., 2003; 2007). http://www.cs.wustl.edu/%7Ezhang/projects/mips.zip

NetPrimer. http://www.premierbiosoft.com/netprimer/index.html

OLIGO. http://www.oligo.net/

PAMPS (Najafabadi et al., 2008). http://www.biomedcentral.com/content/supplementary/1471-2105-9-55-S1.zip

Primer3 (Rozen and Skaletsky, 2000). http://frodo.wi.mit.edu/primer3/

Primer3Plus (Untergasser et al., 2007). http://www.bioinformatics.nl/cgi-bin/primer3plus/primer3plus.cgi

PrimerStation (Yamada et al., 2006) http://ps.cb.k.u-tokyo.ac.jp

Pythia (Mann at al., 2009). http://frodo.wi.mit.edu/primer3/

ThermoBLAST™. http://dnasoftware.com/tabid/110/Default.aspx

UNAFold ($Mfold^{++}$, Markham and Zuker, 2008). http://dinamelt.bioinfo.rpi.edu/download.php http://mfold.bioinfo.rpi.edu/

Vector NTI®. http://www.invitrogen.com/site/us/en/home/LINNEA-Online-Guides/LINNEA-Communities/Vector-NTI-Community/Vector-NTI.html

Visual OMP™. http://dnasoftware.com/tabid/108/Default.aspx

ACKNOWLEDGMENTS

The author thanks Dr. Sara Lopez-Gomollon for making a comprehensive review of the manuscript.

REFERENCES

Referenced Literature

Boyce, R., Chilana, P., & Rose, T. M. (2009). iCODEHOP: A new interactive program for designing COnsensus-DEgenerate Hybrid Oligonucleotide Primers from multiply aligned protein sequences. *Nucleic Acids Research 37(web server issue)*. http://dx.doi.org/10.1093/nar/gkp379.

Chamberlain, J. S., Gibbs, R. A., Rainer, J. E., Nguyen, P. N., & Thomas, C. (1988). Deletion screening of the Duchenne muscular dystrophy locus via multiplex DNA amplification. *Nucleic Acids Research*, *16*, 11141–11156.

Costantino, N., & Court, D. L. (2003). Enhanced levels of λ Red-mediated recombinants in mismatch repair mutants. *Proceedings of the National Academy of Sciences of the United States of America*, *100*, 15748–15753. http://dx.doi.org/10.1073/pnas.2434959100.

Dieffenbach, C. W., Lowe, T. M., & Dveksler, G. S. (1993). General concepts for PCR primer design. *Genome Research*, *3*, S30–S37.

Karsten Tischer, B., von Einem, J., Kaufer, B., & Osterrieder, N. (2006). Two-step Red-mediated recombination for versatile high-efficiency markerless DNA manipulation in *Escherichia coli*. *BioTechniques*, *40*, 191–197. http://dx.doi.org/10.2144/000112096.

Li, K., Brownley, A., Stockwell, T. B., et al. (2008). Novel computational methods for increasing PCR primer design effectiveness in directed sequencing. *BMC Bioinformatics*, *9*, 191. http://dx.doi.org/10.1186/1471-2105-9-191.

Linhart, C., & Shamir, R. (2002). The degenerate primer design problem. *Bioinformatics*, *18*, S172–S180.

Linhart, C., & Shamir, R. (2005). The degenerate primer design problem: Theory and applications. *Journal of Computational Biology*, *12*, 431–456. http://dx.doi.org/10.1089/cmb.2005.12.431.

Liu, Y.-G., & Whittier, R. F. (1995). Thermal asymmetric interlaced PCR: Automatable amplification and sequencing of insert end fragments from PI and YAC clones for chromosome walking. *Genomics*, *25*, 674–681.

Mann, T., Humbert, R., Dorschner, M., Stamatoyannopoulos, J., & Noble, W. S. (2009). A thermodynamic approach to PCR primer design. *Nucleic Acids Research*, *37*(13), e97. http://dx.doi.org/10.1093/nar/gkp443.

Markham, N. R., & Zuker, M. (2008). UNAFold: Software for nucleic acid folding and hybriziation. ch. 1,In J. M. Keith (Ed.), *Methods in Molecular Biology: vol. 453. Bioinformatics, Volume II. Structure, Function and Applications* (pp. 3–31). Totowa, NJ: Humana Press.

Mülhardt, C. (2007). The polymerase chain reaction. In *Molecular Biology and Genomics* (pp. 65–94): Elsevier.

Najafabadi, H. S., Saberi, A., Torabi, N., & Chamankhah, M. (2008). MAD-DPD: Designing highly degenerate primers with maximum amplification specificity. *BioTechniques*, *44*, 519–526.

Najafabadi, H. S., Torabi, N., & Chamankhah, M. (2008). Designing multiple degenerate primers via consecutive pairwise alignments. *BMC Bioinformatics*, *9*, 55. http://dx.doi.org/10.1186/1471-2105-9-55.

Rose, T. M., Henikoff, J. G., & Henikoff, S. (2003). CODEHOP (COnsensus-DEgenerate Hybrid Oligonucleotide Primer) PCR primer design. *Nucleic Acids Research*, *31*, 3763–3766. http://dx.doi.org/10.1093/nar/gkg524.

Rozen, S., & Skaletsky, H. J. (2000). Primer3 on the WWW for general users and for biologist programmers. In S. Krawetz & S. Misener (Eds.), *Bioinformatics Methods and Protocols: Methods in Molecular Biology* (pp. 365–386). Totowa, NJ: Humana Press.
Rychlik, W. (1990). Optimization of the annealing temperature for DNA amplification in vitro. *Nucleic Acids Research, 18*, 6409–6412.
Rychlik, W. (1993). Selection of primers for polymerase chain reaction. In B. A. White (Ed.), *Methods in Molecular Biology: vol. 15. PCR Protocols: Current Methods and Applications* (pp. 31–40).
SantaLucia, J., Jr. (2007). Physical principles and visual-OMP software for optimal PCR design. In A. Yuryev (Ed.), *Methods in Molecular Biology: vol. 402. PCR Primer Design* (pp. 245–267). Totowa, NJ: Humana Press.
SantaLucia, J., Jr. (1998). A unified view of polymer, dumbbell, and oligonucleotide DNA nearest-neighbor thermodynamics. *Proceedings of the National Academy of Sciences of the United States of America, 95*, 1460–1465.
Sawitzke, J. A., Thomason, L. C., Costantino, N., Bubunenko, M., Datta, S., & Court, D. L. (2007). Recombineering: In vivo genetic engineering in *E. coli, S. enterica*, and beyond. In K. T. Hughes & S. R. Maloy (Eds.), *Methods in Enzymology: vol. 421. Advanced Bacterial Genetics: Use of Transposons and Phage for Genomic Engineering* (pp. 171–199). San Diego, CA: Academic Press. 10.1016/S0076-6879(06)21015-2.
Souvenir, R., Buhler, J., Stormo, G., & Zhang, W. (2003). Selecting degenerate multiplex PCR primers. *Proceedings of the 3rd Workshop on Algorithms in Bioinformatics (WABI 2003)* (pp. 512–526).
Souvenir, R., Buhler, J., Stormo, G., & Zhang, W. (2007). PCR primer design. In A. Yuryev (Ed.), *pp: 402. Methods in Molecular Biology* (pp. 245–267). Totowa, NJ: Humana Press.
Untergasser, A., Nijveen, H., Rao, X., Bisseling, T., Geurts, R., & Leunissen, J. A. M. (2007). Primer3Plus, an enhanced web interface to Primer3. *Nucleic Acids Research, 35*, W71–W74. http://dx.doi.org/10.1093/nar/gkm306.
Vallone, P. M., & Butler, J. M. (2004). AutoDimer: A screening tool for primer-dimer and hairpin structures. *BioTechniques, 37*, 226–231.
Viguera, E., Canceill, D., & Ehrlich, S. D. (2001). In vitro replication slippage by DNA polymerases from thermophilic organisms. *Journal of Molecular Biology, 312*, 323–333. http://dx.doi.org/10.1006/jmbi.2001.4943.
Wong, M. L., & Medrano, J. F. (2005). Real-time PCR for mRNA quantitation. *BioTechniques, 39*, 1–11.
Yamada, T., Soma, H., & Morishita, S. (2006). PrimerStation: A highly specific multiplex genomic PCR primer design server for the human genome. *Nucleic Acids Research 34(web server issue):* (pp. W665–W669). http://dx.doi.org/10.1093/nar/gkl297.
Zhou, M.-Y., & Gomez-Sanchez, C. E. (2000). Universal TA cloning. *Current Issues in Molecular Biology, 2*, 1–7.

Related Literature

Primer Design, P. C. R. (2007). In A. Yuryev (Ed.), *Methods in Molecular Biology: vol. 402.* Totowa, NJ: Humana Press.
http://www.ambion.com/techlib/basics/rtpcr/index.html.

Referenced Protocols in Methods Navigator

General PCR
Explanatory Chapter: Troubleshooting PCR
Purification of DNA Oligos by Denaturing Polyacrylamide Gel Electrophoresis (PAGE)
Molecular Cloning
Restrictionless cloning

Reverse-transcription PCR (RT-PCR)
Northern blotting
Explanatory Chapter: Nuclease Protection Assays
Explanatory Chapter: Quantitative PCR
Recombineering: Highly Efficient *in vivo* genetic engineering using single-strand oligos
Yeast - Gene Replacement Using PCR Products

Explanatory Chapter: How Plasmid Preparation Kits Work

Laura Koontz[1]
Department of Molecular Biology and Genetics, Johns Hopkins University School of Medicine, Baltimore, MD, USA
[1]Corresponding author: e-mail address: laurakoontz@gmail.com

Contents

1. Theory 23
2. Equipment 24
3. Materials 24
4. Protocol 24
   4.1 Preparation 24
   4.2 Duration 24
5. Step 1 Pellet Bacteria and Resuspend in Resuspension Buffer 24
6. Step 2 Lyse Bacteria 25
7. Step 3 Neutralize the Solution 26
8. Step 4 Apply Clarified Lysate to Column 26
9. Step 5 Pass Binding Buffer Through Column 27
10. Step 6 Wash the Column 27
11. Step 7 Elute the DNA 27
   11.1 Tip 27
   11.2 Glossary of Buffers (note *x* M denotes unknown concentration) 28
References 28

Abstract

To isolate plasmid DNA from bacteria using a commercial plasmid miniprep kit (if interested, compare this protocol with Isolation of plasmid DNA from bacteria).

1. THEORY

Modern laboratories depend upon high-quality, contaminant-free plasmid DNA preps in order to carry out many standard applications, subcloning, sequencing, transfections, and microinjections being a few examples. Many miniprep kits are commercially available, all of which can be used

Methods in Enzymology, Volume 529
ISSN 0076-6879
http://dx.doi.org/10.1016/B978-0-12-418687-3.00002-1

to quickly and cleanly purify plasmid DNA. These kits all work on the same principles, which are described in detail.

2. EQUIPMENT

Microcentrifuge
Micropipettors
Micropipettor tips
1.5-ml microcentrifuge tubes

3. MATERIALS

Plasmid isolation kit (e.g., miniprep or maxiprep kit, from any commercial supplier).

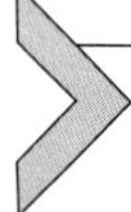

4. PROTOCOL

4.1. Preparation

Grow 3–5 ml overnight cultures of bacteria containing plasmid of interest at 37 °C, with shaking.

Add RNase A to the Resuspension Buffer and ethanol to the Wash Buffer according to the manufacturer's instructions.

4.2. Duration

Preparation	Overnight
Protocol	About 1 h

See Fig. 2.1 for the flowchart of the complete protocol.

5. STEP 1 PELLET BACTERIA AND RESUSPEND IN RESUSPENSION BUFFER

To begin, pellet your bacterial culture by centrifugation, remove the supernatant, and resuspend the cells in resuspension buffer.

Resuspension buffer contains Tris–HCl, pH 8.0, EDTA, and RNase A. Tris–HCl, pH 8.0, acts as a buffering agent. EDTA is a chelator, meaning that it soaks up free divalent cations (such as Mg^{2+} and Ca^{2+}) from the surrounding solution. There are two main effects of removing divalent cations

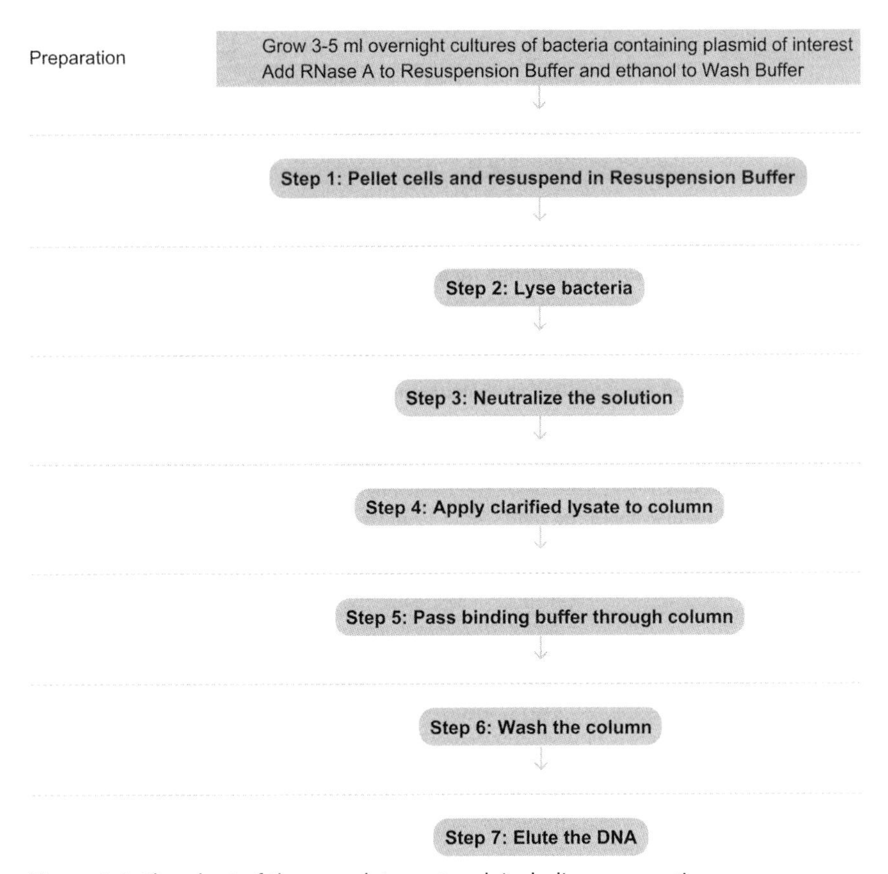

Figure 2.1 Flowchart of the complete protocol, including preparation.

with EDTA: (1) destabilization of bacterial lipid membranes and (2) inhibition of DNases, which often require divalent cations as cofactors. Therefore, EDTA prepares cells for lysis and prevents the degradation of your plasmid DNA of interest. The RNase A in the resuspension buffer enzymatically degrades all RNAs since later purification steps cannot separate RNA from plasmid DNA.

6. STEP 2 LYSE BACTERIA

Lysis buffer is then added to the resuspended cells in order to break them open and release cellular contents, including your plasmid of interest.

(If interested, compare this protocol with a milder one used to extract proteins.) Lysis buffer contains SDS and NaOH, which serve to break bacterial cells open. SDS, a detergent, breaks down the lipid membrane, while NaOH helps break down the bacterial cell wall. The action of these two chemicals releases both plasmid DNA and long genomic DNA into solution. Once the DNAs are in solution, NaOH disrupts the hydrogen bonding between DNA bases, separating the double-stranded DNA into single-stranded DNA.

7. STEP 3 NEUTRALIZE THE SOLUTION

To neutralize the pH and precipitate some contaminants, neutralization buffer is then added to the tube.

This solution contains a high concentration of potassium acetate, pH 5.5, which is in equilibrium with acetic acid and K^+. Neutralization of the alkaline conditions created by the lysis buffer establishes an environment in which hydrogen bonds can reform, allowing the small, single strands of your plasmid DNA to reanneal. Your double-stranded plasmid DNA will now easily dissolve in solution while the long stretches of bacterial genomic DNA will be unable to reanneal and dissolve. The potassium ion from the potassium acetate creates a high-salt environment, which precipitates the cellular debris, the single-stranded genomic DNA, and the SDS (as KDS, the white precipitate that forms in your tube). Your plasmid DNA will be left in the supernatant after centrifugation.

8. STEP 4 APPLY CLARIFIED LYSATE TO COLUMN

After spinning down the lysates, the clear supernatant containing your plasmid is poured over the silica membrane column and centrifuged.

Your plasmid DNA will reversibly bind to the silica in the column, with the help of guanidine HCl, a chaotropic salt, provided in the neutralization buffer from Step 3. Chaotropic salts are large monovalent ions with a low charge density. They weakly interact with H_2O and disrupt the network of hydrogen bonding in the surrounding H_2O. Disruption of this network effectively dehydrates macromolecules, such as your DNA, that are saturated with hydrogen bonded H_2O molecules. Disruption of intra- and intermolecular hydrogen bonding allows your DNA to open up, increasing its surface area and allowing it to bind reversibly to the silica membrane through an unknown mechanism.

9. STEP 5 PASS BINDING BUFFER THROUGH COLUMN

Before beginning the wash step, you have the option of passing a binding buffer through the column.

This buffer contains isopropanol and some guanidine HCl, which again helps to secure your DNA to the silica. The isopropanol dissolves salts and other contaminants. With your plasmid of interest safely bound to the silica in the column, you are now able to wash it to remove contaminants.

10. STEP 6 WASH THE COLUMN

Wash the column with the provided wash buffers.

Wash buffers are a buffered solution containing 80% EtOH. Washing with EtOH removes the chaotropic salts but does not disrupt the bonds between your DNA and the silica membrane.

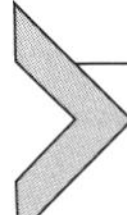

11. STEP 7 ELUTE THE DNA

Elute your DNA.

At this point, your DNA should be the only thing bound to your column. Because your DNA is dehydrated, application of an aqueous buffer will rehydrate it and allow your DNA to be eluted. Depending upon your intended downstream use, you will be able to elute your DNA in your choice of several different aqueous buffers: ddH_2O, TE, or Tris–HCl, pH 8.5.

11.1. Tip

Which elution buffer to choose?

ddH_2O: Water is the best choice if you are going to use your DNA for salt-sensitive applications such as PCR or sequencing.

TE: TE contains both Tris–HCl, (pH 8.0) and EDTA. DNA will be stable in TE over a longer period of time than in water, making this a good choice for long-term plasmid storage.

Tris–HCl, pH 8.5: Typically the elution buffer provided with kits is Tris–HCl, pH 8.5. This slightly basic buffer makes dissolving DNA easier than with ddH_2O. This buffer also does not contain EDTA, which could inhibit downstream reactions.

11.2. Glossary of Buffers (note *x* M denotes unknown concentration)

Resuspension Buffer – 50 mM Tris–HCl, pH 8.0, 10 mM EDTA, and 100 μg μl^{-1} RNase.

Lysis Buffer – 200 mM NaOH, 1% SDS.

Neutralization Buffer (Proprietary) – *x* M potassium acetate, pH 5.5, and *x* M guanidine HCl. (While the concentration of guanidine HCl in the neutralization buffer is not known, it is listed on the MSDS as an ingredient.)

Binding Buffer (Proprietary) – Isopropanol and *x* M guanidine HCl.

Wash Buffer – 10 mM Tris–HCl, pH 7.5, and 80% EtOH.

Elution Buffer Choices (Miniprep) – ddH_2O; Tris–HCl, pH 8.5; and TE (10 mM Tris–HCl, pH 8.0, and 1 mM EDTA).

REFERENCES

Referenced Protocols in Methods Navigator

Isolation of plasmid DNA from bacteria.

CHAPTER THREE

Explanatory Chapter: Introducing Exogenous DNA into Cells

Laura Koontz[1]

Department of Molecular Biology and Genetics, Johns Hopkins University School of Medicine, Baltimore, MD, USA

[1]Corresponding author: e-mail address: laurakoontz@gmail.com

Contents

Abstract

The ability to efficiently introduce DNA into cells is essential for many experiments in biology. This is an explanatory chapter providing an overview of the various methods for introducing DNA into bacteria, yeast, and mammalian cells.

1. THEORY

Transformation, transfection, transduction: each of these words describes a process by which exogenous genetic material is introduced into cells. Care should be taken when using these terms as they have specific meanings and thus are not interchangeable.

The term transformation originated in 1928 when Frederick Griffith noted that innocuous *Streptococcus pneumoniae* bacteria became virulent after being exposed to heat-killed virulent strains. He reasoned that they must have taken up some sort of 'transforming' factor that gave them that new virulence. Although he did not know it at the time, these bacteria must have been able to incorporate the DNA encoding virulence genes from the dead strains. Since then, the term *transformation* has come to mean a process by which cells incorporate and express exogenous genetic material, specifically bacteria taking up nonviral, typically plasmid, DNA.

Methods in Enzymology, Volume 529
ISSN 0076-6879
http://dx.doi.org/10.1016/B978-0-12-418687-3.00003-3

Transfection is a portmanteau of 'trans' and 'infection'; however, this is misleading, as transfected cells are not actually *infected* with anything. The term *transfection* has since evolved to describe the process by which a eukaryotic cell incorporates and expresses any nonviral genetic material (plasmid DNA, siRNA, or dsRNA).

Transduction is the process by which genetic material is transferred to a cell through a viral vector.

2. PROTOCOL

2.1. Transforming bacteria

Bacteria are only able to take up foreign genetic material if they are 'competent' to do so. Typically, bacteria are not naturally competent, but competency can be induced by two different transformation techniques: (1) calcium chloride transformation or (2) electroporation.

1. *Calcium chloride transformation* (sometimes referred to as heat shock) is the cheapest, simplest, and most commonly used method for transforming bacteria with plasmid DNA. Cells are chilled on ice in a solution containing calcium chloride ($CaCl_2$). The presence of divalent cations (Ca^{2+}) prepares the bacterial cell membrane for permeabilization. Bacteria are then briefly heat shocked (typically either 37 °C for 5 min or 42 °C for 30–60 s) and then returned to ice. During the heat shock, the cell membrane becomes permeable to nucleic acids and allows the plasmid DNA to enter the cells. If the plasmid has an origin of replication, it will be retained in the bacteria during divisions and can be selected for based on a selectable marker.

   Note: Competent cells can be kept at −80 °C and only need to be thawed directly before transformation.

2. *Electroporation of bacteria* is typically employed when the plasmid of interest is too large to be taken up through calcium chloride transformation or when high transformation efficiency is required. During electroporation, cells are briefly exposed to an electric field (typically 10–20 kV cm^{-1}), which creates holes in the membrane that plasmids are able to move through. At the termination of the electric current, the holes in the cell membrane are repaired by bacterial machinery. The plasmid can then be selected for by standard means.

   Note: Electroporation competent cells can be kept at −80 °C in a 10% glycerol solution.

2.2. Transforming yeast

There have been many attempts to induce competency in yeast; however, none has been successful. Therefore, cultures of yeast must be grown up prior to transformation each and every time. Typically, log-phase growth yeast cultures are best for transformation; however, the best OD for transformation varies by strain of yeast and by method of transformation. The two most common methods for transformation will be discussed here: (1) lithium acetate transformation and (2) electroporation.

1. *Lithium acetate transformation* is the most common method of transforming yeast. To carry out this protocol, yeast are cultured to log growth and then resuspended in a solution containing lithium acetate. PEG (polyethylene glycol) and DNA to be transformed (+ carrier DNA if needed) are then added. This mixture of yeast and DNA is then heat-shocked for ~15 min at 42 °C. The presence of the lithium cations in the solution destabilizes the membrane and prepares it for permeabilization. It is not entirely understood why PEG and heat shock are required; however, they have both been shown to be necessary for the uptake of DNA. If the concentration of DNA to be transformed is low, it is beneficial to add carrier DNA from salmon testes which stimulates recombination between the yeast genome and the transformed exogenous DNA of interest.
2. If the amount of DNA to be transformed is limiting, *electroporation of yeast* is an easy and fast way to transform small amounts of DNA efficiently. The rationale behind electroporation of yeast cells is identical to that discussed above for bacterial cultures. To electroporate yeast, log growth phase cells are concentrated and resuspended in a solution of 1 M sorbitol. A small amount of DNA is added, without carrier DNA, and the mixture subjected to an electric field. Addition of sorbitol to the solution maintains the osmotic stability of the yeast so that they do not lyse under hypotonic conditions. In the case of electroporation, a carrier DNA is not used because it has been shown to decrease the efficiency of transformation.

2.3. Introducing genetic material into eukaryotic cells

There are several different methods for transfecting genetic material into eukaryotic cell lines depending upon downstream applications. In addition to various methods of transfection, genetic material can either be transfected transiently or stably. During transient transfection, only a percentage of plated cultured cells will express the transfected construct(s). This percentage, the transfection efficiency, varies with the cell line and transfection method being used. For many applications such as co-immunoprecipitations and Western blotting, transient transfection is sufficient. If a homogenous population of cells expressing a gene is required, the cells must be stably transfected instead. To do this, cells are transfected normally with a plasmid containing the gene of interest and an antibiotic resistance gene. The cells are then passaged under antibiotic selection until a population of cells all expressing the gene(s) of interest is obtained. Since it is relatively time consuming to generate a stable cell line, it is only done for experiments in which it is absolutely necessary.

The following sections will give overviews of several different methods for introducing DNA into higher eukaryotes: (1) chemical-based methods; (2) electroporation; (3) viral transduction; (4) particle-based approaches; and (5) microinjection. Exact methods will not be addressed here since they vary widely by purpose, cell type, transfection reagent, and lab.

1. *Chemical-based methods for transfection* have the advantage that they are relatively easy, can transfect large amounts of DNA, are fast, and are cheap to use. For those reasons, they are the most commonly used for routine, every day transfections. While there are many types of kits which employ different methods of chemical delivery, they all work based on the same basic principle: DNA is bundled together with something that will interact with the plasma membrane of cells, that bundle is then deposited onto a layer of adherent cells and a proportion of them will take up the exogenous DNA via endocytosis. The oldest of these methods is calcium phosphate-mediated transfection. Calcium phosphate is mixed with the DNA to be transfected and a fine precipitate is formed. This is added to the cell media and, after it settles on the cells, it is taken up. Most commercially available kits use lipids to mediate these transfections. The DNA of interest is incubated with synthetic cationic lipids, which encase it in a liposome. These liposomes are then bathed over the cells to be transfected. The positively charged liposomes interact with the negatively charged lipids of the cellular plasma membrane and are eventually endocytosed, releasing the DNA into the cell.
2. *Electroporation of eukaryotic cells* follows the same principles as discussed above for bacteria. Electroporation is typically employed when the plasmid to be transfected is quite large; for example, electroporation is sometimes used to transfect mouse embryonic stem cells with the targeting vectors needed for generation of a knockout mouse.

3. *Viral transduction* is used when cell lines, such as primary cell cultures, are difficult to transfect through other mechanisms, when it is necessary to stably introduce a gene into the cell's genome; or when it is necessary to express a protein at high levels. There are several different types of viruses used, the most common of which will be discussed here. For mammalian tissues, it is often useful to use adenovirus transduction. Adenovirus has the advantage that it infects a wide variety of cell types and allows for high expression of transduced genetic material. However, the viral DNA does not integrate into the host cell chromosomes and is not replicated, thus, it is most useful for experiments where efficient, high-level, short-term expression is needed. Retrovirus vectors have long been used for long-term stable expression of genes in mammalian cells; however, they have the drawback of infecting and integrating only in dividing cells. Another mammalian cell specific vector is the lentivirus family of retroviruses. Lentiviruses are similar to retroviruses in that they integrate into the host genome but they have the advantage that they can infect nondividing cells as well as dividing cells. Baculoviruses are able to infect many types of invertebrate cells so they are often used for expressing proteins in insect cells, including Sf9 cells, derived from pupal ovarian tissue of the fall armyworm, *Spodoptera frugiperda*, and S2 cells, derived from late embryonic stages of *Drosophila*.
4. *Particle-based approaches* are often employed when other transfection techniques fail. These approaches have the benefit that they are extremely targeted: the nuclei of single cells are bombarded with DNA attached to a particle (e.g., gold beads, nanoparticles). The particles are shot into a cell using a gene gun or other bolistic delivery system. Nucleic acids associated with magnetic nanoparticles can be introduced into cells by applying a magnetic force. However, this technique has several serious drawbacks. It is expensive, time consuming, and has a low efficiency of transfection.
5. *Microinjection* is frequently used to create genetically engineered or transgenic animals. This technique requires quite a bit of precision and expertise, therefore it is typically only employed in specialized circumstances. To microinject DNA into a cell, the DNA of interest is loaded into a microinjection needle, ~0.5–5 μm in diameter. The needle is then carefully and slowly pushed through both the plasma membrane and nuclear envelope of the cell to be injected. At that time, a small amount of the

DNA is injected into the nucleus, and the needle carefully removed. Injected cells are then allowed to recover and later screened for insertion of the DNA of interest.

SOURCE REFERENCE

Griffith, F. (1928). The significance of pneumococcal types. *Epidemiology and Infection*, *27*(2), 113–159. Cambridge Univ Press. ISSN=0950-2688.

CHAPTER FOUR

Agarose Gel Electrophoresis

Laura Koontz[1]
Department of Molecular Biology and Genetics, Johns Hopkins University School of Medicine, Baltimore, MD, USA
[1]Corresponding author: e-mail address: laurakoontz@gmail.com

Contents

1. Theory 36
2. Equipment 36
3. Materials 36
   3.1 Solutions & buffers 37
4. Protocol 39
   4.1 Preparation 39
   4.2 Duration 40
   4.3 Caution 40
5. Step 1 Casting an Agarose Gel 40
   5.1 Overview 40
   5.2 Duration 40
   5.3 Tip 41
   5.4 Tip 41
   5.5 Tip 41
   5.6 Tip 41
   5.7 Tip 41
6. Step 2 Loading and Running an Agarose Gel 42
   6.1 Overview 42
   6.2 Duration 43
   6.3 Tip 43
   6.4 Tip 44
   6.5 Tip 44
7. Step 3 Visualization of Samples on an Agarose Gel 44
   7.1 Overview 44
   7.2 Duration 44
   7.3 Tip 45
   7.4 Tip 45
   7.5 Tip 45

Abstract

Agarose gel electrophoresis is used to separate DNA or RNA molecules based upon their size.

Methods in Enzymology, Volume 529
ISSN 0076-6879
http://dx.doi.org/10.1016/B978-0-12-418687-3.00004-5

1. THEORY

The separation of nucleic acids based upon their size is required for many common laboratory practices (e.g., subcloning, genotype diagnostics, RT-PCR). Separation of nucleic acids by agarose gel electrophoresis works by harnessing the negative charge of the phosphate backbone of nucleic acids. DNA and RNA molecules have a net negative charge spread evenly over their entire length so they will move through an agarose matrix in an electric field toward the positive pole. Shorter nucleic acids will be able to migrate through the matrix faster than larger ones during a given period of time. Depending upon the percentage of agarose used to make the gel, the range of linear separation will vary.

2. EQUIPMENT

pH Meter and calibration standards
Magnetic stir plate
Microwave oven
Agarose gel casting tray
Agarose gel comb
Agarose gel electrophoresis tank
Power supply
Transilluminator or other UV light source
Micropipettors
Micropipettor tips
Magnetic stir bars
Beaker, 1 l
Erlenmeyer flask, 250 or 500 ml

3. MATERIALS

Agarose
Tris base
EDTA, disodium salt
Glacial acetic acid
Boric acid (H_3BO_3)
Sodium hydroxide (NaOH)

Lithium hydroxide (LiOH)
Ethidium bromide or other fluorescent DNA intercalator (e.g., SYBR-Safe, Invitrogen)
Ficoll
Bromophenol blue
Xylene cyanol FF
DNA ladder

3.1. Solutions & buffers

There are several different buffers that you can choose from for your electrophoresis. The recipes for all of these buffers are included here. Briefly the advantages and disadvantages of each buffer are:

TAE *– The most common of all buffers, TAE has a low buffering capacity that prohibits gels from being run at extremely high voltages. However, it is best at resolving large DNA or RNA fragments.*

TBE *– TBE has been shown to be better than TAE at resolving fragments smaller than 3 kb. Gel purification of DNA bands may not work as well in TBE as in TAE. Refer to the manual of your specific kit for instructions and limitations.*

SB *– Sodium borate is a Tris-free buffer. Sodium borate has low conductivity and therefore gels can be run at higher voltages. SB produces sharp bands and nucleic acids can be purified for all downstream applications. However, SB is not as efficient as Tris-based buffers at resolving bands larger than 5 kb.*

LB *– Lithium borate buffer is similar to SB but has an even lower conductivity. LB can resolve very small size differences if used in conjunction with high-percentage agarose gels. However, lithium borate is somewhat expensive and thus not necessary for most common electrophoretic needs.*

50× TAE

Component	Final concentration	Stock	Amount
Tris base	2 M	N/A	242 g
EDTA, pH 8.0	1×	0.5 M	100 ml
Glacial acetic acid	N/A	N/A	57.1 ml

Dissolve Tris in ~750 ml of deionized water. Add EDTA and acetic acid. Adjust the final volume to 1000 ml with water. There is no need to adjust the pH of this solution.

1× TAE

Add 20 ml of 50× TAE stock to 980 ml of deionized water

10× TBE

Component	Final concentration	Stock	Amount
Tris base	890 mM	N/A	108 g
EDTA, pH 8.0	20 mM	0.5 M	40 ml
Boric acid	890 mM	N/A	55 g

Dissolve Tris and boric acid in approximately 750 ml of deionized water. Add EDTA. Adjust final volume to 1000 ml with water. There is no need to adjust the pH of this solution.

1× TBE

Add 100 ml of 10× TBE stock to 900 ml of deionized water

20× SB

Component	Final concentration	Stock	Amount
Boric Acid	N/A	N/A	45 g
NaOH	N/A	N/A	8 g

Dissolve boric acid and NaOH in 950 ml of deionized water. Adjust the volume to 1000 ml with water. There is no need to adjust the pH of this solution.

1× SB

Add 50 ml of 20× SB stock to 950 ml of deionized water

20× LB

Component	Final concentration	Stock	Amount
Boric acid	N/A	N/A	36 g
Lithium hydroxide	N/A	N/A	8.392 g

Dissolve boric acid and lithium hydroxide in 950 ml of deionized water. Adjust the volume to 1000 ml with water. There is no need to adjust the pH of this solution.

1× LB

Add 50 ml of 20× LB stock to 950 ml of deionized water

1% Agarose Gel Mix

Component	Final concentration	Stock	Amount
Agarose	1% (w/v)	N/A	1 g
Buffer of choice	1×	1×	100 ml
Ethidium bromide	0.5 μg ml^{-1}	10 mg ml^{-1}	5 μl

Mix agarose and buffer in a flask and microwave until dissolved. Add ethidium bromide (or other DNA intercalating dye) **after** removing from the microwave oven, so as not to vaporize the ethidium bromide. NOTE: This is just a template recipe. The concentration of agarose gel that you will need depends upon the size(s) of the nucleic acid(s) you wish to resolve.

Agarose Concentration	Linear Resolution of Gel
0.5%	1000–30 000 bp
0.7%	800–12 00 bp
1.0%	500–10 000 bp
1.2%	400–7000 bp
1.5%	200–3000 bp
2.0%	50–2000 bp

6× DNA loading dye

Component	Final concentration	Stock	Amount
Ficoll	15% (w/v)	N/A	1.5 ml
Bromophenol blue	0.25% (w/v)	N/A	25 mg
Xylene cyanol FF	0.25% (w/v)	N/A	25 mg

Add water to 10 ml

4. PROTOCOL

4.1. Preparation

Prepare the running buffer you wish to use.

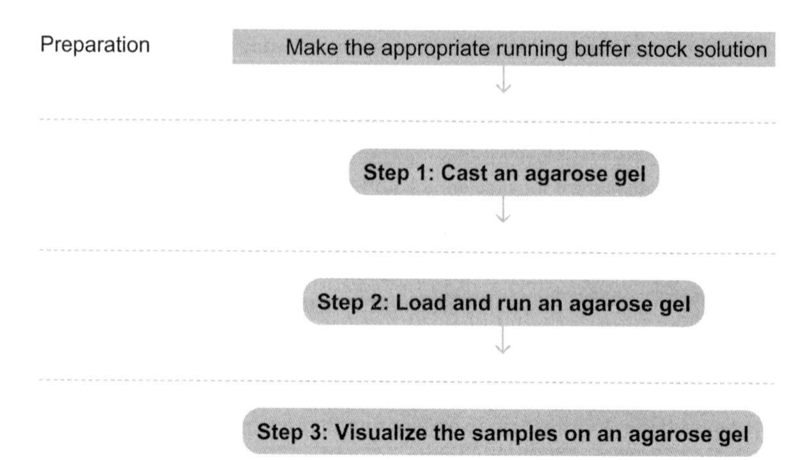

Figure 4.1 Flowchart of the complete protocol, including preparation.

4.2. Duration

Preparation	**About 2 h**
Protocol	About 1 h

4.3. Caution

Ethidium bromide is a known mutagen and carcinogen. It, and all other nucleic acid intercalators, should be handled with care. Wear gloves. Do not microwave any buffer containing ethidium bromide or any other dye. The dyes can vaporize and be inhaled. Instead, wait until the buffer has cooled slightly and then add the dye. Wear gloves when handling the gel and solutions containing ethidium bromide. Dispose of gel and buffer in accordance with local regulations.

See Fig. 4.1 for the flowchart of the complete protocol.

5. STEP 1 CASTING AN AGAROSE GEL

5.1. Overview

Cast an agarose gel for ~30 min before you are ready to run your samples.

5.2. Duration

30 min

1.1 First, refer to the table in (Materials section) and determine what percentage gel you need to resolve all of your samples. For example, if you are digesting DNA from plasmid minipreps to check for the insertion of a 1.5 kb fragment in a vector of 7 kb, you will need a 1% gel in order to linearly resolve all bands.

Once you have determined what percentage gel you need, add the appropriate amount of agarose to a 250 or 500 ml flask. Add 100 ml of 1× buffer.

1.2 Microwave on high for about 2 min, or until all agarose is dissolved.

1.3 Remove from microwave and allow the agarose to cool on the benchtop for a few minutes.

1.4 Once the agarose solution has cooled slightly, add your DNA dye of choice. Mix and pour into your assembled casting tray. Take care to pop any bubbles with a pipette tip. Insert the desired type of comb.

1.5 Allow gel to cool and solidify. This will take ~30 min.

1.6 Remove comb (and tape, if used) and place gel in electrophoresis tank filled with 1× running buffer.

5.3. Tip

Do not make your gel with H_2O! Water does not have the ionic strength to conduct electricity through the gel and your samples will not migrate correctly.

5.4. Tip

You must run your gel in the same buffer that you use to make the gel. Otherwise it will not properly conduct electricity and your gel will not run correctly.

5.5. Tip

Assemble the gel-casting tray while microwaving or cooling your agarose solution. Take care not to move the gaskets or else your gel will leak. If you do not have a setup with a mold, secure lab tape over the ends of the gel-casting tray.

5.6. Tip

Wide combs are typically used for purification, while thin combs are used for diagnostics (Fig. 4.2(a) and *4.2(b))*

5.7. Tip

If you are in a hurry, you can cast your gel at 4 °C and it will solidify faster.

See Fig. 4.3 for the flowchart of Step 1.

Figure 4.2 Loading and running a gel. (a) A typical gel electrophoresis apparatus consists of a gel box full of running buffer connected to a power source (arrow). Here the positive end is at the bottom of the box. Note the wide wells that the DNA is loaded in – these wells are ideal for purification because they can hold a large volume. (b) In contrast, thin wells are ideal for running a large number of diagnostic samples, such as in the case of genotyping experimental animals. (c) In order to ensure that the power source is connected correctly, look for the presence of bubbles emanating from the wire at the negative side of the gel box.

6. STEP 2 LOADING AND RUNNING AN AGAROSE GEL

6.1. Overview

Here you will load your samples on the gel and separate them in an electric current.

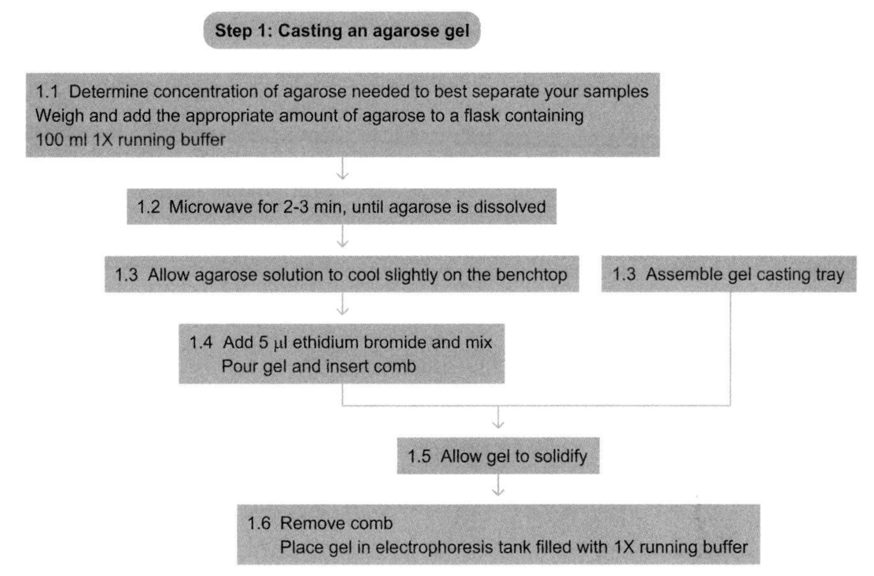

Figure 4.3 Flowchart of Step 1.

6.2. Duration

About 1 h

2.1 Determine the amount of sample you would like to run on the gel and add the appropriate amount of 6× DNA loading buffer. For example, if you are running a genotyping PCR, add 2 μl 6× DNA loading buffer to 10 μl of sample. If you are purifying a fragment for subcloning, you will want to load more.

2.2 Load an appropriate DNA ladder in one well, either 100 bp or 1 kb. Carefully pipette your samples into the wells. Be careful not to damage the wells or accidentally get your sample in adjacent wells.

2.3 Connect your gel tank to the power source. Make sure that you have connected the positive wire from your tank to the positive electrode on the power source, and likewise connected the negative to the negative. Turn your power source on (Fig. 4.2(a)).

2.4 When you are ready to visualize your gel, turn off the power supply and disconnect it from the gel tank. Carefully remove your gel. Do not put your hands in the tank while the gel is still running.

6.3. Tip

If you need to load a larger volume of sample than your well can hold, you can centrifuge it in a Speed-vac to concentrate it. Additionally, using the long thin gel loading pipet tips will allow you to load more sample than regular tips.

Step 2: Loading and running an agarose gel

2.1 Add 6X DNA loading buffer to samples
(e.g., add 2 μl loading buffer to 10 μl sample)

↓

2.2 Add DNA ladder and samples into the wells

↓

2.3 Connect leads and turn on power supply
(e.g., 100 V for 1X TAE)

↓

2.4 Run gel for 30-45 min
Use xylene cyanol and bromophenol blue dyes to judge length of time
Remove gel to UV light box

Figure 4.4 Flowchart of Step 2.

6.4. Tip

You can run your gel at a variety of voltages depending upon the buffer that you choose to use. Keep track of how far your sample has migrated by looking at the dye front on the gel.

6.5. Tip

To ensure that everything is connected correctly and an electric current is being supplied to the tank, check for the presence of bubbles at the positive pole of the tank (Fig. 4.2(c)).

See Fig. 4.4 for the flowchart of Step 2.

7. STEP 3 VISUALIZATION OF SAMPLES ON AN AGAROSE GEL

7.1. Overview

Since your sample is stained with a fluorescent dye, you will be able to visualize it using a UV light source.

7.2. Duration

5 min

3.1 Carefully place your gel on the glass top of a UV light box. Turn the light on. You should now be able to see your DNA light up (Fig. 4.5(a) and 4.5(b).

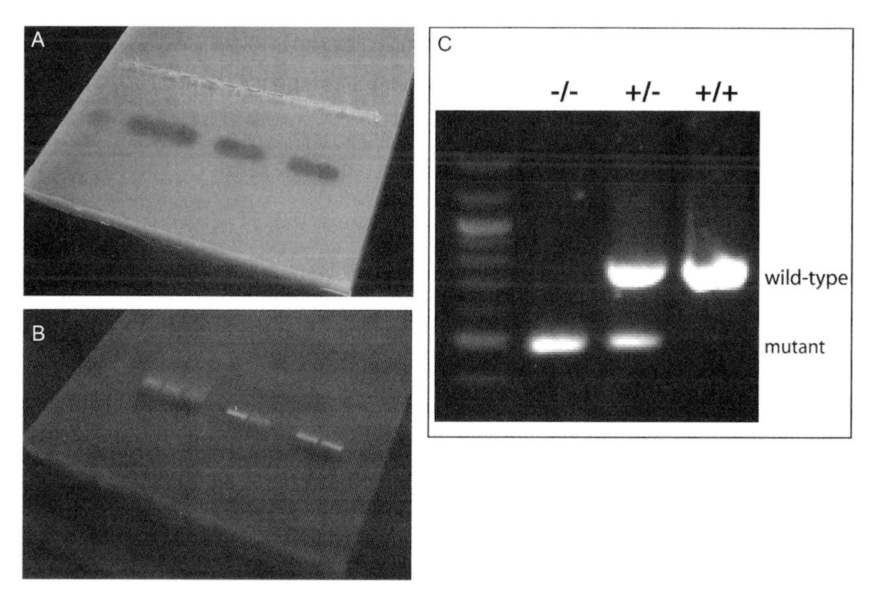

Figure 4.5 Visualization of DNA. (a) Image of an agarose gel illuminated with white light. Note the dye front. (b) Image of the same agarose gel illuminated with UV light. DNA is in green (stained with SYBR-Safe). (c) Gel electrophoresis can be used to genotype experimental animals or diseases. Here wild-type and mutant alleles can be distinguished based on a size difference.

3.2 Take a picture of your gel with a camera mounted to your UV box (Fig. 4.5(c)). If necessary, excise your DNA bands with a razor blade and transfer them to 1.5-ml microcentrifuge tubes for later gel purification and cloning.

7.3. Tip

Do not look directly at UV light. Always wear UV goggles when using UV light. Take care not to expose your skin to UV light too much as well.

7.4. Tip

Note that different DNA stains illuminate different colors – ethidium bromide glows orange and SYBR-Safe glows green, for example.

7.5. Tip

Minimize the amount of time that your sample is exposed to UV light. UV light induces crosslinks in DNA by creating pyrimidine dimers between adjacent cytosine and thymine residues. This is particularly crucial and therefore should be kept in mind if DNA is to be cut from your gel for purification and used for downstream sub-cloning.

CHAPTER FIVE

Analysis of DNA by Southern Blotting

Gary Glenn[1], Lefkothea-Vasiliki Andreou
Ear Institute, University College London, London, United Kingdom
[1]Corresponding author: e-mail address: gary638@gmail.com

Contents

1. Theory 48
2. Equipment 49
3. Materials 49
   3.1 Solutions & Buffers 50
4. Protocol 53
   4.1 Duration 53
   4.2 Preparation 53
   4.3 Caution 53
5. Step 1 Restriction Digestion of Genomic DNA and Agarose Gel Electrophoresis 54
   5.1 Overview 54
   5.2 Duration 54
   5.3 Tip 54
6. Step 2 Denature DNA and Transfer It to a Nylon Membrane 54
   6.1 Overview 54
   6.2 Duration 55
   6.3 Tip 56
   6.4 Tip 56
   6.5 Tip 56
7. Step 3A Label the Probe with [α-^{32}P]-dCTP Using a Random Primed Labeling Kit 56
   7.1 Overview 56
   7.2 Duration 56
   7.3 Caution 58
8. Step 3B Label the Probe with Digoxigenin-11-dUTP for Nonradioactive Detection 59
   8.1 Overview 59
   8.2 Duration 59
   8.3 Tip 59
   8.4 Tip 59
9. Step 4 Hybridize the Labeled Probe to the Membrane 59
   9.1 Overview 59
   9.2 Duration 60
   9.3 Tip 60

Methods in Enzymology, Volume 529
ISSN 0076-6879
http://dx.doi.org/10.1016/B978-0-12-418687-3.00005-7

Abstract

The purpose of this protocol is to detect specific DNA sequences in a complex sample by hybridization to a labeled probe.

1. THEORY

Southern blotting is routinely used to analyze DNA samples for a number of different applications. It was first described by Edwin Southern (Southern, 1975), hence the name, Southern blotting. In most applications, genomic DNA is digested with restriction enzymes and size-fractionated by agarose gel electrophoresis. The two DNA strands are denatured in the gel by alkaline treatment and then transferred to a nylon membrane by capillary action. The DNA fragments are immobilized on the membrane, which is now an exact replica of the size-fractionated DNA molecules in the gel. Specific DNA sequences are detected by hybridizing the membrane with a labeled probe that is complementary to the sequence of interest and then visualized according to the nature of the labeled probe. Originally probes were labeled using [α-^{32}P]-labeled deoxynucleotides; however, a number of nonradioactive labeling and detection techniques have been developed (Cate et al., 1991; Engler-Blum et al., 1993).

In its basic form, Southern blotting is used to determine the size of a particular restriction fragment in a complex mixture of digested genomic DNA. It can be used to study normal gene rearrangements, as in T-cell receptor or immunoglobulin gene rearrangements, or chromosomal rearrangements or translocations of oncogenes in tumor cells. It is relatively quantitative and can be used to determine the number of copies of a gene in the genome. By reducing the stringency of hybridization, it can be used to find related sequences in the genome or find similar sequences found in other species. It can also be used with restriction mapping using different restriction enzymes to characterize a particular region and in some cases to identify single nucleotide polymorphisms that affect a particular restriction enzyme recognition site. It is also used to verify that no gross rearrangements have occurred during the cloning process for a piece of DNA.

2. EQUIPMENT

Water baths (37, 65, 95 °C)
Agarose gel rig
Power supply
UV cross-linker (e.g., Stratalinker)
Plexiglass shields, radioactive waste disposal
Hybridization oven
Hybridization bottles
Trays (for washing membranes)
Film processor
– or –
Phosphorimager
X-ray film
– or –
Phosphor-capture screen
Micropipettors
Micropipettor tips
1.5-ml microcentrifuge tubes
Glass tray
Glass plate
Whatman 3MM chromatography paper
Nylon membrane (e.g., Hybond)
Paper towels
Aluminum foil
Plastic wrap
17-gauge needle
Syringe
0.45-μm disposable cellulose acetate membrane
Adhesive dots with phosphorescent ink
Adhesive dots with radioactive ink

3. MATERIALS

Restriction enzymes
10× restriction enzyme buffer
Agarose
TBE buffer

6× DNA loading buffer
Tris base
Sodium chloride (NaCl)
Sodium hydroxide (NaOH)
Sodium citrate
Random primed DNA labeling kit (e.g., Roche)
[α-^{32}P]-dCTP, 3000 Ci mmol^{-1}, aqueous solution
Digoxigenin-11-dUTP random primed DNA labeling kit (e.g., DIG-High Prime, Roche)
DIG nucleic acid detection kit (e.g., Roche)
Sodium dodecyl sulfate (SDS)
Ficoll 400
Polyvinylpyrrolidone
Bovine serum albumin, Fraction V
Formamide
Poly(A) (optional)
Salmon sperm DNA, type III, sodium salt
Phenol
Phenol:chloroform:isoamyl alcohol

3.1. Solutions & Buffers

Step 2 Denaturation Buffer

Component	Final concentration	Stock	Amount
NaOH	0.5 N	10 N	25 ml
NaCl	1.5 M	5 M	150 ml

Add purified water to 500 ml

Neutralization Buffer

Component	Final concentration	Stock	Amount
Tris–HCl, pH 8.0	1 M	2 M	250 ml
NaCl	1.5 M	5 M	150 ml

20× SSC

Component	Final concentration	Amount
NaCl	3 M	175.3 g
Sodium citrate	0.3 M	88.2 g

Dissolve in 800 ml purified water. Adjust pH to 7.0 using 1 N HCl (a few drops). Add purified water to 1 l. Sterilize by autoclaving

Step 4 Prehybridization Solution

Component	Final concentration	Stock	Amount
SSC	6×	20×	3 ml
Denhardt's solution	5×	50×	1 ml
SDS	0.5% (w/v)	20%	0.25 ml
Salmon sperm DNA	100 μg ml^{-1}	10 mg ml^{-1}	0.1 ml
Formamide	50% (v/v)		5 ml
Poly(A) (optional)	1 μg ml^{-1}	10 mg ml^{-1}	1 μl

Add purified water to 10 ml. Mix and filter through a 0.45-μm disposable cellulose acetate membrane

50× Denhardt's solution

Component	Final concentration	Amount
Ficoll 400	1% (w/v)	1 g
Polyvinylpyrrolidone	1% (w/v)	1 g
Bovine serum albumin	1% (w/v)	1 g

Add purified water to a final volume of 100 ml

Salmon sperm DNA, 10 mg ml^{-1}

Dissolve salmon sperm DNA in purified water at a concentration of 10 mg ml^{-1} in a 1.5-ml microcentrifuge tube (may take 2–4 h, stir if necessary). Adjust concentration of NaCl to 0.1 M. Extract the solution once with phenol and once with phenol/chloroform/isoamyl alcohol. Recover the aqueous phase that contains the DNA. Pass the aqueous phase quickly through a 17-gauge needle 12 times to shear the DNA. Add two volumes of ice-cold ethanol and centrifuge the tube in a microcentrifuge at maximum speed for 10 min. Resuspend the pellet in purified water at a concentration of 10 mg ml^{-1}. Check the concentration by measuring the absorbance at A_{260} and adjust the concentration to 10 mg ml^{-1}. Boil the DNA solution for 10 min. Dispense into 100 μl aliquots and store at −20 °C. Just before use, heat the DNA solution in a boiling water bath for 5 min and immediately place the tube on ice

Poly(A), 10 mg ml^{-1} (optional)

Dissolve 10 mg poly(A) RNA in 1 ml sterile purified water. Store at −20 °C in 50 μl aliquots

Wash Solution #1

Component	Final concentration	Stock	Amount
SSC	2×	20×	50 ml
SDS	0.1% (w/v)	20%	2.5 ml

Add purified water to 500 ml

Wash Solution #2

Component	Final concentration	Stock	Amount
SSC	0.1×	20×	250 ml
SDS	0.1% (w/v)	20%	2.5 ml

Add purified water to 500 ml

Maleic Acid Buffer

Component	Final concentration	Amount
Maleic acid	0.1 M	11.6 g
NaCl	0.15 M	8.77 g

Dissolve in 800 ml purified water (maleic acid will not go into solution). Add NaOH pellets while stirring, adjusting pH to 7.5. Add purified water to 1 l

10× Blocking Reagent

Dissolve The Blocking Reagent powder (Roche) in Maleic Acid Buffer at a concentration of 10% (w/v). Stir, shake, and heat to dissolve, the solution will remain turbid. Sterilize the solution by autoclaving

Washing Buffer

Component	Final concentration	Stock	Amount
Maleic acid	0.1 M		11.6 g
NaCl	0.15 M		8.77 g
Tween-20	0.3%	20%	15 ml

Prepare Maleic Acid Buffer as above. Add Tween-20 after adjusting the pH to 7.5 and then add purified water to 1 l

Detection Buffer

Component	Final concentration	Stock	Amount
Tris–HCl, pH 9.5	0.1 M	1 M	10 ml
NaCl	0.1 M	5 M	2 ml

Mix with 88 ml purified water

4. PROTOCOL

4.1. Duration

Preparation	About 1 day
Protocol	About 4–5 days

4.2. Preparation

Isolate genomic DNA from the appropriate source.

4.3. Caution

Consult your institute Radiation Safety Officer for proper ordering, handling, and disposal of radioactive materials.

See Fig. 5.1 for the flowchart of the complete protocol.

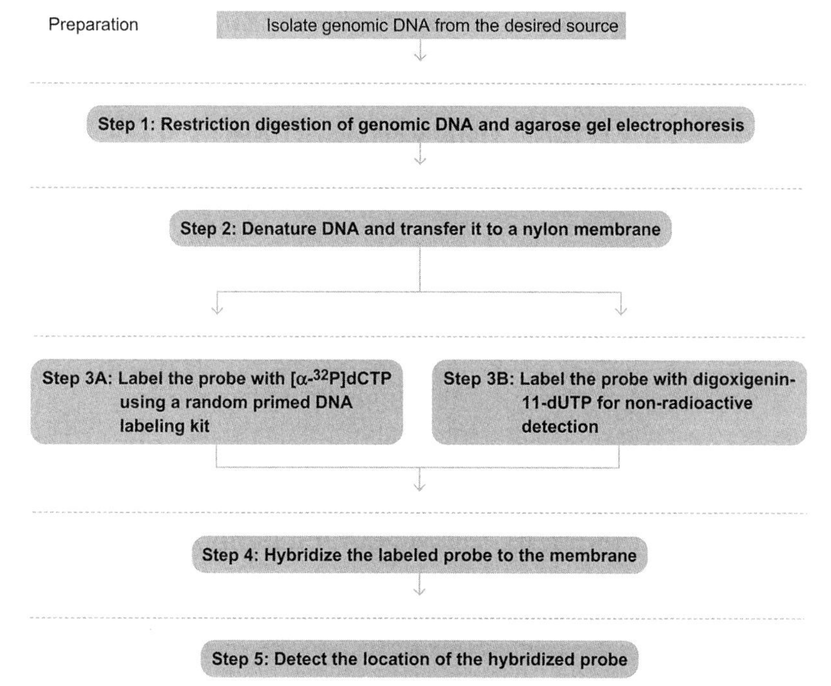
Preparation
Isolate genomic DNA from the desired source
Step 1: Restriction digestion of genomic DNA and agarose gel electrophoresis
Step 2: Denature DNA and transfer it to a nylon membrane
Step 3A: Label the probe with [α-32P]dCTP using a random primed DNA labeling kit
Step 3B: Label the probe with digoxigenin-11-dUTP for non-radioactive detection
Step 4: Hybridize the labeled probe to the membrane
Step 5: Detect the location of the hybridized probe

Figure 5.1 Flowchart of the complete protocol, including preparation.

5. STEP 1 RESTRICTION DIGESTION OF GENOMIC DNA AND AGAROSE GEL ELECTROPHORESIS

5.1. Overview

Genomic DNA is digested to completion using the desired restriction enzyme(s). Next, the DNA fragments are separated by agarose gel electrophoresis according to size.

5.2. Duration

About 24 h

1.1 Digest 10 μg of genomic DNA with the appropriate restriction enzyme(s). Add to a 1.5-ml polypropylene microcentrifuge tube:

Genomic DNA (10 μg)	*X* μl
10× enzyme buffer	5 μl
Restriction enzyme (1–2 U μg^{-1})	1–5 μl
Purified water	to 50 μl

1.2 Incubate reaction at 37 °C, overnight.

1.3 Heat reaction at 65 °C for 20 min to denature the restriction enzyme (optional).

1.4 Add 10 μl of 6× DNA sample buffer.

1.5 Pour an agarose gel and separate the DNA fragments by electrophoresis (see Agarose Gel Electrophoresis).

5.3. Tip

The total amount of restriction enzyme(s) added to the reaction mix should be <10% of the final volume. If too much is added, the higher glycerol concentration may adversely affect the digestion.

See Fig. 5.2 for the flowchart of Step 1.

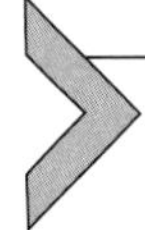

6. STEP 2 DENATURE DNA AND TRANSFER IT TO A NYLON MEMBRANE

6.1. Overview

The DNA will be denatured in the gel by soaking it in alkaline buffer and then transferred to a nylon membrane by capillary action.

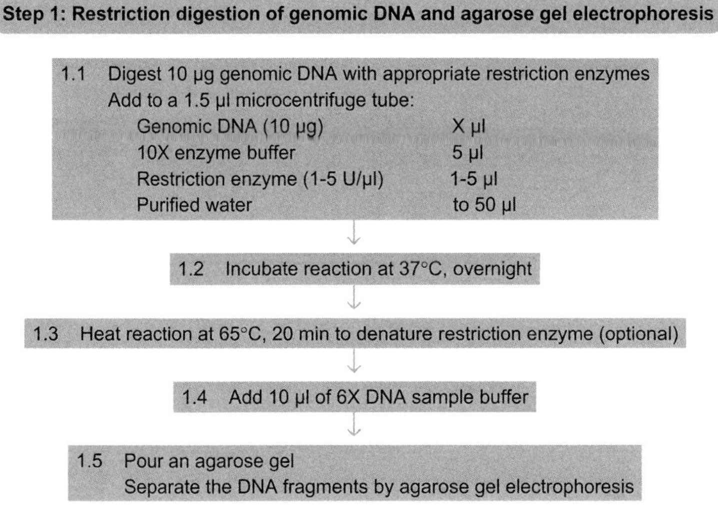

Figure 5.2 Flowchart of Step 1.

6.2. Duration

Overnight

2.1 Remove the gel from the electrophoresis apparatus and place it in a glass tray with 500 ml Denaturation Buffer. Let gel soak for 45 min at room temperature while rotating slowly on a platform rotator.

2.2 Pour off Denaturation Buffer and replace it with 500 ml Neutralization Buffer. Let gel soak for 1 h at room temperature while rotating slowly on a platform rotator.

2.3 Set up the transfer. Place a piece of glass across a larger glass tray so that it rests on the sides of the tray. Cut two pieces of Whatman 3MM chromatography paper so that they cover the piece of glass and hang over the edges to reach the bottom of the glass tray beneath it. Wet the paper thoroughly and fill the glass tray with 10× SSC. Roll a pipette over the paper to remove air bubbles.

2.4 Place the gel upside down on the wetted chromatography paper and remove air bubbles. Cover the parts of the Whatman paper not covered with the gel with aluminum foil.

2.5 Cut a piece of nylon membrane to the size of the gel and wet it thoroughly with 2× SSC. Place the membrane on top of the gel and remove air bubbles as before.

2.6 Cut two pieces of Whatman 3MM chromatography paper the same size as the membrane. Wet them thoroughly in 2× SSC and place them on top of the membrane. Remove air bubbles as before.

2.7 Cut a stack of paper towels about 4–5 inches high, the same size as the membrane, and place on top of the Whatman paper.

2.8 Place a piece of plexiglass on top of the stack of paper towels. Cover the entire setup with plastic wrap and place a weight (about 500 g) on top of the plexiglass. Allow DNA to transfer out of the gel overnight (18–20 h).

2.9 After the transfer is complete, disassemble the transfer setup. Invert the gel with the membrane on a dry surface.

2.10 Mark the positions of the wells on the membrane using a pencil. Discard the gel.

2.11 Immerse the membrane in 6× SSC and wash for 5 min at room temperature with gentle rotating.

2.12 Expose the membrane to UV light (302 nm) for 2–3 min using a UV cross-linker.

6.3. Tip

Add more 10× SSC as needed during the day before wrapping the transfer setup in plastic wrap for overnight transfer.

6.4. Tip

Remove the wet paper towels from the bottom of the stack and replace with dry ones during the day before wrapping the transfer setup in plastic wrap for overnight transfer.

6.5. Tip

In the case that nitrocellulose membranes are used instead of nylon membranes, the DNA is fixated to the membrane by heat.

See Fig. 5.3 for the flowchart of Step 2.

7. STEP 3A LABEL THE PROBE WITH [α-^{32}P]-dCTP USING A RANDOM PRIMED LABELING KIT

7.1. Overview

The probe will be labeled with [α-^{32}P]-dCTP using a random primed DNA labeling kit (e.g., Roche).

7.2. Duration

About 1 h

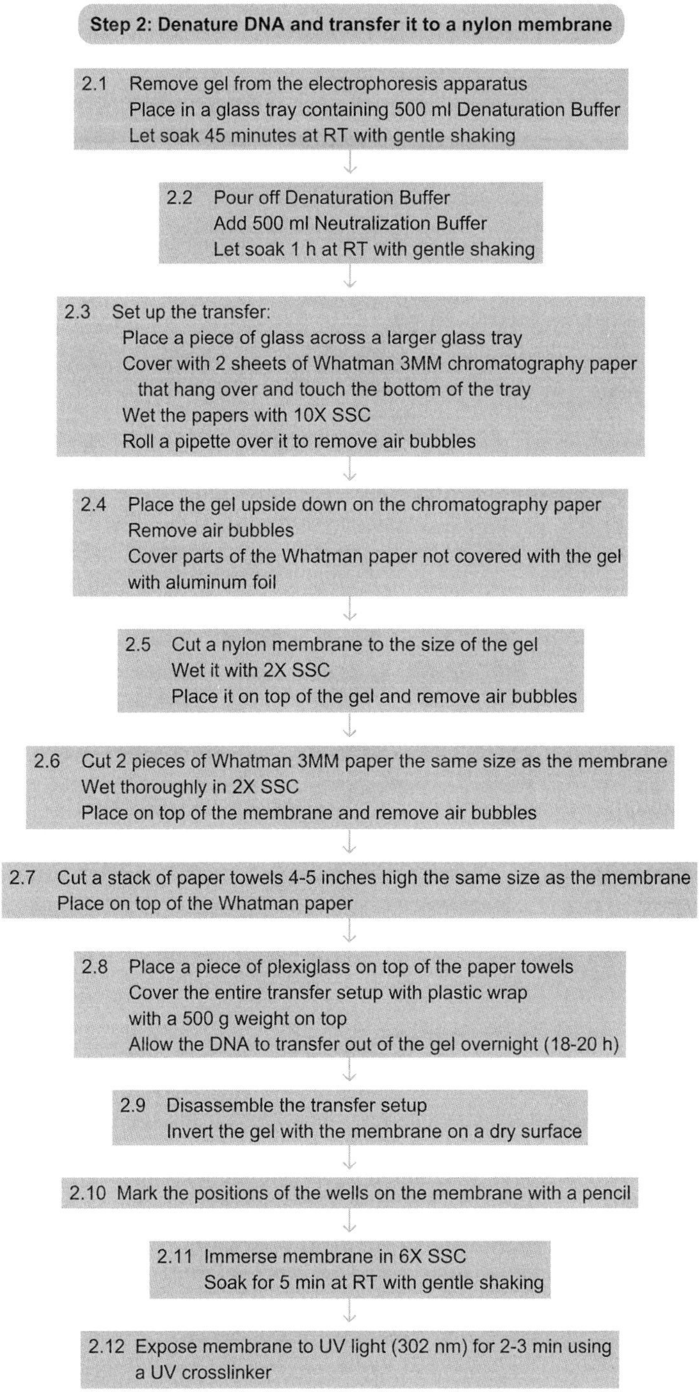

Step 2: Denature DNA and transfer it to a nylon membrane

2.1 Remove gel from the electrophoresis apparatus
Place in a glass tray containing 500 ml Denaturation Buffer
Let soak 45 minutes at RT with gentle shaking

↓

2.2 Pour off Denaturation Buffer
Add 500 ml Neutralization Buffer
Let soak 1 h at RT with gentle shaking

↓

2.3 Set up the transfer:
Place a piece of glass across a larger glass tray
Cover with 2 sheets of Whatman 3MM chromatography paper that hang over and touch the bottom of the tray
Wet the papers with 10X SSC
Roll a pipette over it to remove air bubbles

↓

2.4 Place the gel upside down on the chromatography paper
Remove air bubbles
Cover parts of the Whatman paper not covered with the gel with aluminum foil

↓

2.5 Cut a nylon membrane to the size of the gel
Wet it with 2X SSC
Place it on top of the gel and remove air bubbles

↓

2.6 Cut 2 pieces of Whatman 3MM paper the same size as the membrane
Wet thoroughly in 2X SSC
Place on top of the membrane and remove air bubbles

↓

2.7 Cut a stack of paper towels 4-5 inches high the same size as the membrane
Place on top of the Whatman paper

↓

2.8 Place a piece of plexiglass on top of the paper towels
Cover the entire transfer setup with plastic wrap with a 500 g weight on top
Allow the DNA to transfer out of the gel overnight (18-20 h)

↓

2.9 Disassemble the transfer setup
Invert the gel with the membrane on a dry surface

↓

2.10 Mark the positions of the wells on the membrane with a pencil

↓

2.11 Immerse membrane in 6X SSC
Soak for 5 min at RT with gentle shaking

↓

2.12 Expose membrane to UV light (302 nm) for 2-3 min using a UV crosslinker

Figure 5.3 Flowchart of Step 2.

7.3. Caution

Consult your institute Radiation Safety Officer for proper ordering, handling, and disposal of radioactive materials.

3A.1 Thaw the [α-^{32}P]-dCTP behind proper shielding.

3A.2 Add 25–50 ng of linear probe DNA to purified water in a total volume of 9 μl.

3A.3 Heat the tube in a boiling water bath (>95 °C) for 10 min. Immediately place the tube on ice.

3A.4 Briefly spin the tube of denatured DNA and add in the following order:
3 μl dNTP stock mix
2 μl reaction mixture (containing random hexanucleotides)
5 μl 50 μCi [α-^{32}P]-dCTP
1 μl Klenow polymerase

3A.5 Mix the reaction by pipetting. Centrifuge the tube briefly.

3A.6 Incubate the reaction at 37 °C for 30 min.

3A.7 Add 2 μl of 0.2 M EDTA, pH 8.0 or heat at 65 °C for 10 min.

3A.8 Pass the reaction mix through a QuickSpin Sephadex G-50 column to remove unincorporated nucleotides.

See Fig. 5.4 for the flowchart of Step 3A.

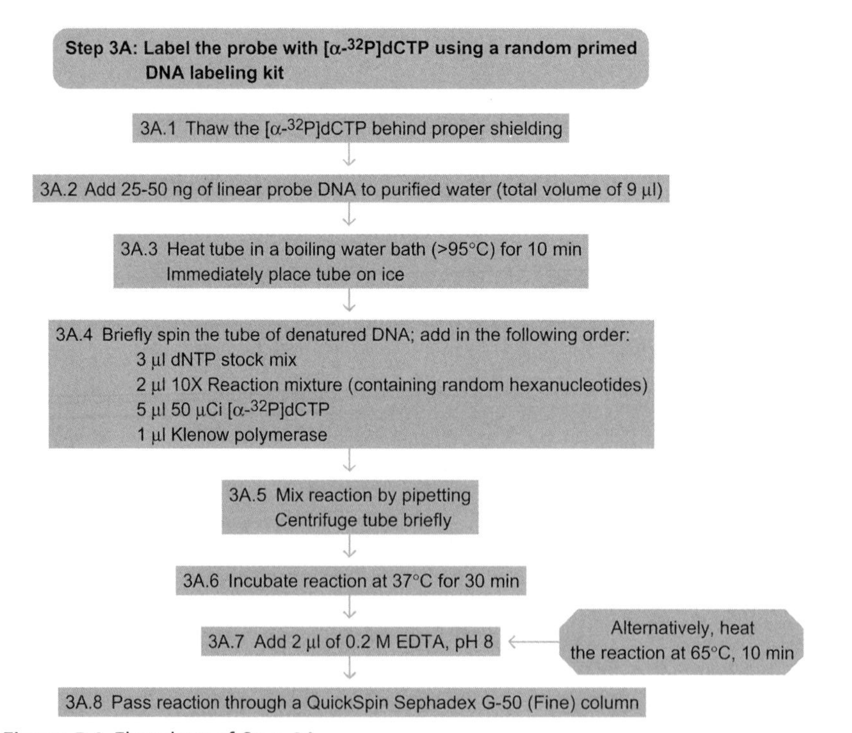

Figure 5.4 Flowchart of Step 3A.

8. STEP 3B LABEL THE PROBE WITH DIGOXIGENIN-11-dUTP FOR NONRADIOACTIVE DETECTION

8.1. Overview

The probe will be labeled with digoxigenin-11-dUTP for nonradioactive detection using a random primed DNA labeling kit (e.g., DIG-High Prime, Roche).

8.2. Duration

Overnight

3B.1 Add 1 μg of linearized probe DNA to purified water in a total volume of 16 μl in a 1.5-ml microcentrifuge tube.

3B.2 Heat the DNA in a boiling water bath for 10 min to completely denature the DNA. Immediately place the tube on ice.

3B.3 Add 4 μl of DIG-labeling mix (5× concentration).

3B.4 Incubate the reaction at 37 °C, overnight.

3B.5 Add 2 μl of 0.2 M EDTA or heat the tube at 65 °C for 10 min to stop the reaction.

8.3. Tip

The 5× DIG-labeling mix contains a mixture of deoxynucleotides plus digoxigenin-11-dUTP, 5× reaction buffer with random hexanucleotides, and Klenow polymerase.

8.4. Tip

The duration of the labeling reaction should be from 1 to 20 h. The longer incubation time will increase the yield of Dig-labeled probe DNA.

See Fig. 5.5 for the flowchart of Step 3B.

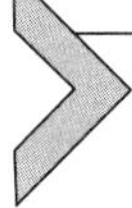

9. STEP 4 HYBRIDIZE THE LABELED PROBE TO THE MEMBRANE

9.1. Overview

The labeled probe will be hybridized to the denatured DNA on the membrane. After hybridization, the membrane will be washed to reduce non-specific background binding.

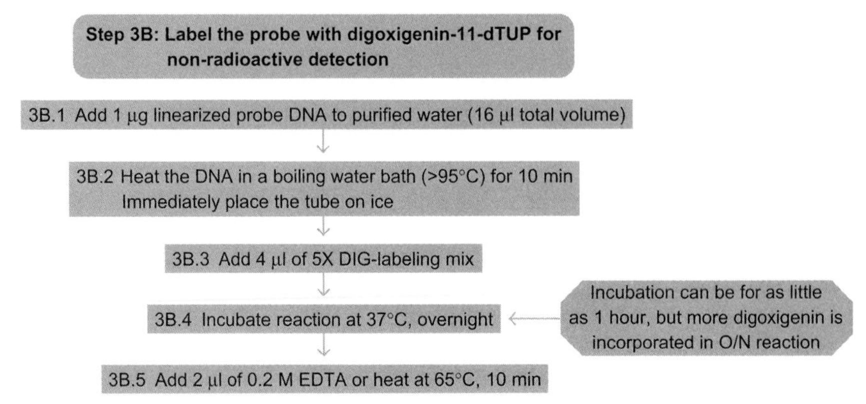

Figure 5.5 Flowchart of Step 3B.

9.2. Duration

Overnight + 2 h

4.1 Place the membrane in a hybridization bottle containing 0.1–0.2 ml prehybridization buffer per square centimeter of membrane.

4.2 Incubate at 42 °C for 1 h, rotating in a hybridization oven.

4.3 Denature the probe at 95 °C for 10 min. Immediately place the tube on ice.

4.4 Add the denatured probe to the prehybridization buffer.

4.5 Hybridize it to the membrane at 42 °C, overnight.

4.6 Remove the membrane from the hybridization bottle and place it in a tray with 100 ml Wash Solution #1. Discard the hybridization solution (in radioactive waste, if appropriate, otherwise, in hazardous waste due to the formamide).

4.7 Incubate with shaking for 15 min at room temperature. Discard the waste (as appropriate for radioactive waste).

4.8 Repeat wash using Wash Solution #1.

4.9 Wash the membrane with 100 ml Wash Solution #2 at 65 °C for 30 min, with shaking. Discard wash solution.

4.10 Repeat wash using Wash Solution #2.

4.11 Wash the membrane with 6× SSC for 5 min at room temperature.

9.3. Tip

Do not add the denatured probe directly on top of the membrane. Instead add it to the prehybridization buffer and mix it before letting it contact the membrane.

See Fig. 5.6 for the flowchart of Step 4.

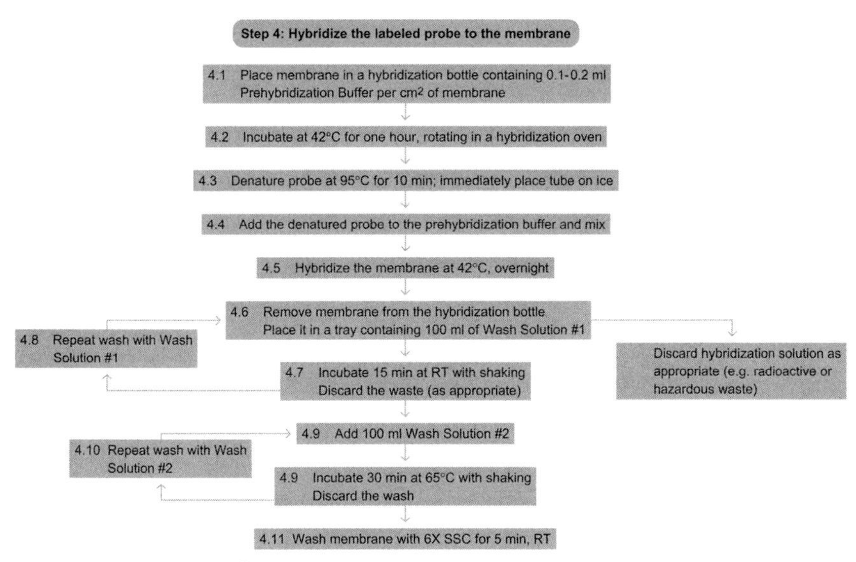

Figure 5.6 Flowchart of Step 4.

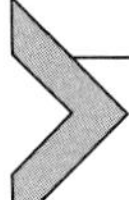

10. STEP 5 DETECT THE LOCATION OF THE HYBRIDIZED PROBE

10.1. Overview

The radiolabeled probe will be detected by autoradiography. The hybridized digoxigenin-labeled probe will be detected using an alkaline phosphatase-conjugated antibody that recognizes digoxigenin. Reactivity will be visualized by colorimetric assay using NBT/BCIP as a substrate.

10.2. Duration

2 h – overnight, depending on the strength of the signal

5.1 For detecting a radiolabeled probe, wrap the membrane in plastic wrap, and expose to X-ray film or a phosphor-capture screen.

5.2 Develop the film or scan the phosphor-capture screen to see the bands. Fig. 5.7 shows an autoradiograph of a Southern blot probed with a different probe in each of the six panels.

5.3 For detecting a digoxigenin-labeled probe, rinse the membrane in Washing Buffer. Dilute the 10× Blocking Reagent 1:10 in Maleic Acid Buffer to give a 1× Blocking Solution.

5.4 Incubate the membrane in 100 ml 1× Blocking Solution for 30 min at room temperature with gentle shaking.

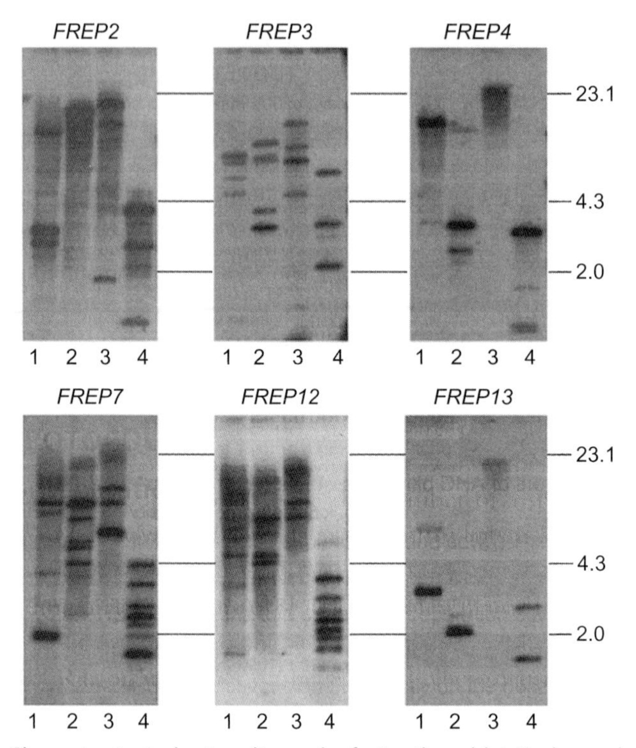

Figure 5.7 Shown is a typical autoradiograph of a Southern blot. Each panel was hybridized with a different probe. **Reprinted with permission from Zhang and Loker (2004).

5.5 Dilute the anti-digoxigenin-AP conjugated antibody 1:5000 in 20 ml 1× Blocking Solution (4 μl in 20 ml).

5.6 Incubate the membrane in the diluted antibody conjugate for 30 min at room temperature with gentle shaking.

5.7 Wash twice with 100 ml Washing Buffer for 15 min each wash with gentle shaking.

5.8 Equilibrate the membrane in 20-ml Detection Buffer for 2–5 min.

5.9 Add 200 μl of NBT/BCIP stock solution to 10-ml Detection Buffer in a foil-wrapped tube.

5.10 Incubate the membrane in Detection Buffer containing the NBT/BCIP substrate in the dark without shaking. Check periodically to monitor color development.

5.11 When it reaches the desired color, place the membrane in 50 ml purified water to stop the reaction.

5.12 Photograph or photocopy the membrane to record the results.

See Fig. 5.8 for the flowchart of Step 5.

Step 5: Detect the location of the hybridized probe

For radiolabeled probes:

5.1 Wrap the membrane in plastic wrap Expose it X-ray film or a phosphor capture screen

↓

5.2 Develop the film or scan the phosphor capture screen

For digoxigenin-labeled probes:

5.3 Dilute 10X Blocking Reagent in maleic acid buffer to give a 1X Blocking Solution

↓

5.4 Incubate the membrane in 100 ml 1X Blocking Solution, 30 min, RT, with gentle shaking

↓

5.5 Dilute 4 μl anti-digoxigenin-AP conjugate in 20 ml 1X Blocking Solution

↓

5.6 Incubate membrane in diluted antibody for 30 min, RT, with gentle shaking

↓

5.7 Wash membrane twice with 100 ml Washing Buffer for 15 min (each wash) at RT with gentle shaking

↓

5.8 Equilibrate membrane in 20 ml Detection Buffer for 2-5 minutes

↓

5.9 Add 200 μl NBT/BCIP stock solution to 10 ml Detection Buffer in a foil-wrapped tube

↓

5.10 Incubate membrane in Detection Buffer + NBT/BCIP in the dark without shaking Check periodically to monitor color development

↓

5.11 When it reaches the desired color, place membrane in 50 ml purified water

↓

5.12 Photograph or photocopy the wet membrane

Figure 5.8 Flowchart of Step 5.

REFERENCES

Referenced Literature

Cate, R. L., Ehrenfels, C. W., Wysk, M., et al. (1991). Genomic Southern analysis with alkaline-phosphatase-conjugated oligonucleotide probes and the chemiluminescent substrate AMPPD. *Genetic Analysis, Techniques and Applications*, *8*, 102–106.

Engler-Blum, G., Meier, M., Frank, J., & Muller, G. A. (1993). Reduction of background problems in nonradioactive northern and Southern blot analyses enables higher sensitivity than ^{32}P-based hybridizations. *Analytical Biochemistry*, *210*, 234–244.

Southern, E. M. (1975). Detection of specific sequences among DNA fragments separated by gel electrophoresis. *Journal of Molecular Biology*, *98*, 503–517.

Zhang, S. M., & Loker, E. S. (2004). Representation of an immune responsive gene family encoding fibrinogen-related proteins in the freshwater mollusc *Biomphalaria glabrata*, an intermediate host for *Schistosoma mansoni*. *Gene*, *341*, 255–266.

Referenced Protocols in Methods Navigator

Agarose Gel Electrophoresis.

CHAPTER SIX

Purification of DNA Oligos by Denaturing Polyacrylamide Gel Electrophoresis (PAGE)

Sara Lopez-Gomollon[1], Francisco Esteban Nicolas
University of East Anglia, School of Biological Sciences, Norwich, United Kingdom
[1]Corresponding author: e-mail address: s.lopez-gomollon@uea.ac.uk

Contents

1. Theory 66
2. Equipment 69
3. Materials 70
   3.1 Solutions & buffers 71
4. Protocol 72
   4.1 Preparation 72
   4.2 Duration 72
5. Step 1 Butanol Precipitation of Crude Oligonucleotide 72
   5.1 Overview 72
   5.2 Duration 73
   5.3 Tip 74
   5.4 Tip 74
6. Step 2 Preparative Polyacrylamide Gel Electrophoresis 75
   6.1 Overview 75
   6.2 Duration 75
   6.3 Tip 76
   6.4 Tip 76
   6.5 Tip 76
   6.6 Tip 76
7. Step 3 Isolation of Oligonucleotides from Polyacrylamide Gels 76
   7.1 Overview 76
   7.2 Duration 77
   7.3 Caution 77
   7.4 Tip 77
   7.5 Tip 78
8. Step 4A Elution of Oligonucleotide from Polyacrylamide Gel by Diffusion 78
   8.1 Overview 78
   8.2 Duration 78
   8.3 Tip 79

Methods in Enzymology, Volume 529
ISSN 0076-6879
http://dx.doi.org/10.1016/B978-0-12-418687-3.00006-9

9. Step 4B Elution of Oligonucleotide from Polyacrylamide Gel by Electroelution 80
 9.1 Overview 80
 9.2 Duration 80
 9.3 Tip 80
 9.4 Tip 80
 9.5 Tip 81
 9.6 Tip 81
10. Step 5 Ethanol Precipitation 82
 10.1 Overview 82
 10.2 Duration 82
 10.3 Tip 82
References 83

Abstract

After chemical synthesis, the oligonucleotide preparation contains the desired full-length oligonucleotide but also all of the DNA molecules that were aborted during each cycle in the synthesis, and the by-products generated during the chemical reactions. The purification of oligonucleotides is a critical step for demanding applications where the exact length or sequence of the oligonucleotide is important, or for oligonucleotides longer than 50 bases. There are several methods of increasing oligonucleotide purity, the choice of which will depend on modifications of the oligonucleotides and their intended use. Polyacrylamide gel purification (PAGE purification) is the method of choice when the highest percentage of full-length oligonucleotide is desired. This chapter describes a protocol for oligonucleotide purification using denaturing polyacrylamide gel electrophoresis, and includes oligonucleotide preparation, polyacrylamide gel electrophoresis, and purification from the gel slice by two different methods: by diffusion or by electroelution. This chapter also includes recommendations as well as protocol advice.

1. THEORY

Oligonucleotides are synthesized by the addition of one base at a time, extending from the 3′ to the 5′-end, attached to a solid-phase support. When the chemical synthesis of the oligonucleotide is complete, the DNA chain is released from the solid support by incubation with ammonium hydroxide. Ammonium also acts as a deprotective agent, cleaving off the groups protecting the phosphates in the phosphodiester bonds. After this, hot ammonia is added to hydrolyze the protecting groups from the exocyclic amino groups of deoxyadenosine, deoxycytosine, and deoxyguanosine. The oligonucleotide could be supplied to the user in the ammonia solution as a crude preparation. In this case, before use, it may be necessary to carry out an additional step to remove the salts and by-products from the oligonucleotide preparation generated during deprotection.

After deprotection, the oligonucleotide preparation is usually desalted, quantified, evaporated to dryness in a lyophilizer, and supplied to the user in the form of a powder. The desalted oligonucleotide solution contains the full-length oligonucleotide but also contains a mixture of truncated products that were accumulated during the chemical synthesis. Truncation occurs when the specified base fails to add to the chain in the corresponding cycle (Hecker & Rill, 1998). Thus, the percentage of truncated products accumulated in the preparation is determined by the synthesis efficiency (the fraction of full-length product after synthesis) and the length of the molecule. Long oligonucleotides, which require more cycles to be synthesized, are less pure than short oligonucleotides. These failures may compete with the full-length product in some applications and may need removal before the oligonucleotide can be used. However, desalted oligonucleotide preparations are usually suitable for many applications, such as standard PCR or microarrays, especially if the oligonucleotide is <20 nucleotides in length.

Further purification for oligonucleotides longer than 50 bases is recommended because the percentage of full-length product in the preparation is typically no higher than 80%. For demanding applications where the exact length of the oligonucleotide is important, such as gel-shift assays, site-directed mutagenesis, sequencing, production of cloning adapters, or first-strand cDNA synthesis for the generation of libraries, additional purification is recommended, even for shorter oligonucleotides (see more information on the techniques above on Standard in vitro Assays for Protein-Nucleic Acid Interactions - Gel shift Assays for RNA and DNA Binding and Site-Directed Mutagenesis).

Purification may be required after synthesis or after enzymatic modification of the oligonucleotide. There are several methods for achieving this, and the choice depends on the nature of the oligonucleotide and the application that it is intended for.

Reverse-phase HPLC purification is based on hydrophobicity. It uses a nonpolar stationary phase, and aqueous buffers and organic solvents as the mobile phase to elute the analytes; polar compounds are eluted first, while nonpolar compounds are retained. This is the best method for purifying oligonucleotides modified with a hydrophobic group, such as biotin, fluorescent dye labels, or NHS-ester conjugations. The mass recovery is higher than when using PAGE purification and it is amenable to the purification of greater amounts of oligonucleotides (mmole scale), although it does not remove incomplete products so effectively. Oligonucleotide purification cartridges, based on reverse-phase chromatography, provide a fast method of desalting and purifying oligonucleotides.

Polyacrylamide gel electrophoresis (PAGE) resolves oligonucleotides according to their length and thereby offers enough resolution to differentiate full-length products from failed molecules (Atkinson and Smith, 1984). Polyacrylamide gels are more effective for separating small fragments of DNA than agarose gels (see Analysis of RNA by analytical polyacrylamide gel electrophoresis and Agarose Gel Electrophoresis). The sole disadvantage of polyacrylamide gels is that they are more difficult to prepare and handle than agarose gels. Polyacrylamide gels are run in a vertical configuration in a constant electric field. After electrophoresis, the oligonucleotide is eluted from the gel and concentrated, using ethanol precipitation (find more on precipitation of nucleic acids on RNA purification – precipitation methods) or reverse-phase chromatography. This is the method of choice when the highest percentage of full-length oligonucleotides is desired for demanding applications. It is also strongly recommended for oligonucleotides longer than 30 bases. Also, several samples can be run simultaneously and no expensive equipment is required. The only drawback of this technique is the small amount of oligonucleotide obtained per purification. Typically, oligonucleotides are used at the scale of 1 μmol or less, and the mass recovery after PAGE purification is <50% and even lower in the case of modified oligonucleotides. Nevertheless, although the yield is low, it is satisfactory for most biochemical applications.

The oligonucleotide preparation is loaded onto a denaturing polyacrylamide gel at an amount of at least 1 mg per lane. The range of resolution of the gel depends on the concentration of polyacrylamide. For short oligonucleotides (from 15 to 35 bases), 13–15% polyacrylamide gels are recommended; for longer oligonucleotides (from 35 to 70 bases), 8–13% polyacrylamide gels are recommended. The percentage of the gel should be adapted to the running system. We usually use the Bio-Rad Mini Protean II system and run 15% gels to purify oligonucleotides of 25 bases in length. After identification of the correct band by UV visualization, the band is excised and extracted from the gel.

Two different methods of purifying oligonucleotides from polyacrylamide gels are described in this chapter. The simplest method is elution by means of diffusion. This is based on the motion of molecules from a region of higher concentration to one of a lower concentration. The result of diffusion is a gradual mixing of material. This method is time consuming but it does not require too much labor or any equipment. Basically, the polyacrylamide band containing the oligonucleotide of interest is sliced into small pieces and covered with elution buffer. After incubation, the DNA is purified from the supernatant by ethanol precipitation. The yield is limited, between 20% and 70%, depending on the length of the oligonucleotide. The greater the amount of elution buffer used, the more oligonucleotide is diffused from the gel.

The second method is electroelution into a dialysis bag (McDonell et al., 1977). The gel piece containing the oligonucleotide is excised and placed into a dialysis bag with a buffer. The application of an electric current causes the DNA to migrate out of the gel into the dialysis bag buffer. The oligonucleotide is recovered from this buffer and purified. Using this method, the oligonucleotide can be isolated with a good yield, although it is time consuming if a large number of samples are purified at the same time, because it requires the insertion of each gel band into individual dialysis bags. This method also works well for larger DNA fragments and can even be used to extract DNA from agarose gels.

2. EQUIPMENT

Chemical fume hood
Microcentrifuge
Speed-vac
Vortex mixer
Heating block
Spectrophotometer
Rocking shaker
Power supply
UV lamp or transilluminator
PAGE gel system (e.g., Bio-Rad Mini Protean II system (Bio-Rad)), including glass plates, comb, and spacers
Horizontal electrophoresis system
Micropipettors
Micropipettor tips
Gloves
2.0-ml microcentrifuge tubes
1.5-ml microcentrifuge tubes
0.5-ml microcentrifuge tubes
Costar® Spin-X® centrifuge tube filter (Corning)
Scalpel/razor blade

Plastic wrap
Syringe
20-gauge needle
21-gauge needle
Dialysis tubing
Dialysis tubing clips

3. MATERIALS

n-Butanol
Ethidium bromide
Sodium acetate (NaOAc)
Ethanol
GlycoBlue™ (Ambion)
Molecular weight marker (e.g., 10 bp DNA ladder, invitrogen)
Tris base
Boric acid (H_3BO_3)
EDTA, disodium
Bromophenol blue
Xylene cyanol
Formamide
Ammonium acetate (NH_4OAc)
Magnesium acetate (MgOAc)
40% acrylamide:bisacrylamide (19:1)
Urea
Ammonium persulfate (APS)
N,*N*,*N*′,*N*′-tetramethylethylenediamine (TEMED)
Ultrapure water (e.g., purified through a Milli-Q system)

Caution	*N*-Butanol is irritating to the mucous membranes, upper respiratory tract, skin and especially the eyes. Avoid breathing the vapors. Wear appropriate gloves and safety glasses. Use in a chemical fume hood. Keep away from heat, sparks, and open flames.
	Ethidium bromide is a powerful mutagen. Consult label for specific handling and disposal procedures. Avoid breathing the dust. Wear appropriate gloves when working with solutions that contain this molecule.
	Formamide is teratogenic. The vapor is irritating to the eyes, skin, mucous membranes, and the upper respiratory tract. It may be harmful

(*Continued*)

by inhalation, ingestion, or skin absorption. Wear appropriate gloves and safety glasses. Always use in a chemical fume hood when working with concentrated solutions of formamide. Keep working solutions covered as much as possible.

Unpolymerized acrylamide is a potent neurotoxin and is absorbed through the skin. Polyacrylamide is considered to be nontoxic but it should be handled with care because it might contain small quantities of unpolymerized acrylamide.

APS and TEMED are extremely destructive to tissue of the mucous membranes and upper respiratory tract, eyes, and the skin. Wear appropriate gloves, glasses, and protective clothing. Always use in a chemical fume hood.

3.1. Solutions & buffers

Step 2 5×TBE

Component	Final concentration	Stock	Amount
Tris base	890 mM		54 g
Boric acid	890 mM		27.5 g
EDTA, pH 8.0	20 mM	0.5 M	20 ml

Add water to 1 l

1×TBE

Add 200 ml of 5×TBE to 800 ml water

2×Formamide loading buffer

Component	Final concentration	Amount
Formamide	95%	95 ml
H_2O		5 ml
EDTA	5 mM	2.7 g
Bromophenol blue	0.1%	1 g
Xylene cyanol	0.1%	1 g

Ethidium bromide (10 mg ml^{-1})
Dissolve 10 mg of ethidium bromide in 1 ml water

Step 4.A Elution buffer

Component	Final concentration	Amount
Ammonium acetate	0.5 M	3.85 g
Magnesium acetate	10 mM	0.14 g
H_2O		100 ml

Step 4.B 0.25×TBE

Add 50 ml of 5×TBE to 800 ml water

Step 5 3M NaOAc, pH5.2

Dissolve 12.3 g of sodium acetate in 40 ml of water. Adjust pH to 5.2. Add water to a volume of 50 ml

75% ethanol

Mix 75 ml ethanol with 25 ml water

4. PROTOCOL

4.1. Preparation

Order or synthesize the desired oligonucleotide.

4.2. Duration

Preparation	1–2 days (variable)
Protocol	About 8–10 h

See Fig. 6.1 for the flowchart of the complete protocol.

5. STEP 1 BUTANOL PRECIPITATION OF CRUDE OLIGONUCLEOTIDE

5.1. Overview

When the oligonucleotide is supplied as a crude oligonucleotide, it may be necessary to perform a butanol precipitation. This procedure removes ammonia and the by-products produced during deprotection (Sawadogo

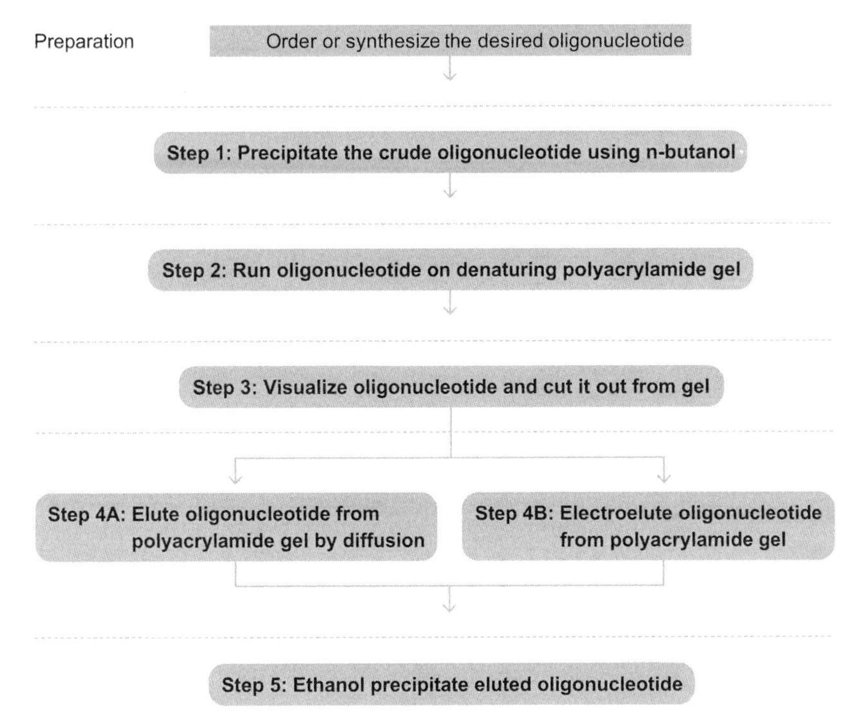
Preparation
Order or synthesize the desired oligonucleotide
Step 1: Precipitate the crude oligonucleotide using n-butanol
Step 2: Run oligonucleotide on denaturing polyacrylamide gel
Step 3: Visualize oligonucleotide and cut it out from gel
Step 4A: Elute oligonucleotide from polyacrylamide gel by diffusion
Step 4B: Electroelute oligonucleotide from polyacrylamide gel
Step 5: Ethanol precipitate eluted oligonucleotide

Figure 6.1 Flowchart of the complete protocol, including preparation.

and Van Dyke, 1991). When opening a tube of crude oligonucleotide for the first time, spin the sample and vent the tube in a chemical fume hood by opening it slowly to allow the ammonia gas to escape.

5.2. Duration

About 4 h

1.1 Dissolve the lyophilized deprotected oligonucleotide in 100 µl of ultrapure water to give a 10 µM solution. Mix by vortexing.

1.2 Centrifuge at 10 000 × g for 5 min at room temperature and transfer the supernatant to a new 1.5-ml microcentrifuge tube.

1.3 Extract the oligonucleotide by adding 1 ml of *n*-butanol (not water saturated). Mix by vortexing and centrifuge at 12 000 × g for 1 min at room temperature.

1.4 Remove the water-containing butanol and discard it. Redissolve the pellet in 100 µl ultrapure water.

1.5 Repeat the *n*-butanol extraction.

1.6 Evaporate the solution to dryness in a speed-vac. The pellet, if it is visible, should be yellowish.

1.7 Resuspend the pellet in ultrapure water and quantify the oligonucleotide. Dilute 2 μl of the oligonucleotide into 200 μl water to estimate the concentration of the oligonucleotide by measuring the absorbance at 260 nm, or use 1 μl in a NanoDrop spectrophotometer (see Explanatory Chapter: Nucleic Acid Concentration Determination).

5.3. Tip

If the oligonucleotide is supplied as an ammonia solution, it can be evaporated to dryness in a speed-vac at room temperature. The ammonia will also be removed during butanol precipitation, so another possibility is to do the protocol without lyophilization, starting at Step 1.3.

5.4. Tip

When initially dissolving the crude oligonucleotide, the solution may turn slightly cloudy after vortexing because of the presence of insoluble benzamides.

See Fig. 6.2 for the flowchart of Step 1.

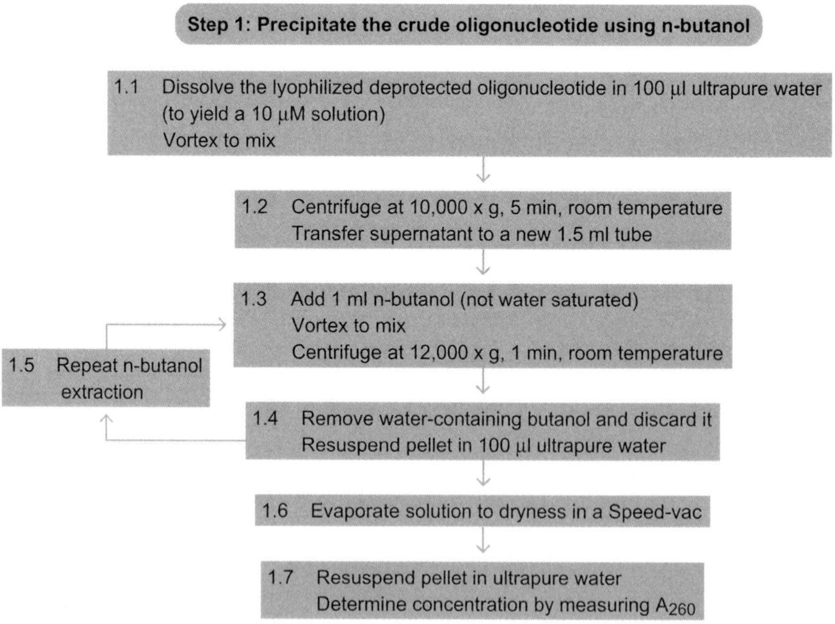

Step 1: Precipitate the crude oligonucleotide using n-butanol

1.1 Dissolve the lyophilized deprotected oligonucleotide in 100 μl ultrapure water (to yield a 10 μM solution)
Vortex to mix

1.2 Centrifuge at 10,000 x g, 5 min, room temperature
Transfer supernatant to a new 1.5 ml tube

1.3 Add 1 ml n-butanol (not water saturated)
Vortex to mix
Centrifuge at 12,000 x g, 1 min, room temperature

1.4 Remove water-containing butanol and discard it
Resuspend pellet in 100 μl ultrapure water

1.5 Repeat n-butanol extraction

1.6 Evaporate solution to dryness in a Speed-vac

1.7 Resuspend pellet in ultrapure water
Determine concentration by measuring A_{260}

Figure 6.2 Flowchart of Step 1.

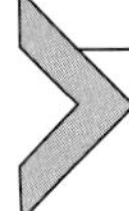

6. STEP 2 PREPARATIVE POLYACRYLAMIDE GEL ELECTROPHORESIS

6.1. Overview

Oligonucleotides are heat denatured and loaded into a preparative denaturing polyacrylamide gel (for more information, see RNA purification by preparative polyacrylamide gel electrophoresis) to separate the full-length oligonucleotide from the truncated products produced as impurities during the chemical synthesis of the oligonucleotide.

6.2. Duration

About 2 h

2.1 Prepare a 15% denaturing polyacrylamide gel:

Component	Final concentration	Stock	Amount
Urea	0.42 g ml^{-1}		2.1 g
H_2O			1.25 ml
TBE	0.5×	5×	500 µl
Acrylamide–bisacrylamide (19:1)	15%	40%	1.87 ml
APS	0.1%	10%	50 µl
TEMED	0.01×		5 µl

Dissolve the urea in the water by heating in a microwave oven, but do not let it boil. When the urea is completely dissolved, add the 5×TBE, and acrylamide–bisacrylamide; swirl to mix. Add sterile water to a final volume of 5 ml. Just before pouring the gel, add 10% APS and TEMED, mix and pour the gel immediately into the cassette made with the two glass plates and the spacers, and then insert the comb. Allow gel to polymerize for about 30 min; 5 ml is enough to prepare one gel for the Bio-Rad Mini Protean II system.

2.2 Prepare the sample by adding an equal volume of 2×formamide loading buffer. Mix well by vortexing and incubate for 5 min at 65 °C to denature the DNA. Keep on ice until ready to load the gel. Load at least 1 mg of oligonucleotide per lane.

2.3 Place the gel into the tank. Be sure that the short plate faces toward the inner reservoir.

2.4 Add 1×TBE buffer to both reservoirs covering the wells, and rinse out the wells with a syringe and a 20-gauge needle.

2.5 Prerun the gel in 1×TBE at 100 V for 30 min. Rinse out the wells again with a syringe and a 20-gauge needle.

2.6 Load the samples.

2.7 Run the gel at 200 V until the bromophenol blue is two-thirds down the gel.

2.8 Pry apart the glass plates and stain the gel for 5 min in 10 μg ml^{-1} of ethidium bromide in 1×TBE.

6.3. Tip

Try to keep the volume of the sample loaded onto the gel down to a minimum (10 μl or less) to obtain a better resolution. If necessary, dry down the oligonucleotide in a speed-vac and resuspend in ultrapure water. Determine the concentration of the oligonucleotide by measuring the absorbance at 260 nm.

6.4. Tip

In a 15% polyacrylamide gel, bromophenol blue runs around 15 nucleotides and xylene cyanol around 60 nucleotides. If the oligonucleotide migrates at the same rate as the tracking dyes, this will interfere with detection of the oligonucleotide by UV. The loading buffer can be prepared without the tracking dyes, and instead, load 5 μl of formamide tracking dye mixture into an unused well. Alternatively, the tracking dyes can be substituted by 0.2% Orange G, which migrates with the buffer front and does not interfere with UV detection.

6.5. Tip

During the prerunning, the ammonium persulfate migrates from the wells and the gel will become warmed up, which helps to keep the DNA denatured during electrophoresis, resulting in better resolution.

6.6. Tip

Load a molecular weight marker into a separate lane and keep an empty lane between the marker and the oligonucleotide to avoid contamination.

See Fig. 6.3 for the flowchart of Step 2.

7. STEP 3 ISOLATION OF OLIGONUCLEOTIDES FROM POLYACRYLAMIDE GELS

7.1. Overview

After the oligonucleotide samples are run on a polyacrylamide gel, the fragment of interest is identified and cut out from the gel.

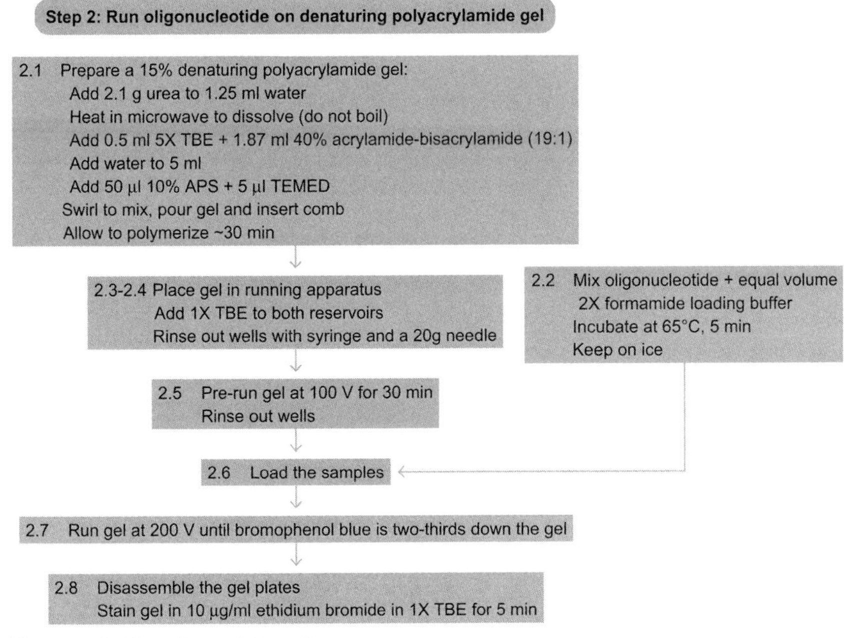

Figure 6.3 Flowchart of Step 2.

7.2. Duration

10 min

3.1 Place the gel on a piece of plastic wrap.

3.2 Examine the gel using a UV lamp. Identify the band of interest, which should be the slowest migrating band.

3.3 Cut out the region of the gel containing the full-length oligonucleotide. Use a clean, sharp scalpel or razor blade. Keep the size of the polyacrylamide slice as small as possible to avoid contamination.

7.3. Caution

Protect yourself from exposure to UV by wearing face and eye protection, as well as a lab coat.

7.4. Tip

Preparations of oligonucleotides that are up to 20 nucleotides in length generally contain only one band because the efficiency is more than 99%. Long oligonucleotides are less pure than short oligonucleotides, so the preparations of very long oligonucleotides may contain several bands of prematurely truncated products. The slowest migrating

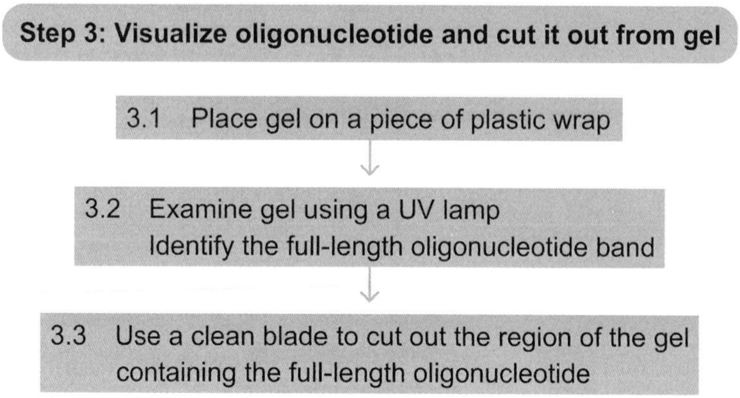

Step 3: Visualize oligonucleotide and cut it out from gel

3.1 Place gel on a piece of plastic wrap

↓

3.2 Examine gel using a UV lamp
Identify the full-length oligonucleotide band

↓

3.3 Use a clean blade to cut out the region of the gel containing the full-length oligonucleotide

Figure 6.4 Flowchart of Step 3.

band usually corresponds to the desired product. Very rarely, however, the oligonucleotide may migrate not according to its molecular weight.

7.5. Tip

If there is not enough oligonucleotide to be detected by UV, consider radiolabeling a portion of the sample by phosphorylating it to use as a marker.

See Fig. 6.4 for the flowchart of Step 3.

8. STEP 4A ELUTION OF OLIGONUCLEOTIDE FROM POLYACRYLAMIDE GEL BY DIFFUSION

8.1. Overview

After the oligonucleotide of interest is identified and isolated from the gel, the band is sliced into small pieces and the oligonucleotide is recovered from the gel by diffusion.

8.2. Duration

3 h

4A.1 Puncture the bottom of a sterile 0.5-ml microcentrifuge tube 4–5 times using a 21-gauge needle.

4A.2 Place the polyacrylamide gel fragment into the 0.5-ml microcentrifuge tube.

4A.3 Place the 0.5-ml microcentrifuge tube into a sterile 2-ml microcentrifuge tube.

4A.4 Centrifuge the stacked tubes at 10 000 × g in a microcentrifuge for 2 min at room temperature to crush the gel by forcing it through the holes.

4A.5 Add 300 μl of elution buffer to the gel debris and elute the DNA by rocking the tube gently at room temperature for 2–3 h.

4A.6 Transfer the eluate and the gel debris to a Costar Spin-X filter.

4A.7 Centrifuge the filters at 10 000 × g for 2 min at room temperature.

4A.8 Add 100 μl of elution buffer to the gel debris in the filter. Centrifuge at 10 000 × g for 2 min to collect the extra eluate in the same collection tube.

8.3. Tip

It is also possible to grind up the gel band in a tissue homogenizer for 10 min after addition of the elution buffer, or with a disposable pipette tip using a circular motion and pressing the fragments of the gel against the sides of the tube.

See Fig. 6.5 for the flowchart of Step 4A.

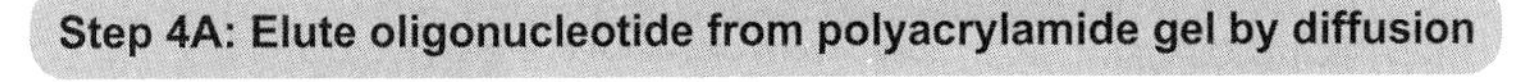

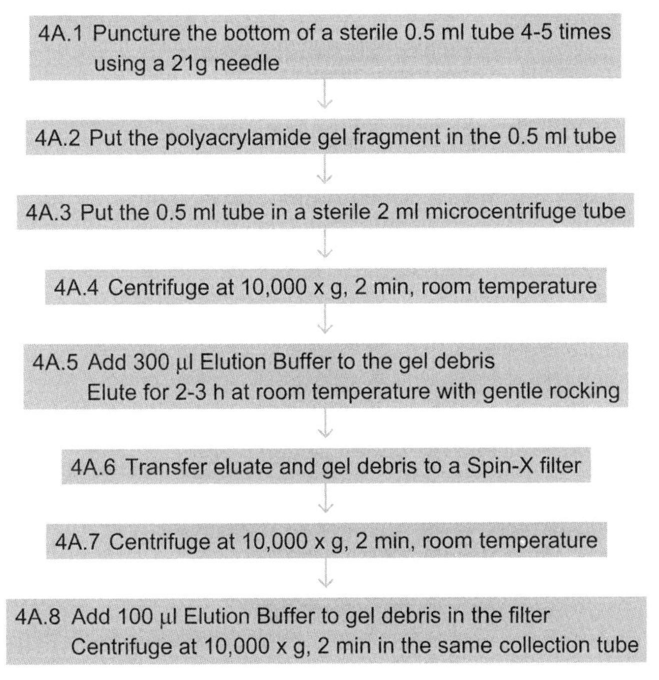

Figure 6.5 Flowchart of Step 4A.

9. STEP 4B ELUTION OF OLIGONUCLEOTIDE FROM POLYACRYLAMIDE GEL BY ELECTROELUTION

9.1. Overview

After the oligonucleotide of interest is identified and isolated from the gel, the band is placed in a small dialysis tube containing TBE buffer. The tube is placed in a horizontal electrophoresis tank and the DNA is electroeluted out of the gel band.

9.2. Duration

About 2 h

4B.1 Seal one end of a piece of dialysis tubing with a secure knot. Fill the dialysis bag with 0.25×TBE and transfer the gel slice into it.

4B.2 Squeeze out most of the buffer, leaving just enough to keep the gel slice in constant contact with the buffer.

4B.3 Place a dialysis tubing clip above the gel slice to seal the bag. Seal above the clip with a secure knot.

4B.4 Immerse the bag in a shallow layer of 0.25×TBE buffer in a horizontal electrophoresis tank. Prevent the dialysis bag from floating and keep the gel fragment parallel to the electrodes.

4B.5 Apply an electric current through the bag at 7.5 V cm^{-1} for 45–60 min.

4B.6 Reverse the polarity of the electric current for 20 s to release the DNA from the wall of the dialysis bag.

4B.7 Recover the bag from the electrophoresis tank and gently mix the eluted DNA into the buffer.

4B.8 Transfer the buffer to a 1.5-ml microcentrifuge tube. Wash out the empty bag with a small volume of 0.25×TBE and transfer to the same tube.

9.3. Tip

Wear gloves when handling the dialysis tubing.

9.4. Tip

Avoid trapping air bubbles or clipping the gel slice itself.

9.5. Tip

Under these conditions, the yield of a DNA fragment 0.1–2.0 kb in length is typically about 85%. The electroelution conditions should be adapted to each situation. If the electroelution is too short, the DNA will remain in the gel slice and the yield will be low. If the electroelution is too long, the DNA will become attached to the wall of the dialysis tubing.

9.6. Tip

To confirm that the oligonucleotide has been eluted, remove the gel slice from the bag and stain it with ethidium bromide (10 µg ml^{-1}) in 0.25× TBE buffer for 20 min and visualize by UV illumination.

See Fig. 6.6 for the flowchart of Step 4B.

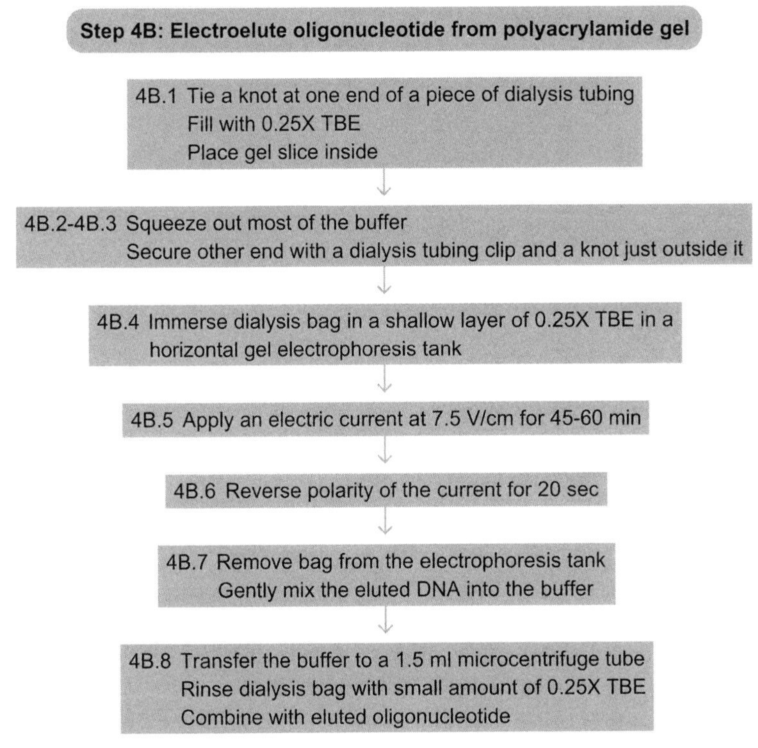

Figure 6.6 Flowchart of Step 4B.

10. STEP 5 ETHANOL PRECIPITATION

10.1. Overview

After elution of the oligonucleotide from the gel slice, it may be concentrated by ethanol precipitation. It is also possible to purify it using commercial resins.

10.2. Duration

1 h

5.1 Add 1/10 volume of 3 M NaOAc, pH 5.2, and 2.5 volumes 100% ethanol. Incubate at −20 °C for 30 min.

5.2 Centrifuge at 6000 × g for 30 min at 4 °C.

5.3 Remove the supernatant. Wash the pellet by adding 500 μl of 70% ethanol.

5.4 Centrifuge at 10 000 × g for 5 min at room temperature.

5.5 Remove the supernatant. Air dry for 5 min and resuspend in ultrapure water.

10.3. Tip

Add 3 μl of GlycoBlue™ to increase the size and visibility of the DNA pellet. It consists of a blue dye covalently linked to glycogen, a branched chain carbohydrate.

See Fig. 6.7 for the flowchart of Step 5.

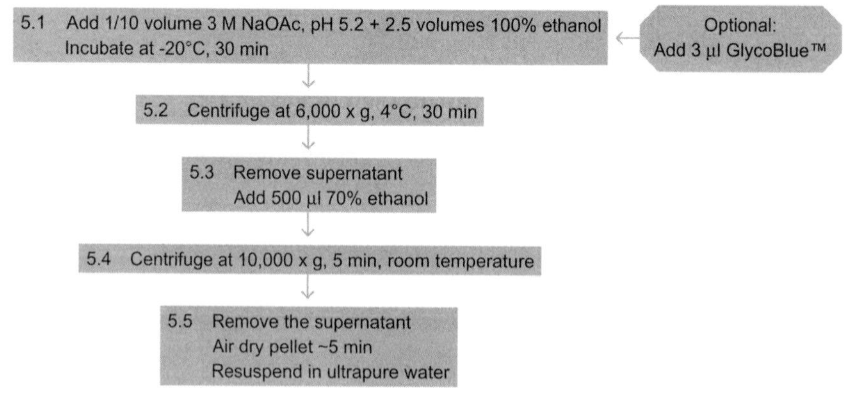

Figure 6.7 Flowchart of Step 5.

REFERENCES

Referenced Literature

Atkinson, T., & Smith, M. (1984). Solid-phase synthesis of oligodeoxyribonucleotides by the phosphite-triester method. In M. J. Cait (Ed.), *Oligonucleotide Synthesis: A Practical Approach* (pp. 35–82). Oxford, UK: IRL Press.

Hecker, K. H., & Rill, R. L. (1998). Error analysis of chemically synthesized polynucleotides. *BioTechniques*, *24*(2), 256–260.

McDonell, M. W., Simon, M. N., & Studier, E. W. (1977). Analysis of restriction fragments of T7 DNA and determination of molecular weights by electrophoresis in neutral and alkaline gels. *Journal of Molecular Biology*, *110*, 119–146.

Sawadogo, M., & Van Dyke, M. W. (1991). A rapid method for the purification of deprotected oligodeoxynucleotides. *Nucleic Acids Research*, *19*(3), 674.

Related Literature

Sambrook, J., & Russell, D. W. (2001). *Molecular Cloning: A Laboratory Manual*. Cold Spring Harbor, NY: Cold Spring Harbor lab Press.

www.idtdna.com.

www.invitrogen.com.

Referenced Protocols in Methods Navigator

Standard in vitro Assays for Protein-Nucleic Acid Interactions - Gel shift Assays for RNA and DNA Binding.

Site-Directed Mutagenesis.

Analysis of RNA by analytical polyacrylamide gel electrophoresis.

Agarose Gel Electrophoresis.

RNA purification – precipitation methods.

Explanatory Chapter: Nucleic Acid Concentration Determination.

RNA purification by preparative polyacrylamide gel electrophoresis.

CHAPTER SEVEN

Molecular Cloning

Juliane C. Lessard[1]

Department of Biochemistry and Molecular Biology, Johns Hopkins School of Public Health, Baltimore, MD, USA

[1]Corresponding author: e-mail address: jkellne2@jhmi.edu

Contents

1. Theory 86
2. Equipment 88
3. Materials 88
   3.1 Solutions & Buffers 89
4. Protocol 90
   4.1 Preparation 90
   4.2 Duration 91
5. Step 1 Restriction Digests of Vector and Insert 91
   5.1 Overview 91
   5.2 Duration 91
   5.3 Tip 92
   5.4 Tip 92
   5.5 Tip 92
   5.6 Tip 93
6. Step 2 Ligation 93
   6.1 Overview 93
   6.2 Duration 94
   6.3 Tip 94
   6.4 Tip 94
   6.5 Tip 94
7. Step 3 Transformation into Chemically Competent *E. coli* 94
   7.1 Overview 94
   7.2 Duration 95
   7.3 Tip 95
   7.4 Tip 96
   7.5 Tip 96
   7.6 Tip 96
   7.7 Tip 96
8. Step 4 Identify Successful Ligation Events 96
   8.1 Overview 96
   8.2 Duration 96
   8.3 Tip 97
   8.4 Tip 97
References 98

Methods in Enzymology, Volume 529
ISSN 0076-6879
http://dx.doi.org/10.1016/B978-0-12-418687-3.00007-0

Abstract

This protocol describes the basic steps involved in conventional plasmid-based cloning. The goals are to insert a DNA fragment of interest into a receiving vector plasmid, transform the plasmid into *E. coli*, recover the plasmid DNA, and check for correct insertion events.

1. THEORY

Molecular cloning is an essential technique to create DNA-based experimental tools for expression in bacterial or mammalian cells. Examples of such DNA constructs include a promoter element fused to a reporter gene or a cDNA sequence under the control of a ubiquitous promoter. Molecular cloning entails the preparation of the vector and insert DNAs, ligation of the insert into the vector, transformation of competent *E. coli*, and identification of positive clones (Fig. 7.1).

Traditionally, molecular cloning is defined as the isolation and amplification of a specific DNA fragment. Most of these fragments are created either by digesting an existing piece of DNA with restriction enzymes or by targeting it via PCR. Short inserts of ~100 bp can also be commercially synthesized as complementary single-stranded oligos, which are subsequently annealed to form a double-stranded fragment.

After successful isolation, the DNA of interest is ligated into a vector plasmid, a double-stranded circular piece of DNA that can be propagated in *E. coli*. Vectors used in the laboratory represent a smaller version of naturally occurring plasmids that include several basic features: a replication origin, a drug-resistance gene, and unique restriction sites to facilitate the insertion of DNA fragments. Often, several different restriction sites are clustered together in so-called 'polylinker regions' or 'multiple cloning sites,' making it easier to choose convenient and unique restriction enzyme combinations for a variety of inserts.

The choice of restriction enzymes is critical when designing a cloning strategy. While some sever the double-stranded DNA in one place, creating 'blunt' ends, others leave an overhang of a few bases at the cut site. These complementary 'sticky' ends find one another easily, increasing the efficiency of the ligation reaction and thus the chances for a successful cloning event. Thoughtful combination of restriction enzymes can also help to control the directionality of the insert, which is critical to many applications.

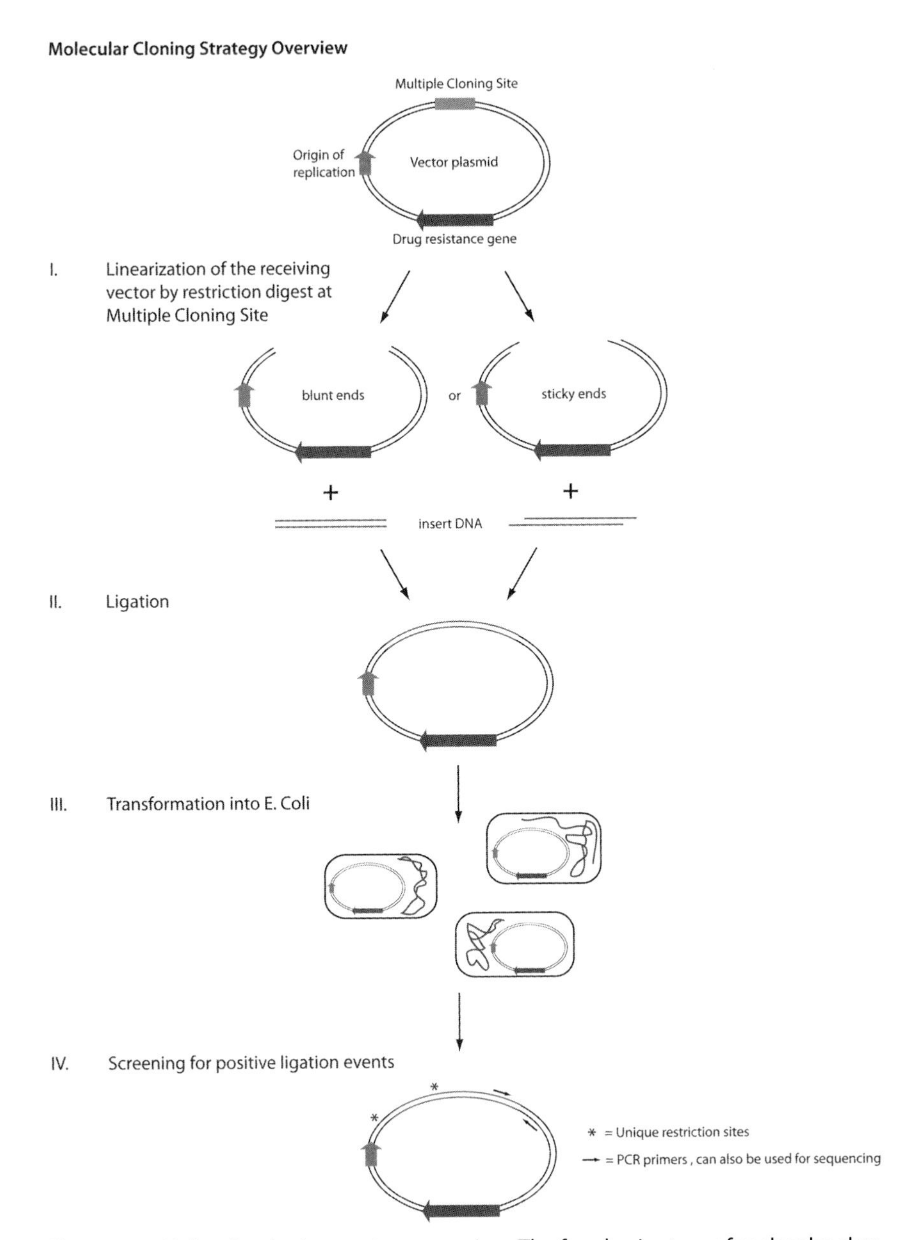

Figure 7.1 Molecular cloning strategy overview. The four basic steps of molecular cloning are outlined.

2. EQUIPMENT

Microcentrifuge
Shaking incubator (37 °C)
Incubator (37 °C)
Heating block or water bath (42 °C)
Heating block (65 °C) (optional)
Bunsen burner
agarose gel electrophoresis equipment
1.5-ml microcentrifuge tubes
15-ml polypropylene tubes
Micropipettors
Micropipettor tips (autoclaved)

3. MATERIALS

Vector DNA (receiving vector, insert vector or insert PCR product)
Insert DNA
Restriction enzymes
10× Restriction enzyme buffers
10× BSA
Agarose
Tris base
EDTA
Glacial acetic acid
Ficoll
Bromophenol blue
Xylene cyanol
Ethidium bromide
Sodium chloride (NaCl)
Bacto tryptone
Yeast extract
Bacto/Difco agar
10× DNA T4 ligase buffer
T4 DNA ligase
Chemically competent *E. coli*
Antibiotic (e.g., ampicillin, kanamycin)

DNA Gel purification kit
PCR purification kit
optional:
10× Restriction enzyme buffer #2 (NEB2)
2.5 mM dNTPs
Klenow fragment
0.5 M EDTA

3.1. Solutions & Buffers

Step 1 0.5 M EDTA, pH 8.0

Dissolve 14.6 g EDTA (MW 292.24 g mol^{-1}) in 95 ml water. Adjust pH to 8.0 using NaOH and add water to 100 ml.

50× TAE

Component	Final concentration	Stock	Amount
Tris base	2 M	N/A	242 g
EDTA, pH 8.0	1×	0.5 M	100 ml
Glacial acetic acid	N/A	N/A	57.1 ml

Dissolve Tris in ~750 ml of deionized water. Add EDTA and acetic acid. Adjust the final volume to 1000 ml with water. There is no need to adjust the pH of this solution.

6× DNA loading dye

Component	Final concentration	Stock	Amount
Ficoll	15% (w/v)	N/A	1.5 ml
Bromophenol blue	0.25% (w/v)	N/A	25 mg
Xylene cyanol FF	0.25% (w/v)	N/A	25 mg

Add water to 10 ml

Step 3 LB medium

Component	Amount
NaCl	10 g
Bacto tryptone	10 g
Yeast extract	5 g

Add water to 1 l and autoclave. For LB plates, add 15 g Bacto/Difco agar, autoclave and cool to ~55 °C before adding the antibiotic and pouring the plates

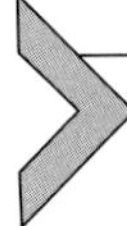

4. PROTOCOL

4.1. Preparation

Ensure that good-quality DNA preps are available for the receiving vector and any vectors used to retrieve the insert (see Isolation of plasmid DNA from bacteria).

Identify the restriction enzymes needed and their buffer requirements for single and double digests.

For PCR product inserts:

Add the sequence of any desired restriction site(s) to the 5′ end of each primer (see Explanatory chapter: PCR -Primer design). It is a good idea to precede these sites with a few random nucleotides, such as GATC, since some restriction enzymes can cut only at sites surrounded by double-stranded DNA. Note that these regions should be excluded from annealing temperature calculations since they do not contain any homology to the target sequence (Fig. 7.2).

Once you obtain the primers, run the PCR reaction and gel-purify the product, eluting in 30 μl of ddH_2O.

Design one or more strategies to test for vectors that carry the correct insert (in the appropriate orientation).

Examples:

1. Identify restriction enzymes that cut inside (and outside) of the new insert, producing distinct bands.
2. Design PCR primers, one outside and one inside the insert, to use for colony PCR.
3. Identify or design primers that can be used to sequence all or part of the insert.

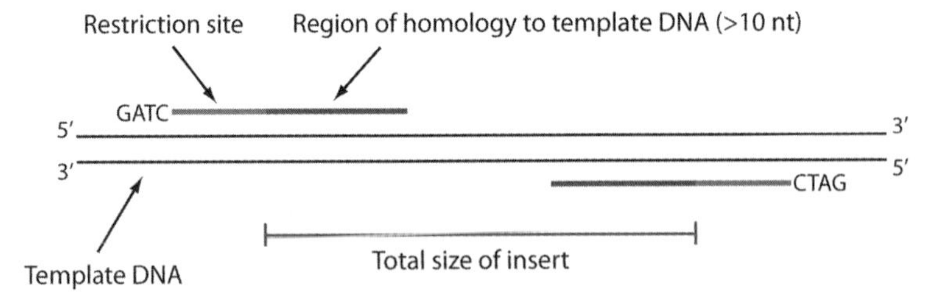

Figure 7.2 Primer design for molecular cloning. The schematic shows the elements of PCR primers to amplify a fragment of DNA to be cloned.

4.2. Duration

Preparation	variable
Protocol	4 days

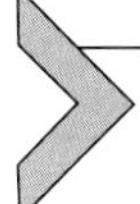

5. STEP 1 RESTRICTION DIGESTS OF VECTOR AND INSERT

5.1. Overview

Linearize the receiving vector by restriction digest and cut out the insert from another vector or digest a gel-purified PCR product to create sticky ends.

If using DNA oligos, anneal the two single-stranded pieces to form a double-stranded insert with sticky ends according to the manufacturer's directions and go directly to Step 2.

5.2. Duration

2.5–3 h

1.1 Set up the restriction digests for both the vector and the insert in 1.5-ml microcentrifuge tubes:

DNA	3 μg for vector DNA, 30 μl for PCR product
10× μ Buffer	5 μl
10× μ BSA	5 μl
Enzyme 1	1 μl
Enzyme 2	1 μl (if required)
ddH2O	to 50 μl

1.2 Incubate at 37 °C for 1 h.

1.3 Pour a 0.8% agarose gel in 1× TAE buffer containing ethidium bromide (see Agarose Gel Electrophoresis).

1.4 Add 10 μl of 6× DNA gel loading dye and load onto the agarose gel. Run the gel at 100 V for 30–45 min.

1.5 Gel-purify the receiving vector and the insert, using a DNA Gel Purification kit. This purification step is essential as it removes any enzymatic activity that might interfere with the ligation reaction. When running the gel, include an uncut vector control to confirm that the digests are complete. Any residual uncut plasmid will produce a high background of negative colonies.

1.6 **Optional:** To create blunt ends, fill in any overhangs left by restriction enzymes. Set up the following reaction in a 1.5-ml microcentrifuge tube:

DNA	2.5 μg
10 ×μ NEB Buffer 2	2.5 μl
dNTPs (2.5 mM)	1 μl
Klenow fragment (5000 U μl μ1)	0.5 μl
ddH_2O	to 25 μl

1.7 Incubate at room temperature for 30 min.
1.8 Stop the reaction with 1 μl of 0.5 M EDTA.
1.9 Purify using a PCR purification kit or by gel-extraction.

5.3. Tip

To help prevent unwanted self-ligation of the vector, add 1 μl of a phosphatase, for example, Calf Intestinal Phosphatase (CIP) or antarctic phosphatase, to the vector digest ONLY. This is especially important when using just one enzyme to create sticky ends or enzymes that create blunt ends.

5.4. Tip

Some restriction enzyme combinations require a sequential digest. If this is the case, add the first enzyme for 1 h, then heat-inactivate it by incubating the reaction at 65 °C for 20 min. Add the second enzyme (adjusting the buffer conditions if necessary) and incubate for another hour at 37 °C.

5.5. Tip

If the insert is a purified PCR product, gel purification is not necessary at this point. Heat-inactivation of the enzyme(s) through incubation at 65 °C for 20 min is sufficient. However, using a PCR purification kit will improve the quality of the insert DNA by removing all enzymatic activity and is highly recommended.

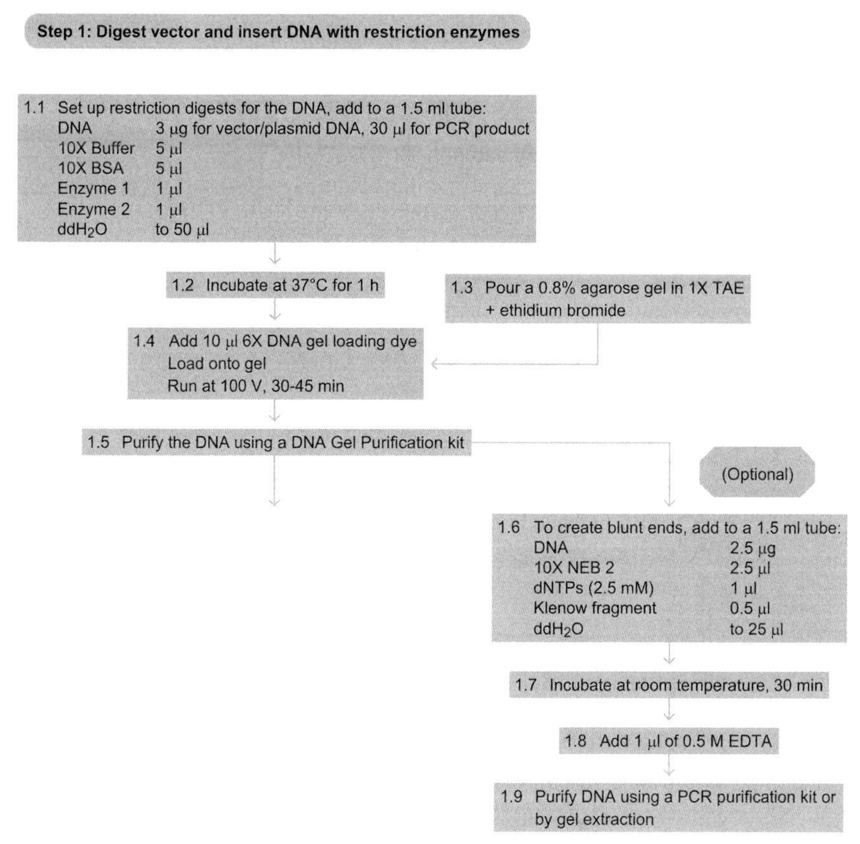

Figure 7.3 Flowchart of Step 1.

5.6. Tip

After digesting and gel purifying the DNA, run 2–5 μl on an agarose gel to verify recovery and determine the amounts to use in the ligation.

See Fig. 7.3 for the flowchart of Step 1.

6. STEP 2 LIGATION

6.1. Overview

Ligate the insert into the linearized receiving vector.

6.2. Duration

Overnight

2.1 Set up the ligation reactions, including a negative control reaction using ddH_2O instead of the insert DNA. Add to a 1.5-ml microcentrifuge tube:

Vector DNA	100 ng
Insert DNA	≥400 ng
10× T4 DNA ligase buffer	2 μl
T4 DNA ligase	1 μl
ddH_2O	to 20 μl

2.2 Incubate overnight at room temperature.

2.3 Place the ligation at −20 °C until ready to transform competent *E. coli*.

6.3. Tip

After thawing, T4 ligase buffer often still contains small white precipitates. Prior to adding the buffer to the ligation reaction, vortex it thoroughly until the solution is completely clear. Incomplete resuspension may alter the concentration of buffer ingredients and interfere with the ligation reaction.

6.4. Tip

For sticky end ligations, a T4 ligase stock of 1 U μl^{-1} *usually works well. For more complicated ligations, including blunt-end ligations, use a more concentrated T4 ligase at 400 U* μl^{-1}.

6.5. Tip

The overnight ligation described here usually produces reliable and efficient results, especially for difficult (including blunt-end) ligations. However, manufacturer-recommended incubation times and temperatures for their ligases vary widely and most eliminate the overnight step. In general, increasing the incubation time will increase the efficiency of the reaction.

See Fig. 7.4 for the flowchart of Step 2.

7. STEP 3 TRANSFORMATION INTO CHEMICALLY COMPETENT *E. COLI*

7.1. Overview

The ligated plasmid is transferred into chemically competent *E. coli* via heat-shock (see Transformation of Chemically Competent *E. coli*, or alternatively

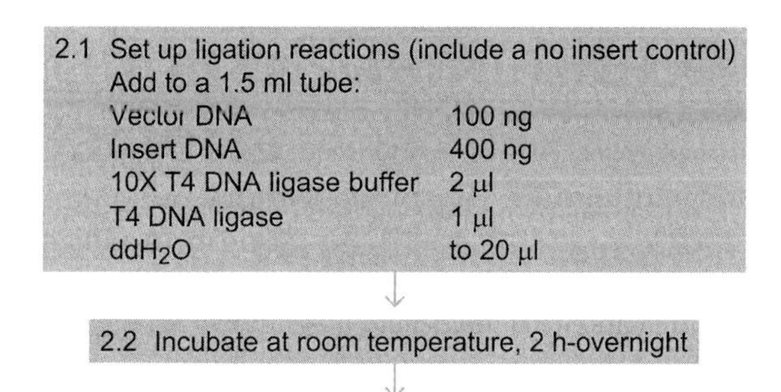

Figure 7.4 Flowchart of Step 2.

see Transformation of *E. coli* via electroporation if you are using electrocompetent cells) to select against linear (nonligated) pieces of DNA. The transformed bacteria are then grown on media containing an antibiotic corresponding to the resistance gene in the receiving vector.

7.2. Duration

2 h + overnight incubation

3.1 Let competent cells thaw on ice (~30 min).

3.2 Add 20 μl of the ligation reaction and or negative control to about 200 μl of cells (the ratio of the ligation reaction to *E. coli* should be at least 1:10).

3.3 Mix gently to avoid damage to the cells (do not pipette up and down).

3.4 Incubate on ice for 30 min.

3.5 Heat shock at 42 °C for 1 min.

3.6 Immediately add 1 ml of room temperature LB (or other medium, e.g., SOC).

3.7 Incubate at 37 °C for 1 h with shaking (100–150 rpm).

3.8 Centrifuge the bacteria at a low speed (~7500 rpm) for 2 min.

3.9 Aspirate all but 100 μl of the LB.

3.10 Carefully resuspend the pellet and spread all 100 μl evenly onto a selective plate.

3.11 Incubate overnight at 37 °C with the plates turned upside down.

7.3. Tip

Use autoclaved pipette tips for each step. Alternatively, sterilize tips before touching any bacteria by passing them briefly through a Bunsen burner flame.

7.4. Tip

The amounts of the competent cells and ligation mix can be reduced to 50 and 5 µl, respectively.

7.5. Tip

If time is limited, Steps 3.4 and 3.7 may be shortened to 15 and 30 min respectively without compromising the transformation efficiency too much.

7.6. Tip

During the incubation, let selective plates warm up at 37 °C. Cold plates lead to lower transformation efficiency.

7.7. Tip

Alternately spread the transformed E. coli onto two plates, with 80% on one plate and 20% on the other, to ensure the growth of well isolated colonies.

See Fig. 7.5 for the flowchart of Step 3.

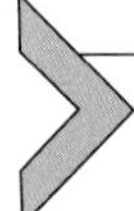

8. STEP 4 IDENTIFY SUCCESSFUL LIGATION EVENTS

8.1. Overview

Colonies appearing on the selective plates after transformation represent individual, unique ligation outcomes. The plasmid DNA is extracted from several colonies and checked for successful ligation events using one or more of the previously determined screening strategies (see Protocol Preparation).

8.2. Duration

Day 1: 5 min or 3–5 h (if colonies are screened by PCR; see Colony PCR).

Day 2: about 1 h for the mini-prep (see Isolation of plasmid DNA from bacteria), otherwise the duration depends on the screening strategy used to identify positive clones.

4.1 Use a sterile pipette tip to pick individual, round, medium-sized colonies from the selective plate containing transformed cells from the ligation reaction with the insert.

4.2 Transfer each colony into a 14-ml polypropylene round-bottom tube containing 3–5 ml LB plus the appropriate antibiotic.

4.3 Incubate overnight at 37 °C, shaking at ~250 rpm. For optimal growth, tilt the tubes slightly in the shaking rack.

4.4 The next day, isolate the plasmid DNA, using a miniprep kit.

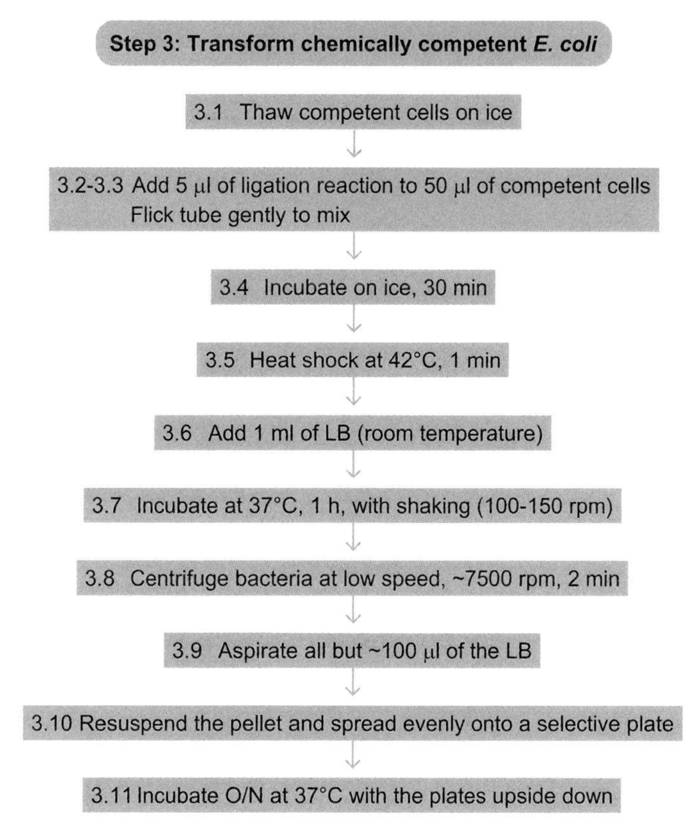

Figure 7.5 Flowchart of Step 3.

4.5 Confirm successful ligation events using one or more of the previously worked out screening strategies.

8.3. Tip

After transformation, the selective plate for the ligation should contain a lot more colonies than the negative control plate. If not, this may be an indication that the restriction digest of the receiving vector was incomplete or that the receiving vector was able to self-ligate too easily. Try increasing the time of the receiving vector digestion and add a phosphatase to limit self-ligation.

8.4. Tip

If primers are available, perform a colony PCR at this point (see Colony PCR). Draw a grid on the back of a selective plate, numbering each field. Pick one colony at a time

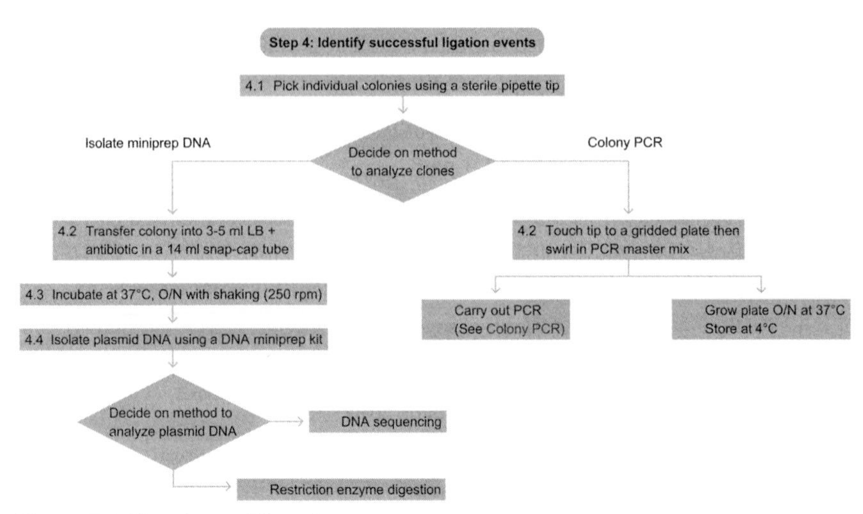

Figure 7.6 Flowchart of Step 4.

using a sterile pipette tip. Lightly touch the selective plate in one of the numbered fields, and then transfer the rest of the colony into a PCR tube with the same number, containing a PCR master mix. Repeat for several colonies. Incubate the plate upside down at 37 °C for a few hours while performing the PCR. Positive clones identified by the colony PCR can be picked directly from the corresponding numbered field on the selective plate.

See Fig. 7.6 for the flowchart of Step 4.

REFERENCES

Allison, L. A. (2010). *Fundamental Molecular Biology,* Chapter 8.

Lodish, H., Berk, A., Kaiser, C. A., Krieger, M., Scott, M. P., Bretscher, A., et al. (2007). In *Molecular Cell Biology* (6th ed.), Chapter 5.

Referenced Protocols in Methods Navigator

Isolation of plasmid DNA from bacteria.
Explanatory chapter: PCR -Primer design.
Agarose Gel Electrophoresis.
Transformation of Chemically Competent *E. coli.*
Transformation of *E. coli* via electroporation.
Colony PCR.

CHAPTER EIGHT

Rapid Creation of Stable Mammalian Cell Lines for Regulated Expression of Proteins Using the Gateway® Recombination Cloning Technology and Flp-In T-REx® Lines

Jessica Spitzer*, Markus Landthaler†, Thomas Tuschl*,[1]

*Howard Hughes Medical Institute, Laboratory of RNA Molecular Biology, The Rockefeller University, New York, NY, USA

†Berlin Institute for Medical Systems Biology, Max-Delbruck-Center for Molecular Medicine, Berlin, Germany

[1]Corresponding author: e-mail address: ttuschl@rockefeller.edu

Contents

1. Theory 100
2. Equipment 104
3. Materials 105
   3.1 Solutions & buffers 108
4. Protocol 110
   4.1 Preparation 110
   4.2 Duration 111
   4.3 Tip 111
5. Step 1 Molecular Cloning of the Gene of Interest into Gateway Expression Vectors 112
   5.1 Overview 112
   5.2 Duration 112
   5.3 Tip 116
6. Step 2 Establishing Stable Cell Lines Expressing the Protein of Interest 118
   6.1 Overview 118
   6.2 Duration 118
   6.3 Tip 122
References 123
Source References 123

Methods in Enzymology, Volume 529
ISSN 0076-6879
http://dx.doi.org/10.1016/B978-0-12-418687-3.00008-2

Abstract

The biochemical analysis of cellular processes in mammalian cells is often facilitated by the creation of cell lines coexpressing or overexpressing an affinity-tagged wild-type or mutant protein of interest in an inducible or noninducible stable manner (Malik and Roeder, 2003). The affinity tag allows for standardization of purification protocols to characterize interacting proteins or nucleic acids and minimizes the need for generating protein-specific antibodies at the early stages of analysis (for more information on affinity tags, see Purification of His-tagged proteins, Affinity purification of a recombinant protein expressed as a fusion with the maltose-binding protein (MBP) tag, Purification of GST-tagged proteins, Protein Affinity Purification using Intein/Chitin Binding Protein Tags, Immunoaffinity purification of proteins or Strep-tagged protein purification). The establishment of stable cell lines with inducible expression is critical to studying proteins that reduce cell growth and/or viability upon overexpression.

Over the past several years, our laboratory has developed an expression platform for analyzing RNA-interacting proteins, including the establishment of stable mammalian cell lines expressing proteins of interest using a recombination-based cloning technology (Landthaler et al., 2008; Hafner et al., 2010). Our aim is to determine the mRNA targets of the hundreds of RNA-binding proteins encoded in the human genome by the isolation and molecular characterization of their ribonucleoprotein complexes (RNPs).

1. THEORY

Invitrogen commercially introduced recombination-based Gateway® Cloning Technology (Walhout et al., 2000). It represents a fast and easy way to transfer a gene of interest (GOI) from one vector into another within a multitude of vectors already available. The reactions are performed in less than 1 h at room temperature without the need for sequencing the new vector.

Here we describe how the process of generating stable cell lines can be shortened to just under a month by adapting conventional cloning-based Flp-In recombination vectors, a separate Invitrogen technology, to the Gateway technology.

Following the recommendations of Malik and Roeder (2003), we use a 17-amino acid N-terminal FLAGHA-double affinity tag. Excellent antibody products are commercially available for these affinity tags. Cell lines expressing FLAGHA-affinity-tagged proteins also represent an excellent source for characterizing antibodies raised against the protein of interest.

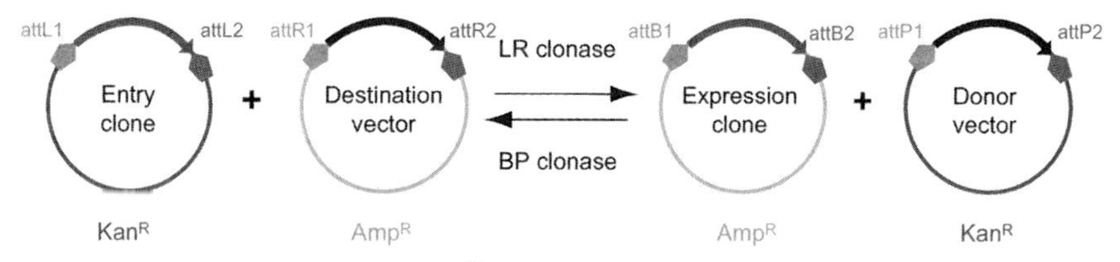

Figure 8.1 The Invitrogen Gateway® Recombination Cloning Technology.

In brief, a Gateway recombination reaction mimics either the excision (LR reaction) or the integration (BP reaction) of phage lambda into or from the *E. coli* genome (Fig. 8.1).

- LR reaction: The phage DNA is excised from the bacterial genome by recombination between the attL and attR sites. This excision reaction requires two vectors, the entry clone (with attL sites) and the destination vector (with attR sites). The LR reaction requires three enzymes, phage protein integrase (Int), bacterial protein integration host factor (IHF), and phage excisionase (Xis). These three enzymes are collectively referred to as 'LR Clonase.' The attL sites give rise to attB sites and the attR sites to attP sites.
- BP reaction: The phage DNA is integrated into the bacterial genome. It requires two vectors, the expression clone with attB sites and the donor vector with attP sites. The BP reaction requires Int and IHF only. These two enzymes are collectively referred to as 'BP Clonase.' Since the attB sites give rise to attL sites and the attP sites to attR sites the cycle completes itself enabling a new round of cloning.

Since both reactions are reversible, this allows for a very flexible approach in shuttling the insert DNA from one expression vector to another.

We adapted two regular Invitrogen vectors pcDNA5/FRT and pcDNA5/FRT/TO for their use as Gateway vectors and added an N-terminal FLAGHA tag, naming the resulting vectors pFRT_DESTFLAGHA and pFRT_TO_DESTFLAGHA (Fig. 8.2). These plasmids were deposited at www.addgene.com, together with many destination vectors our laboratory generated for stably expressing human RNA-binding proteins or components of the RNAi machinery. Plasmids pFRT_DESTFLAGHA_GOI (cDNA of gene of interest) are prepared for stable nonregulated (co-)expression of proteins, whereas the pFRT_TO_DESTFLAGHA_GOI vectors are prepared for inducible expression of otherwise toxic proteins or protein levels. Protein synthesis is induced by addition of tetracycline derivatives to the culture medium of stable GOI-containing cell lines.

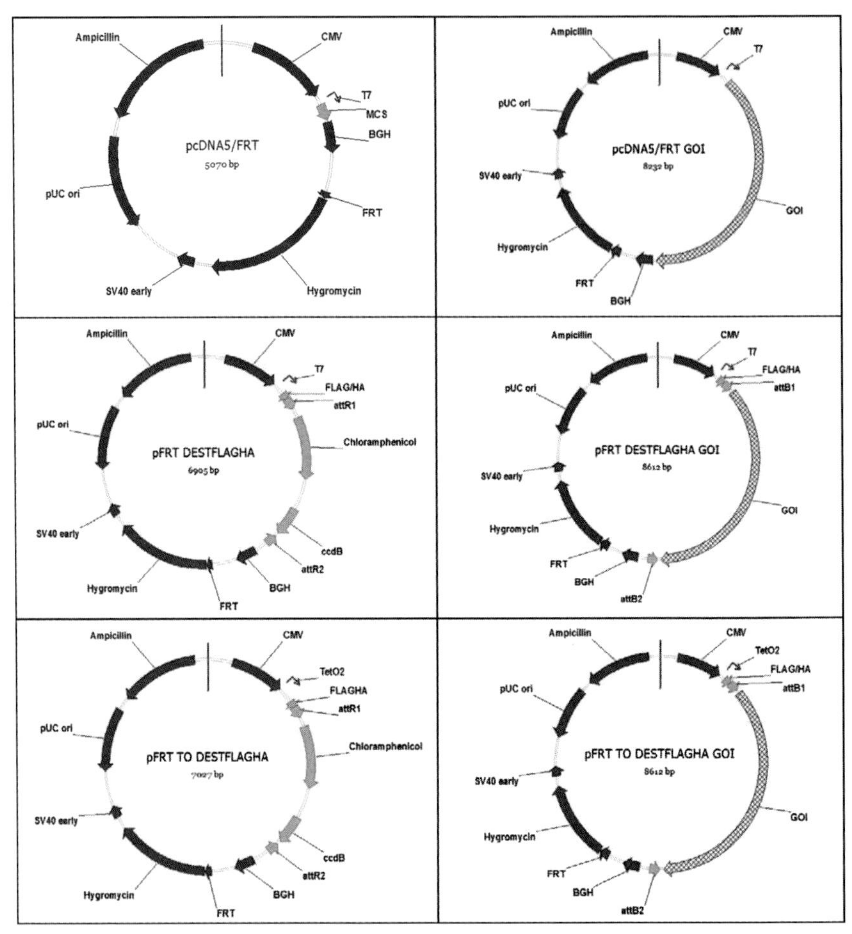

Figure 8.2 Vector maps of pcDNA5/FRT, pcDNA5/FRT_GOI, pFRT_DESTFLAGHA, pFRT_DESTFLAGHA_GOI, pFRT_TO_DESTFLAGHA and pFRT_TO_DESTFLAGHA_GOI vectors. Shown are individual vector properties including replication origins, antibiotic resistance and recombination sites.

For transfer of the GOI into the pFRT vectors, we first establish a Gateway entry vector pENTR_GOI by cloning the cDNA PCR product of the GOI in the correct reading frame into commercial pENTR4, although any of the other available entry vectors can equally be used (Fig. 8.3).

The pENTR4_GOI also allows access to a multitude of different Gateway expression vectors for recombinant protein purification. For bacterial expression (see Small-scale Expression of Proteins in *E. coli*), there is a choice between untagged (pDEST14) and tag-expressing vectors, such as

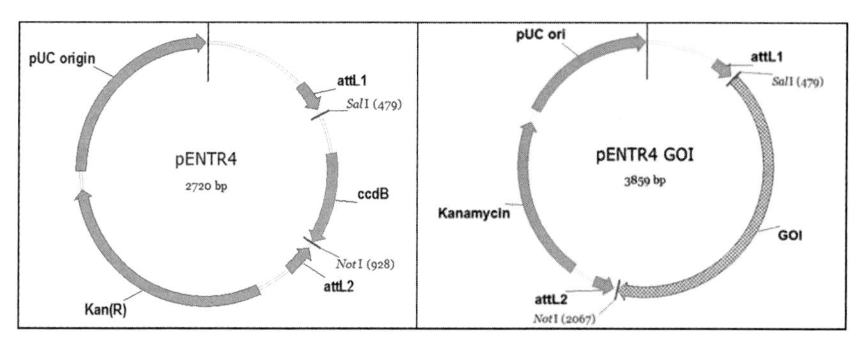

Figure 8.3 pENTR4 and pENTR4_GOI, which was cloned using SalI and NotI restriction sites.

glutathione *S*-transferase (GST, N-terminal, pDEST15; C-terminal, pDEST24) and 6xHis (N-terminal, pDEST17). Analogous vectors exist for baculovirus expression (see Recombinant protein expression in baculovirus-infected insect cells), untagged (pDEST8), N-terminal 6xHis tag (pDEST10), and N-terminal GST (pDEST20). (For purification of these tagged proteins see Purification of GST-tagged proteins or Purification of His-tagged proteins).

Different mammalian Flp-In cell lines are available from Invitrogen, only one of which allows for tetracycline-regulated expression of the GOI:

- Flp-In T-REx 293 Cell Line (human embryonic kidney cell line that also expresses high levels of the Tet repressor protein)
- Flp-In-293 Cell Line (human embryonic kidney cell line)
- Flp-In-CHO Cell Line (Chinese hamster ovary cell line)
- Flp-In-BHK Cell Line (baby hamster kidney cell line)
- Flp-In-CV-1 Cell Line (African Green Monkey kidney cell line)
- Flp-In-3 T3 Cell Line (mouse embryonic fibroblast cell line)
- Flp-In-Jurkat Cell Line (human T lymphocyte cell line)

The Flp-In T-REx 293 cell line contains a single integrated FRT site and was also selected to stably express the Tet repressor protein (TetR). Vectors equipped with two tetracycline operator 2 ($TetO_2$) sites (Hillen and Berens, 1994) in their CMV promoter region (Andersson et al., 1989) are subjected to TetR-mediated repression until TetR is released from the $TetO_2$ sites by the addition of tetracycline (or doxycycline) to the tissue culture medium. Examples of $TetO_2$ site-containing expression vectors (generally characterized by the suffix 'TO') include pcDNA4/TO (Invitrogen) or our pFRT_TO_DESTFLAGHA. To also obtain TetR-expressing stable lines for the Flp-In lines listed above, Invitrogen provides the plasmid pcDNA6/TR, which can be transfected into cells followed by selection of stable clones.

The following additional TetR-expressing cell lines are available for generation of 'Flp-In T-REx' cell lines by transfection of pFRT/lacZeo vector (Invitrogen) and subsequent colony selection and Southern blot screening for FRT site integration (see Invitrogen's user manuals 'pFRT/lacZeo' and 'T-REx System'):

- T-REx-HeLa Cell Line (low levels of TetR expression resulting in limited range of induction)
- T-REx-Jurkat Cell Line
- T-REx-CHO Cell Line

Finally, the integration of the GOI within the Flp-In host cell line is accomplished by cotransfection of the destination vector containing GOI flanked by the FRT sites, pFRT_DESTFLAGHA_GOI or pFRT_TO_DESTFLAGHA_GOI, together with the plasmid pOG44, encoding Flp recombinase, and followed by antibiotic selection of cell clones.

2. EQUIPMENT

Major equipment	365-nm UV-transilluminator
	Agarose gel chambers
	Balances (e.g., 0.1 mg to 64 g and 0.1 g to 4.2 kg)
	Microcentrifuge
	CO_2 incubator (for mammalian cell culture)
	Hemacytometer
	High speed floor centrifuge (capable of at least 13 000 × *g*)
	Incubator (for work with bacteria)
	Shaking incubator (37 °C)
	Microscope with fluorescence filter
	Multichannel pipet
	pH-meter
	Power supply
	PCR thermocycler
	Thermomixer

	Vortex mixer
	Water bath (42 °C)
	Water filter (e.g., MilliQBiocel water purification system)
Consumables	96-well plate, flat bottom
	96-well plate, conical bottom
	24-well plate
	12-well plate
	6-well plate
	10-cm tissue culture plate
	15-cm tissue culture plate
	0.2-ml PCR strips (Thermo Scientific, AB-0266)
	1.5-ml polypropylene tubes
	Cryogenic vials (e.g., Corning 2-ml internal threaded polypropylene cryogenic vial, 430488)
	50-ml conical tubes (e.g., Falcon)
	Scalpels (or razor blades)
	5-μm Supor membrane syringe filter (Pall Acrodisc)
	Syringes

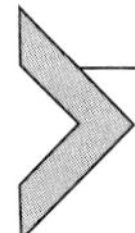

3. MATERIALS

Reagents & chemicals	
	Agarose, electrophoresis grade
	Boric acid (H_3BO_3)
	Bromophenol blue
	Calf intestinal alkaline phosphatase (CIP)
	Dimethyl sulfoxide (DMSO; Sigma, D2438)
	DNA ladder (1 kb)
	EDTA disodium salt dihydrate (Sigma, E5134)
	Ethidium bromide

	Ficoll type 400
	Agar
	Sodium chloride (NaCl)
	Sodium hydroxide (NaOH)
	Phosphate buffered saline (PBS; 10×, commercially available)
	Tris base (Fisher Scientific, BP152-1)
	Tris–HCl (Promega, PR-H5121)
	Tryptone
	Yeast extract
Antibiotics	Ampicillin sodium salt (Sigma, A9518-25G)
	Blasticidin S, cell culture grade (InvivoGen, ant-bl-1)
	Doxycycline, monohydrochloride hemiethanolate hemihydrate (Sigma, D9891-1G)
	Hygromycin B, cell culture grade (InvivoGen, ant-hg-1)
	Kanamycin sulfate (Sigma, K4000-25G)
	Penicillin–Streptomycin, cell culture grade (Invitrogen, 15140-122; 100×, contains 10 000 units of penicillin (base) and 10 000 μg of streptomycin (base)/ml)
	Zeocin, cell culture grade (InvivoGen, ant-zn-1)
Cell culture media	D-MEM, Dulbecco's Modified Eagle Medium high glucose (1×), with L-glutamine, without sodium pyruvate (Invitrogen, 11965-092)
	Fetal bovine serum heat-inactivated (FBS, heat-inactivated)
	Opti-MEM, reduced-serum medium (1×), liquid (with GlutaMAX-I) (Invitrogen, 51985-034)
Commercial kits	QIAfilter Plasmid Midi Kit (Qiagen, 12243)
	QIAprep Spin Miniprep Kit (Qiagen, 27106)
	QIAquick PCR Purification Kit (Qiagen, 28104)
	QIAquick Gel Extraction Kit (Qiagen, 28704)

Competent cells	One Shot® ccdB Survival™ 2 T1R Competent Cells (Invitrogen, A10460)
	One Shot® TOP10 Chemically Competent *E. coli* (Invitrogen, C4040-10)
Vectors	pENTR4 Dual Selection Vector (Invitrogen, A10465)
	pOG44 (part of the Flp-In™ Core System by Invitrogen)
	pFRT_DESTFLAGHA (www.addgene.com)
	pFRT_TO_DESTFLAGHA (www.addgene.com)
	pFRT_DESTFLAGHA_eGFP (www.addgene.com)
Other reagents	AccuPrime Pfx DNA Polymerase (Invitrogen, 12344-024)
	Flp-In™ Core System (Invitrogen, K6010-02)
	Flp-In™ T-REx™ 293 Cell Line (Invitrogen, R780-07)
	Gateway® LR Clonase® II enzyme mix (Invitrogen, 11791-100)
	Lipofectamine 2000 (Invitrogen, 11668-019)
	Protein-specific or anti-HA (Covance, MMS-101P) antibody
	Restriction endonucleases (e.g., SalI and NotI from NEB)
	T4 DNA Ligase (NEB, M0202M)
	Trypsin-EDTA, 0.05% (Gibco, 25300)

DNA oligonucleotides for sequencing	
pENTR4 forward primer	5′ CTGTTAGTTAGTTACTTAAGCTC
pENTR4 reverse primer	5′ AGACACGGGCCAGAGCTG
pFRT_DESTFLAGHA forward primer	5′ CGCAAATGGGCGGTAGGCGTG
pFRT_DESTFLAGHA reverse primer	5′ TAGAAGGCACAGTCGAGG
pFRT_TO_DESTFLAGHA forward primer	Same as pFRT_DESTFLAGHA forward primer
pFRT_TO_DESTFLAGHA reverse primer	Same as pFRT_DESTFLAGHA reverse primer

3.1. Solutions & buffers

Step 1.2.2.1 We use water purified by a Millipore water purification system

DNA loading dye (5×)

Component	Final concentration	Stock	Amount
EDTA–NaOH, pH 8.0	50 mM	0.5 M	1 ml
Bromophenol blue	0.2% (w/v)	n/a	20 mg
Ficoll type 400	20% (w/v)	n/a	2 g
H_2O to 10 ml			

10× TBE

Component	Final concentration	Stock	Amount
Tris base	445 mM	n/a	53.9 g
Boric acid	445 mM	n/a	27.5 g
EDTA–NaOH, pH 8.0	10 mM	0.5 M	20 ml
H_2O to 1 l			

Step 1.5 LB medium

Component	Amount
Yeast extract	5 g
Tryptone	10 g
NaCl	10 g
H_2O to 1 l	

Autoclave for 15 min at 121 °C. Prepare stock solutions of the selection antibiotics (100 mg ml^{-1}) and filter them through a 5-μm syringe filter. Add them fresh before each inoculation to a final concentration of 50 μg ml^{-1} (kanamycin, 1:2000 dilution of stock solution) or 100 μg ml^{-1} (ampicillin, 1:1000 dilution of stock solution).

LB agar plates

Component	Final concentration	Stock	Amount
Yeast extract	n/a	n/a	5 g
Tryptone	n/a	n/a	10 g
NaCl	n/a	n/a	10 g
Agar	n/a	n/a	15 g

NaOH	1 mM	1 M	1 ml
H_2O to 1 l			

Autoclave for 30 min at 121 °C. After cooling it down to 50 °C, add the desired selection antibiotic (for concentrations see above) and pour the plates (about 35 ml per plate).

Step 1.7 TE buffer, pH 8.0 at room temperature

Component	Final concentration	Stock	Amount
Tris–HCl, pH 8.0	10 mM	1 M	5 ml
EDTA–NaOH, pH 8.0	2 mM	0.5 M	2 ml
H_2O			493 ml

Step 2 Maintenance medium for parental Flp-In T-REx HEK 293 cell line

DMEM supplemented with 10% FBS (final concentration), 100 U ml^{-1} of penicillin and 100 μg ml^{-1} streptomycin, 15 μg ml^{-1} blasticidin, and 100 μg ml^{-1} zeocin.

Antibiotic free medium for transfection

DMEM supplemented with 10% FBS (final concentration).

Stable cell line selection medium

D-MEM supplemented with 10% FBS (final concentration), 100 U ml^{-1} of penicillin and 100 μg ml^{-1} streptomycin, 100 μg ml^{-1} hygromycin, and 15 μg ml^{-1} blasticidin. Blasticidin is only added to maintain TetR expression for generating inducible constructs.

Blasticidin	Selects for the TetR gene-expressing cell, only in inducible constructs (stock concentration 10 mg ml^{-1}; diluted to 15 μg ml^{-1} final concentration in medium)
Hygromycin	Selects for the pFRT vector-containing cell (stock concentration 100 mg ml^{-1}; diluted to 100 μg ml^{-1} final concentration in medium)
Zeocin	Selects for the FRT recipient site-containing cell (stock concentration 100 mg ml^{-1}; diluted to 100 μg ml^{-1} final concentration in medium)

Freezing medium for Flp-In T-REx HEK 293 cell lines
90% FBS, 10% DMSO. Cool to 4 °C before using.

1× PBS

Component	Final concentration	Stock	Amount
PBS	1×	10×	100 ml
H_2O			900 ml

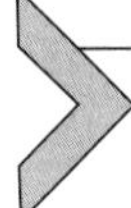

4. PROTOCOL

4.1. Preparation

I. Isolate the following plasmids using a midiprep or maxiprep kit (alternatively see Isolation of plasmid DNA from bacteria). Depending on the number of reactions, you will need 10–50 μg of each of them:

pENTR4 (or any other of the pENTR vector family): entry vector for the Gateway System

pOG44: expression vector for Flp recombinase

pFRT_DESTFLAGHA: plasmid with FRT site, hygromycin resistance, and the N-terminal double-epitope tag (FLAG and HA); receives the cDNA of the gene from the pENTR4 vector using LR recombination sites; used for stable expression of the protein of interest in either Flp-In cell lines or T-REx Flp-In cell lines

pFRT_TO_DESTFLAGHA: plasmid with FRT site, hygromycin resistance, and the N-terminal double-epitope tag (FLAG and HA); receives the cDNA of the gene of interest from the pENTR4 vector using LR recombination sites; used for tetracycline inducible expression of the protein of interest in T-REx Flp-In cell lines

pFRT_DESTFLAGHA_eGFP: plasmid for the stable expression of eGFP and hygromycin resistance; will be used to assess transfection efficiency

II. Maintain Flp-In T-REx HEK 293 cell line:

Culture Flp-In T-REx HEK 293 cells on plates in DMEM with 10% FBS, 1% penicillin/streptomycin, 15 μg ml^{-1} blasticidin, and 100 μg ml^{-1} zeocin. Cells are split 1:5 every 2–4 days.

III. Generate vector maps (e.g., pENTR4_GOI) in a suitable software (Vector NTI or MacVector):

This will simplify the choice of restriction enzymes in Step 1.5 and the sequence analysis in Step 1.6. The programs developed and licensed by Invitrogen are also useful for performing a virtual LR reaction and for a basic analysis of protein properties.

4.2. Duration

Step	Action	Duration
0	Preparation	1 day
1.1	Primer design	2 h + time of primer synthesis
1.2	PCR + PCR product purification	3.5 h
1.3	Restriction digestion (PCR product + pENTR4)	4 h
1.4	Ligation + transformation	3 h
1.5	Inoculation, miniprep and restriction digestion	Overnight + 4 h
1.6	Sequencing + sequence analysis	Overnight + 1 h
1.7	LR reaction + transformation	4 h
1.8	Inoculation, miniprep and restriction digestion	Overnight + 4 h
2.1	Preparation of transfection	1 h + overnight
2.2	Transfection	1 h
2.3–2.8	Colony isolation	12–14 days
2.9	Selection of individual colonies	2 h
2.10–2.11	Expansion of individual clones for cryopreservation	11–14 days
2.12–2.14	Protein detection by Western blotting, cryopreservation of stable cell lines for long-term storage	1–2 days

4.3. Tip

Remember to use ccdB survival cells (One Shot® ccdB Survival™ 2 T1R Competent Cells by Invitrogen) for pENTR4, pFRT_DESTFLAGHA, and pFRT_TO_DESTFLAGHA, since regular bacterial strains such as TOP10 are not suitable

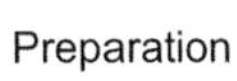

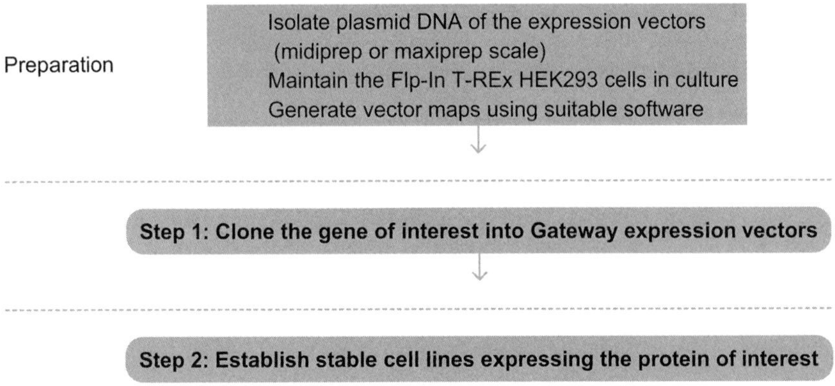

Isolate plasmid DNA of the expression vectors (midiprep or maxiprep scale)
Maintain the Flp-In T-REx HEK293 cells in culture
Generate vector maps using suitable software

Step 1: Clone the gene of interest into Gateway expression vectors

Step 2: Establish stable cell lines expressing the protein of interest

Figure 8.4 Flowchart of the complete protocol, including preparation.

for propagation of plasmids containing the ccdB (coupled cell division B; inhibits cell division) gene.

See Fig. 8.4 for the flowchart of the complete protocol, including preparation.

5. STEP 1 MOLECULAR CLONING OF THE GENE OF INTEREST INTO GATEWAY EXPRESSION VECTORS

5.1. Overview

In the first step the cDNA of the GOI is cloned into the entry vector pENTR4 and subsequently transferred via LR cloning into the destination vector for mammalian expression.

5.2. Duration

Step	Action	Duration
1.1	Primer design	2 h + time of primer synthesis
1.2	PCR + PCR product purification	3.5 h
1.3	Restriction digestion (PCR product + pENTR4)	4 h
1.4	Ligation + transformation	3 h
1.5	Inoculation, miniprep and restriction digestion	Overnight + 4 h
1.6	Sequencing + sequence analysis	Overnight + 1 h
1.7	LR reaction + transformation	4 h
1.8	Inoculation, miniprep and restriction digestion	Overnight + 4 h

1.1 Design primers specific to chosen cDNA of your gene of interest

1.1.1 PCR primers

If you already have the cDNA of your GOI cloned in frame in one of the Invitrogen entry vectors, continue with Step 1.7. If not, look up the multiple cloning site of pENTR4 regarding your choice of restriction endonucleases (http://tools.invitrogen.com/content/sfs/vectors/pentr4_dual_selection_vector_mcs.pdf for cloning your GOI cDNA into this vector. Make sure that your GOI does not contain any internal restriction sites recognized by the chosen restriction endonucleases; use the program supported at http://www.restrictionmapper.org/.

We routinely use SalI/NotI for this purpose. For SalI/NotI cloning, adapt the following PCR primers:

Forward PCR primer:

5′ ACGC**GTCGAC**xxxx

The SalI recognition site is in bold. xxxx represents the first 10–15 bases of your GOI starting from the ATG. The ATG start codon must follow the SalI site (at least in pENTR4) to maintain the reading frame after LR recombination into destination vectors.

Reverse PCR primer

5′ ATAGTTTA**GCGGCCGC**xxxx

The NotI recognition site is in bold. xxxx represents the reverse complement of the last 10–15 bases of your gene of interest, ending with the stop codon.

Refer to the primer design chapter in this book series for general guidelines (see Explanatory chapter: PCR -Primer design). If you choose to use either a different entry vector or a different set of restriction endonucleases make sure to clone your GOI in the correct reading frame.

1.1.2 Sequencing primers

pENTR4 forward primer: 5′ CTGTTAGTTAGTTACTTAA GCTC

pENTR4 reverse primer: 5′ AGACACGGGCCAGAGCTG

pFRT_DESTFLAGHA forward primer: 5′ CGCAAATGGG CGGTAGGCGTG

pFRT_DESTFLAGHA reverse primer: 5′ TAGAAGGCACA GTCGAGG

pFRT_TO_DESTFLAGHA forward primer is the same as pFRT_DESTFLAGHA forward primer pFRT_TO_DESTFLAGHA reverse primer is the same as pFRT_DESTFLAGHA reverse primer

Also design gene-specific internal sequencing primers such that they are about 700 bases apart from each other (assuming a sequencing read length of 900 bases). This is not needed if your GOI is smaller than 1200 bases (again assuming a sequencing read length of 900 bases).

1.2 PCR and PCR product purification

1.2.1 PCR to amplify GOI for subcloning (see Reverse-transcription PCR (RT-PCR))

The number of PCR cycles and the amount of input cDNA depend on whether you amplify from an existing plasmid containing your GOI or from cDNA transcribed from total RNA.

To amplify a GOI from cDNA (obtained from total RNA extracted with TRIzol from Flp-In T-REx HEK 293 cells and reverse transcribed using SuperScript® III, both by Invitrogen), we use the following PCR conditions (using AccuPrime Pfx Polymerase, Invitrogen):

Process	Step	Temperature	Duration
Heat activation	1	95 °C	2 min
Denaturation	2	95 °C	45 s
Annealing	3	55–64 °C	30 s
Extension	4	68 °C	1 min per kb
Final extension	5	68 °C	5 min

Repeat Steps 2–4 20–35 times, depending on the expression level of your GOI and the amount of input cDNA (we usually use around 100 ng cDNA and 30 cycles of PCR). Include a negative control (blank).

1.2.2 PCR purification

1.2.2.1 Cast a 1% agarose gel in 1× TBE containing 0.4 μg ml^{-1} of ethidium bromide (see Agarose Gel Electrophoresis).

1.2.2.2 Take an aliquot of 4 μl of the PCR product and add 1 μl 5× DNA loading dye.

1.2.2.3 Load this mix next to a 1 kb DNA ladder on the agarose gel and run the gel in 1× TBE at 150 V for about 45 min.

1.2.2.4 Visualize PCR products using a UV 365 nm transilluminator.

1.2.2.5 If a single band is visible at the predicted size, purify the remainder of the PCR reaction using the QIAquick PCR Purification kit. If there are multiple bands, gel-purify the remainder of the PCR reaction by cutting only the band at the predicted size. Follow the manufacturer's instructions using the QIAquick Gel Extraction kit.

1.3 Restriction digestion of PCR product and pENTR4 vector

1.3.1 Digest the PCR product and the pENTR4 vector (about 1 μg) using the restriction endonucleases present in your PCR primer sequence in the appropriate enzyme buffer. Include controls for single cuts and a negative control (uncut vector). NotI and SalI can be used simultaneously in buffer 3 (NEB). We usually use about 50% of the initial PCR product in this step in a 30 μl reaction volume with 1 μl of each enzyme.

1.3.2 Incubate for 2 h at 37 °C (check specific enzyme temperature requirements).

1.3.3 If the restriction endonucleases you are using create blunt ends on both sides of the cut vector, dephosphorylate the vector to avoid religation of the vector (e.g., using calf intestinal alkaline phosphatase, NEB).

1.3.4 In the meantime cast a 1% agarose gel in 1× TBE containing 0.4 μg ml^{-1} of ethidium bromide.

1.3.5 Add the appropriate volume of 5× DNA loading dye to the reaction and gel-purify the reaction by running the gel in 1× TBE at 150 V for 1 h until a good separation of the digested fragments is achieved.

1.3.6 Cut fragments of the predicted size from the gel with a scalpel and proceed with the gel extraction according to the manufacturer's instructions (QIAquick Gel Extraction kit).

1.4 Ligation and transformation (pENTR4_GOI)

1.4.1 Choose the overall concentration of vector and insert such that it is between 1 and 10 ng μl^{-1} for efficient ligation. Molar ratios of insert to vector between 2 and 6 are optimal for single insertions. Use 1 μl T4 DNA ligase in a reaction volume of 20 μl. Incubate for 1 h at room temperature. Check for vector religation efficiency by including a reaction where no insert is added to the ligation reaction.

1.4.2 We routinely use One Shot® TOP10 chemically competent *E. coli* (Invitrogen) for transformations (alternatively see Transformation of Chemically Competent *E. coli* or Transformation of *E. coli* via electroporation).

To 50 μl of competent cell suspension, add 1 μl of the ligation reaction and incubate on ice for 30 min. Heat-shock the cells for 30 s at 42 °C without shaking. Place the vial on ice for 2 min. Add 250 μl of prewarmed S.O.C. medium (supplied by Invitrogen) to each vial and incubate horizontally at 37 °C for 1 h, shaking at 225 rpm. Spread 20 and 200 μl from each transformation on a prewarmed selective plate containing 50 μg ml^{-1} kanamycin and incubate overnight at 37 °C.

5.3. Tip

You can split the 50 μl cell suspension into 2 × 25 μl to save competent cells (but remember to adjust the volumes accordingly).

1.5 Miniprep and restriction digestion (pENTR4_GOI)

1.5.1 Inoculate a single colony in 5 ml LB medium containing 50 μg ml^{-1} kanamycin. We typically process —two to three colonies per selective plate. Incubate overnight at 37 °C, shaking at 225 rpm. Isolate plasmid following the manufacturer's instructions for the plasmid DNA extraction (include the optional 0.5 ml buffer PB wash step if using Qiagen miniprep kit. Alternatively, see Isolation of plasmid DNA from bacteria). Elute in 50 μl Buffer EB.

1.5.2 Determine the DNA concentration by measuring UV absorbance at 260 nm (see Explanatory Chapter: Nucleic Acid Concentration Determination).

1.5.3 Digest 200 ng of each plasmid sample selecting one or two restriction endonucleases that only cut once in the vector and once in the GOI.

1.5.4 Analyze the restriction pattern on a 1% agarose gel.

1.6 Sequencing reaction and sequence analysis (pENTR4_GOI)

1.6.1 Prepare the sequencing sample according to the guidelines of your sequencing facility. Sequence one or two of the colonies that showed the expected restriction pattern.

1.6.2 Verify the sequences.

1.7 LR reaction and second transformation (pFRT_DESTFLAGHA_GOI, pFRT_TO_DESTFLAGHA_GOI)

1.7.1 LR reaction

1.7.1.1 Set up the following reaction mix (we routinely scale the LR reaction down by 50% compared to manufacturer's recommendations) in a 1.5-ml polypropylene tube at room temperature (RT):

pENTR4_GOI (75 ng)	× μl
pFRT_DESTFLAGHA (75 ng) or pFRT_TO_DESTFLAGHA (75 ng)	× μl
TE buffer, pH 8.0 (at RT)	to 4 μl

1.7.1.2 Thaw the LR Clonase II enzyme mix on ice for 2 min. Vortex the enzyme briefly twice.

1.7.1.3 Add 1 μl enzyme mix to each sample and mix by vortexing twice. Microcentrifuge briefly.

1.7.1.4 Incubate reactions at 25 °C for 2 h.

1.7.1.5 Add 0.5 μl proteinase K solution (supplied by Invitrogen) and vortex briefly. Incubate for 10 min at 37 °C.

1.7.2 Transformation

1.7.2.1 To 50 μl of competent cell suspension, add 1.5 μl of the LR reaction (e.g., One Shot® TOP10 chemically competent *E. coli*, Invitrogen) and incubate on ice for 30 min.

1.7.2.2 Heat-shock the cells for 30 s at 42 °C without shaking. Place the vial on ice for 2 min.

1.7.2.3 Add 250 μl of prewarmed S.O.C. medium to each vial and incubate horizontally at 37 °C for 1 h, shaking at 225 rpm.

1.7.2.4 Spread 20 and 200 μl from each transformation on a prewarmed selective plate containing 100 μg ml^{-1} ampicillin and incubate overnight at 37 °C.

1.8 Minipreps and restriction digestion (pFRT_DESTFLAGHA_GOI, pFRT_TO_DESTFLAGHA_GOI)

1.8.1 Inoculate a single colony in 5 ml LB medium containing 100 μg ml^{-1} ampicillin. We typically process two to three colonies per selective plate. Incubate overnight at 37 °C, shaking at 225 rpm. Follow the manufacturer's instructions for the plasmid DNA extraction (include the optional 0.5 ml buffer

PB wash step if using Qiagen miniprep kit). Elute in 50 μl buffer EB.

1.8.2 Determine the DNA concentration by measuring UV absorbance at 260 nm.

1.8.3 Digest 200 ng of each plasmid sample with a restriction endonuclease that ideally cuts once in the vector and once in your GOI (or perform a double digest to the same effect).

1.8.4 Analyze the restriction pattern on a 1% agarose gel.

1.8.5 Optional: Check the plasmid once more by sequencing. We do not routinely perform this step if the restriction digestion of the miniprep shows the expected pattern on the agarose gel.

See Fig. 8.5 for the flowchart of Step 1.

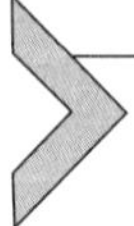

6. STEP 2 ESTABLISHING STABLE CELL LINES EXPRESSING THE PROTEIN OF INTEREST

6.1. Overview

The destination vector is transfected into Flp-In T-Rex HEK 293 cells to generate cells stably expressing the protein or gene of interest (GOI). After the initial transfection, the cells are subjected to antibiotic selection until the appearance of individual clones. Cells from these clones are then isolated and expanded and their protein expression is tested by Western blot analysis (see Western Blotting using Chemiluminescent Substrates). Cells are then cryopreserved.

6.2. Duration

Step	Action	Duration
2.1	Preparation of transfection	1 h and overnight
2.2	Transfection	1 h
2.3–2.8	Colony isolation	12–14 days
2.9	Selection of individual colonies	2 h
2.10–2.11	Expansion of individual clones for cryopreservation	11–14 days
2.12–2.14	Protein detection by Western blotting, cryopreservation of stable cell lines for long-term storage	1–2 days

Step 1: Clone the gene of interest into Gateway expression vectors

1.1 Design PCR primers and sequencing primers for your chosen gene

↓

1.2 Amplify gene of interest by PCR
Purify PCR product

↓

1.3 Digest 1-2 μg PCR product and pENTR4 DNA with chosen restriction enzymes
Add to a 1.5 ml tube:

DNA	
10X restriction enzyme buffer	3 μl
10X BSA	3 μl
Restriction enzyme(s)	1 μl
Water	to 30 μl

Incubate at 37°C, 2 h
Purify the DNA on an agarose gel

↓

1.4 Ligate digested pENTR4 + GOI
Add to a 1.5 ml tube:

10X T4 DNA ligase buffer	2 μl
Digested pENTR4	~50 ng
Digested PCR product	~150 ng
T4 DNA ligase	1 μl
Water	up to 20 μl

Incubate at room temperature, 1 h
Transform competent *E.coli* and grow O/N (kanamycin plates)

↓

1.5 Pick single colonies into 5 ml LB + kanamycin
Grow O/N at 37°C, with shaking
Isolate miniprep DNA
Confirm presence of insert by restriction enzyme digestion

↓

1.6 Sequence positive clones

↓

1.7 Set up LR reaction, add to a 1.5 ml tube:

pENTR4_GOI	75 ng
pFRT_DESTFLAGHA	75 ng
TE buffer	up to 4 μl

Add 1 μl LR Clonase II enzyme mix
Incubate at 25°C, 2 h
Add 0.5 μl Proteinase K solution
Incubate at 37°C, 10 min
Transform competent *E. coli* and grow O/N at 37°C (ampicillin plates)

↓

1.8 Pick single colonies into 5 ml LB + ampicillin
Grow O/N at 37°C, with shaking
Isolate miniprep DNA
Confirm presence of insert by restriction enzyme digestion

Figure 8.5 Flowchart of Step 1.

2.1 Aliquot Flp-In T-REx HEK 293 cells from the growing stock into antibiotic-free medium at a density of 50 000 cells per 1 ml medium per well of a 12-well plate. You will need one well per destination vector as well as one for the negative (pOG44 only) and one for the positive control (pFRT_DESTFLAGHA_eGFP). Make sure to spread the cells evenly by gentle swirling.

2.2 24 h later cotransfect your Flp-In T-REx HEK 293 cells with a 9:1 ratio of pOG44: pFRT_DESTFLAGHA_GOI or pFRT_TO_DESTFLAGHA_GOI destination vector. As suggested by the Invitrogen manual we use 0.9 μg pOG44 (maximum of 3 μl) and 0.1 μg of the destination vector (maximum of 3 μl).

- **2.2.1** In a 1.5-ml polypropylene tube combine the destination vectors for each transfection with 50 μl of Opti-MEM. Pipette the mixture up and down twice to ensure proper mixing.
- **2.2.2** Perform pOG44 only and pFRT_DESTFLAGHA_eGFP + pOG44 transfections to serve as a negative and positive control, respectively.
- **2.2.3** In a separate 1.5-ml polypropylene tube combine 2 μl lipofectamine 2000 with 50 μl of Opti-MEM for each transfection and incubate for 5 min at room temperature. Invert the tube twice to ensure proper mixing.
- **2.2.4** Add the Lipofectamine 2000/Opti-MEM mixture to the destination vector/Opti-MEM mixture. Mix combined solutions and incubate the mixture at room temperature for 20 min. Add the mixture to the cells and move the plate from side to side to distribute evenly across the wells.

2.3 24 h after transfection, aspirate the medium and replace it with fresh antibiotic-free medium.

2.4 Check for the eGFP expression to assess the transfection efficiency.

2.5 48 h after transfection split the cells into a 10-cm plate containing antibiotic-free medium. Incubate the cells at 37 °C for 12 h until they have completely attached to the culture dish.

2.6 Remove the medium and add selective medium containing the appropriate antibiotics (see above; use hygromycin alone for the eGFP cells).

2.7 Change the selective medium every 2–3 days until distinct foci can be identified. After about a week, there will be a noticeable selection of cells.

2.8 Continue changing the medium for a total of 10–14 days.

2.9 Mark 8–12 well separated colonies while looking at them under the microscope.

2.9.1 Prepare the following:
96-well plate (conical bottom) containing 20 μl of trypsin per well
96-well plate (flat bottom) containing 80 μl selection medium per well
Multichannel pipet
1× PBS

2.9.2 Aspirate the medium and replace with 3 ml of 1X PBS. Set a P20 micropipettor to 3 μl and gently scrape one colony off the plate. Once the colony is dislodged take it up into the pipet tip.

2.9.3 Place the cell suspension into 20 μl of trypsin in a conical bottom 96-well plate to prepare a single cell suspension. Pipette up and down several times. Incubate at room temperature for 5 min.

2.9.4 Add 20 μl of selective medium into each well and pipette up and down again several times to ensure that no cell aggregates remain. Avoid introducing bubbles.

2.9.5 Use a multichannel pipettor to transfer these 40 μl into a flat bottom 96-well plate already containing 80 μl/well of stable cell line selective medium.

2.9.6 Check the individual wells under the microscope to make sure that there are no cell clumps visible. If there are, pipette up and down several times to achieve a single cell suspension.

2.9.7 After picking the colonies, remove the PBS and add fresh media to maintain the original plates for a few days in case none of the newly isolated colonies survives. After you expand the colonies into a 24-well plate, induce the saved original plates with 1 μg ml^{-1} doxycycline for 12 h, wash the plates with 3 ml PBS, collect the remaining colonies in a 1.5-ml polypropylene tube, and solubilize in SDS sample buffer. Analyze the samples by Western blotting to ensure that your protein of interest is expressed.

2.10 After 2–3 days when the cells reach 80% confluence, split into a 24-well plate. Avoid letting the cells become overconfluent (usually the cells are ready to be split into the next larger plate after 2–3 days). After the cells have grown in the 24-well plate, sequentially expand them into a 12-well plate, a 6-well plate, and then a 10-cm plate. We usually expand up to five clones to the 10-cm plate stage and up to three clones to the 15-cm plate stage.

2.11 Once cells reach confluence on a 10-cm plate and are ready to be expanded onto a 15-cm plate, keep a fraction of the cells for Western blot analysis (if the cell line is constitutively expressing your protein of interest) or plate cells from each clone into separate wells of a 12-well plate and induce them with 1 μg ml^{-1} doxycycline overnight. Wash the cells with PBS, collect them in a 1.5-ml microcentrifuge tube, and solubilize them in SDS sample buffer. Make sure to assess protein expression in the individual colonies by Western blotting. If no protein-specific antibody is available, we routinely use an anti-HA antibody (see above).

2.12 Continue to split and expand cells until you have 3–5 15-cm plates of each cell line.

2.13 Once protein expression has been confirmed and the cells have reached the desired level of confluence, they are ready to cryopreserve.

2.14 We usually collect one 15-cm plate of 90% confluent cells per vial for cryopreservation.

2.14.1 Remove culture medium from cells.

2.14.2 Wash the cells with 1× PBS.

2.14.3 Add 3 ml trypsin to each plate and incubate for 5 min.

2.14.4 Detach the cells by pipetting up and down.

2.14.5 Add 15 ml of antibiotic-free culture medium to each plate to inactivate trypsin.

2.14.6 Harvest the cells in 50-ml centrifuge tubes by centrifugation at 250 × *g* for 7 min.

2.14.7 Resuspend the cells in 1.5 ml cold freezing medium per initial 15-cm plate.

2.14.8 Aliquot 1–1.5 ml of the cell suspension into cryogenic vials.

2.14.9 Start the freezing process by placing the cryogenic vials at −20 °C for 2 h. Transfer the vials to −80 °C overnight. The next morning transfer the vials to liquid nitrogen storage. Check for cell viability by thawing one vial per batch some days after freezing.

6.3. Tip

Should you not be able to obtain colonies from inducible cell lines expressing a toxic protein of interest, you might wish to use Tet system-approved fetal bovine serum (e.g., Clontech, 631106). Regular FBS usually contains very low levels of tetracycline, which might be sufficient to induce the expression of the protein of interest.

See Fig. 8.6 for the flowchart of Step 2.

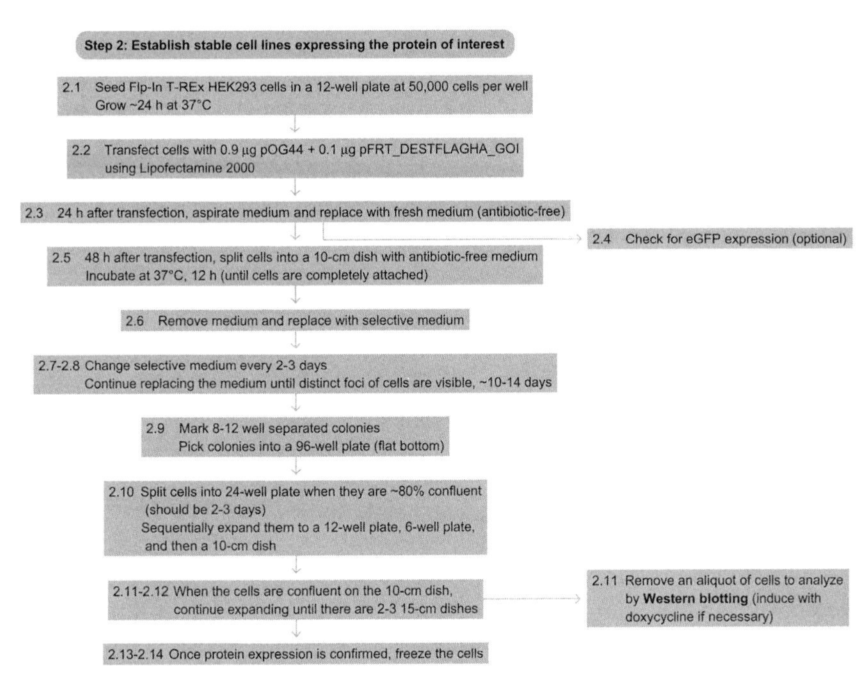

Figure 8.6 Flowchart of Step 2.

REFERENCES

Referenced Literature

Andersson, S., Davis, D. L., Dahlback, H., Jornvall, H., & Russell, D. W. (1989). Cloning, structure, and expression of the mitochondrial cytochrome P-450 sterol 26-hydroxylase, a bile acid biosynthetic enzyme. *Journal of Biological Chemistry*, *264*(14), 8222–8229.

Hafner, M., Landthaler, M., Burger, L., et al. (2010). PAR-CLIP – transcriptome-wide identification of RNA targets and binding sites of RNA-binding proteins. *Cell*, *141*(1), 129–141.

Hillen, W., & Berens, C. (1994). Mechanisms underlying expression of Tn10 encoded tetracycline resistance. *Annual Review of Microbiology*, *48*, 345–369.

Landthaler, M., Gaidatzis, D., Rothballer, A., et al. (2008). Molecular characterization of human Argonaute-containing ribonucleoprotein complexes and their bound target mRNAs. *RNA*, *14*(12), 2580–2596.

Malik, S., & Roeder, R. G. (2003). Isolation and functional characterization of the TRAP/mediator complex. *Methods in Enzymology*, *364*, 257–284.

Walhout, A. J., Temple, G. F., Brasch, M. A., et al. (2000). GATEWAY recombinational cloning: Application to the cloning of large numbers of open reading frames or ORFeomes. *Methods in Enzymology*, *328*, 575–592.

SOURCE REFERENCES

Hafner, M., Landthaler, M., Burger, L., et al. (2010). PAR-CLIP – transcriptome-wide identification of RNA targets and binding sites of RNA-binding proteins. *Cell*, *141*(1), 129–141.

Landthaler, M., Gaidatzis, D., Rothballer, A., et al. (2008). Molecular characterization of human Argonaute-containing ribonucleoprotein complexes and their bound target mRNAs. *RNA*, *14*(12), 2580–2596.

Walhout, A. J., Temple, G. F., Brasch, M. A., et al. (2000). GATEWAY recombinational cloning: Application to the cloning of large numbers of open reading frames or ORFeomes. *Methods in Enzymology*, *328*, 575–592.

Referenced Protocols in Methods Navigator

Purification of His-tagged proteins.

Affinity purification of a recombinant protein expressed as a fusion with the maltose-binding protein (MBP) tag

Purification of GST-tagged proteins

Protein Affinity Purification using Intein/Chitin Binding Protein Tags

Immunoaffinity purification of proteins

Strep-tagged protein purification

Small-scale Expression of Proteins in *E. coli*

Recombinant protein expression in baculovirus-infected insect cells

Isolation of plasmid DNA from bacteria

Explanatory chapter: PCR -Primer design

Reverse-transcription PCR (RT-PCR)

Agarose Gel Electrophoresis

Transformation of Chemically Competent *E. coli*

Transformation of *E. coli* via electroporation

Explanatory Chapter: Nucleic Acid Concentration Determination

Western Blotting using Chemiluminescent Substrates

CHAPTER NINE

Restrictionless Cloning

Mikkel A. Algire[1]
J. Craig Venter Institute, Synthetic Biology Group, Rockville, MD, USA
[1]Corresponding author: e-mail address: malgire@jcvi.org

Contents

1. Theory 126
2. Equipment 126
3. Materials 126
   3.1 Solutions & buffers 127
4. Protocol 127
   4.1 Preparation 127
   4.2 Duration 127
5. Step 1 Cloning Vector Preparation 128
   5.1 Overview 128
   5.2 Duration 128
   5.3 Tip 129
   5.4 Tip 129
   5.5 Caution 129
6. Step 2 PCR Primer Design and PCR 130
   6.1 Overview 130
   6.2 Duration 130
   6.3 Tip 132
   6.4 Tip 132
7. Step 3 Insert Denaturation and Annealing 132
   7.1 Overview 132
   7.2 Duration 132
8. Step 4 Ligation and Transformation 133
   8.1 Overview 133
   8.2 Duration 133
   8.3 Tip 133

References 134

Abstract

Many methods have been developed for the cloning of PCR products. These methods include blunt-end cloning, TA cloning, and using restriction sites incorporated into the PCR primers. The restrictionless cloning technique allows efficient directional cloning of PCR products into any cloning site within a vector regardless of whether the sites are contained within the insert to be cloned.

Methods in Enzymology, Volume 529
ISSN 0076-6879
http://dx.doi.org/10.1016/B978-0-12-418687-3.00009-4

1. THEORY

A standard technique for cloning PCR inserts is to incorporate the restriction sites to be used into the amplification primers. Following amplification the PCR product is digested with the specific restriction enzymes to create the compatible ends for ligation into the digested vector. This method can be difficult because of the variable efficiency that restriction enzymes have when cleaving near the ends of a DNA molecule. Inefficient cleavage results in many inserts not containing the cohesive ends for ligation into the vector and therefore poor cloning efficiencies. Also, this standard method cannot be used if the PCR product to be cloned contains a recognition sequence for the restriction enzymes used to create the cohesive ends. The restrictionless cloning technique utilizes two PCR products to create almost identical inserts that when heat-denatured and annealed, result in a DNA insert containing the desired overhang sequences for ligation into a prepared vector. This protocol shows a specific example of restrictionless cloning although the logic behind the primer design is the same for all situations. Your unique circumstances would require specific modifications to the primer sequences to create your desired overhangs.

2. EQUIPMENT

PCR thermocycler
Benchtop microcentrifuge
Gel electrophoresis equipment
UV/vis spectrophotometer
Micropipettors
Micropipettor tips
1.5-ml microcentrifuge tubes
0.2-ml thin-walled PCR tubes
Gel extraction kit (Qiagen)

3. MATERIALS

Phusion DNA polymerase and buffer (New England Biolabs)
10 mM dNTP mix (New England Biolabs)
T4 DNA ligase and buffer (New England Biolabs)
Calf Intestinal Alkaline Phosphatase (CIP) (New England Biolabs)

5′ phosphorylated PCR primers
Nonphosphorylated PCR primers
Vector DNA
Agarose
Tris base
Glacial acetic acid
EDTA
Ethidium bromide
Competent bacteria of choice (e.g., DH5α)

3.1. Solutions & buffers

Step 1 50 × TAE agarose gel running buffer

Component	Amount
Tris base	242 g
Glacial acetic acid	57.1 ml
EDTA	18.6 g

Add water to 1 l

1 × TAE agarose gel running buffer

Component	Amount
50 × TAE buffer	10 ml
Water	490 ml

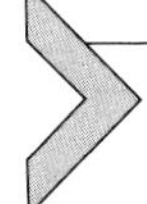

4. PROTOCOL

4.1. Preparation

Design and order oligonucleotides for PCR (see Step 2.2 for PCR primer design).

4.2. Duration

Preparation	About 2 days
Protocol	About 3–5 days

See Fig. 9.1 for the flowchart of the complete protocol.

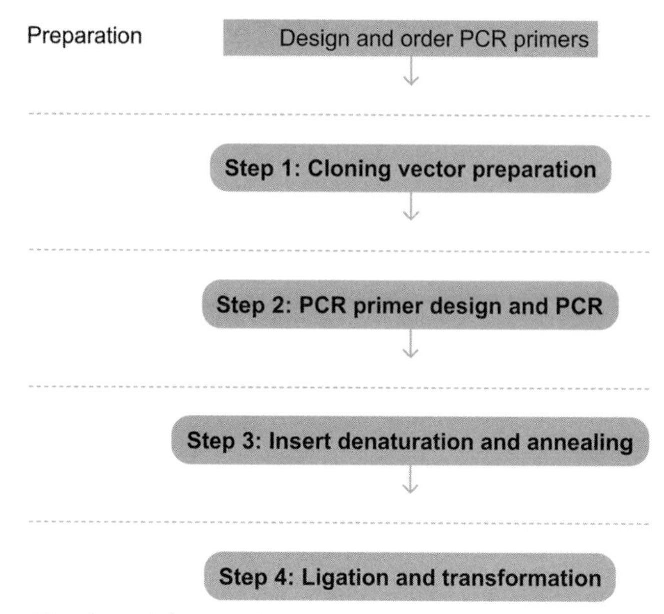
Preparation
Design and order PCR primers
Step 1: Cloning vector preparation
Step 2: PCR primer design and PCR
Step 3: Insert denaturation and annealing
Step 4: Ligation and transformation

Figure 9.1 Flowchart of the complete protocol, including preparation.

5. STEP 1 CLONING VECTOR PREPARATION

5.1. Overview

Prepare vector by digesting with appropriate restriction enzymes to create cohesive ends for cloning. For example, this protocol discusses using pTYB1 (a bacterial expression vector from NEB) and NdeI and SapI as the cloning sites. Your specific cloning project would require modifications to the described protocols. Always follow the manufacturer's specific instructions for restriction enzyme use.

5.2. Duration

3–5 h

1.1 Choose cloning or expression vector and determine cloning sites within vector to be used. In this case the vector is pTYB1 and the cloning sites are NdeI and SapI.

1.2 Digest and dephosphorylate 1 μg of pTYB1 with NdeI, SapI, and alkaline phosphatase.

1.3 Incubate for 1 h at 37 °C.

1.4 Prepare a 1% agarose gel in 1× TAE with ethidium bromide. Run reaction on the 1% TAE agarose gel. Excise digested band and purify DNA with Gel Extraction kit (Qiagen) following recommended directions (see Agarose Gel Electrophoresis).

5.3. Tip

The restriction and dephosphorylation reactions can be done simultaneously in this situation. Check enzyme and buffer compatibility when using different restriction enzymes. Antarctic Phosphatase (NEB) can be used instead of CIP. Antarctic Phosphatase is active in NEB buffers 1–4. CIP is active in NEB buffers 2 and 4.

5.4. Tip

Dephosphorylation and gel purification of the vector are important for significantly reducing background colonies after ligation and transformation.

5.5. Caution

Ethidium bromide is a mutagen. Wear gloves and handle with care. Dispose gel and buffer in accordance with local regulations.

See Fig. 9.2 for the flowchart of Step 1.

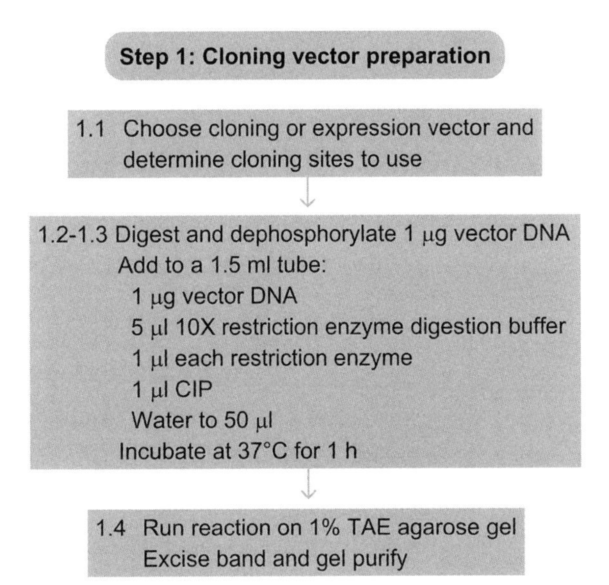

Figure 9.2 Flowchart of Step 1.

6. STEP 2 PCR PRIMER DESIGN AND PCR

6.1. Overview

Two sets of primers (and two PCRs) are required to produce the correct cohesive ends on the final DNA insert that match the prepared vector (see Explanatory chapter: PCR -Primer design). The first PCR will create the top strand for the final insert and the second PCR will create the bottom strand for the final insert.

6.2. Duration

4–5 h

2.1 Determine the correct overhangs needed on the insert to match the ends of the prepared vector. Figure 9.3 shows an overview of expected overhang sequences on the example digested vector and final insert.

2.2 Design two sets of PCR primers that incorporate the desired overhang sequences. The first set of primers is used to create the top strand of the final insert. In this case the forward primer for PCR #1 will include 5′-phosphate-TATG followed by insert sequence to generate the NdeI overhang. The 5′ phosphate is required for ligation into the dephosphorylated vector. Since no overhang needs to be created on the top strand of the insert, the reverse primer for PCR #1 is simply the reverse complement of the desired sequence to amplify. The second set of primers is used to create the bottom strand of the final insert. The forward primer for PCR #2 will include 5′-TG (part of NdeI site) followed by insert sequence. The reverse primer for PCR #2 will have 5′-phosphate-GCA followed by the reverse complement of the insert sequence to generate the SapI overhang.

2.3 After obtaining both sets of primers, amplify insert of interest (see General PCR) with Phusion DNA polymerase following recommended instructions.

2.4 Gel-purify both reactions as performed earlier with the prepared vector.

2.5 Quantitate DNAs after purification using a spectrophotometer at A_{260} (see Explanatory Chapter: Nucleic Acid Concentration Determination).

i. Cloning sites

NdeI 5′ CATATG 3′ / 3′ GTATAC 5′

SapI 5′ GCTCTTCNNNN 3′ / 3′ CGAGAAGNNNN 5′

SapI site in pTYB1 5′ GCTCTTCCTGC 3′ / 3′ CGAGAAGGACG 5′

ii. Overhangs generated on vector after restriction digestion.

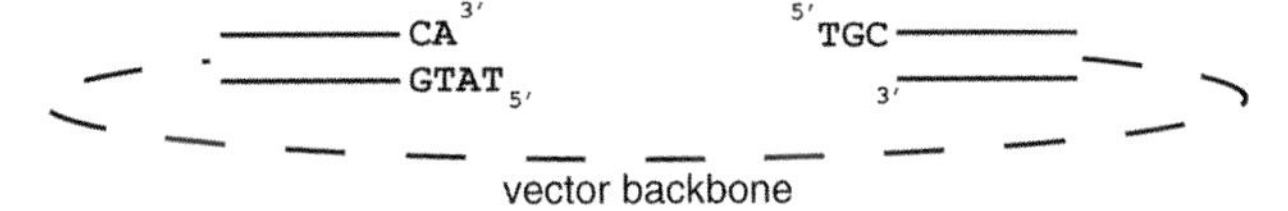

iii. Overhangs needed for insert to match vector.

5′ p-TATG 3′
3′ AC ACG-p 5′
insert

iv. PCR products of insert that produce desired top and bottom strands.

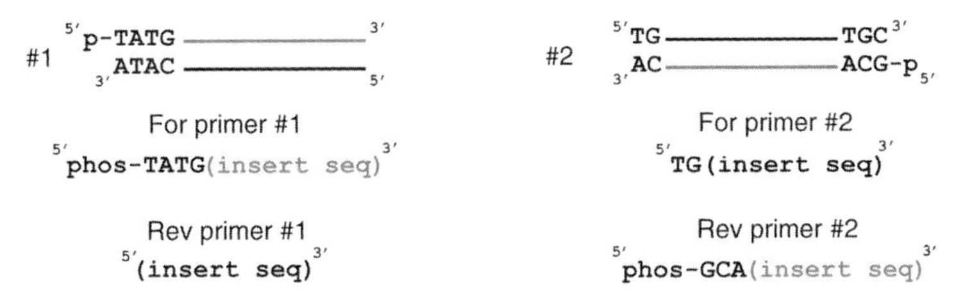

v. Inserts produced after denaturing and annealing

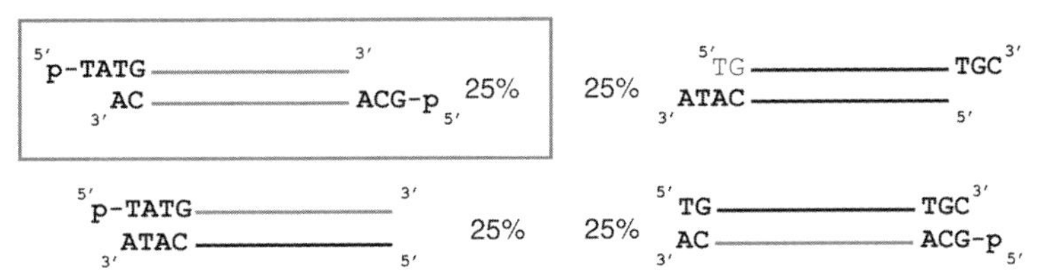

Figure 9.3 Overview of restrictionless cloning primer design and PCRs. The example cloning sites are shown in (i). The overhangs of the cloning vector after digestion are shown in (ii). In order to match the cohesive ends of the vector the insert will need to have the overhangs indicated in (iii). Two PCRs will be performed to generate the strands required to produce the desired overhangs for the insert (colored in red). PCR #1 will generate the top strand of the final insert and PCR #2 will generate the bottom strand of the final insert. The PCR products from these reactions and the primers used to generate them are shown in (iv). Following heat denaturing and annealing of the two PCR products ~25% of the dsDNA molecules will be the desired insert product (boxed) ready for ligation into the prepared vector (v).

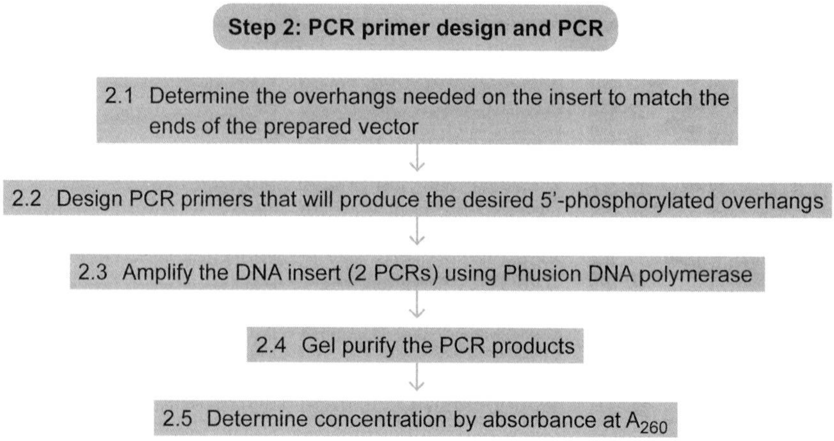

Step 2: PCR primer design and PCR

2.1 Determine the overhangs needed on the insert to match the ends of the prepared vector

↓

2.2 Design PCR primers that will produce the desired 5'-phosphorylated overhangs

↓

2.3 Amplify the DNA insert (2 PCRs) using Phusion DNA polymerase

↓

2.4 Gel purify the PCR products

↓

2.5 Determine concentration by absorbance at A_{260}

Figure 9.4 Flowchart of Step 2.

6.3. Tip

The primers not used to create the strands in the final insert do not need to be phosphorylated.

6.4. Tip

Do not use a DNA polymerase that adds bases to the ends of DNA such as Taq DNA polymerase as this will produce incorrect overhangs.

See Fig. 9.4 for the flowchart of Step 2.

7. STEP 3 INSERT DENATURATION AND ANNEALING

7.1. Overview

Heat denaturation will be used to separate the strands of DNA from one another in a mixture of the two insert PCRs. Slow cooling allows the formation of the correct final insert.

7.2. Duration

15 min

3.1 Mix equal amounts (~500 ng each) of insert PCR #1 and insert PCR #2 in a 0.2-ml PCR tube.

3.2 Heat mixture tube at 95 °C for 5 min in thermocycler.

3.3 Allow tube to cool in thermocycler for 10 min.

8. STEP 4 LIGATION AND TRANSFORMATION

8.1. Overview

The prepared vector and annealed final insert will be ligated together and transformed into the bacterial host of choice **(see Transformation of Chemically Competent *E. coli* or Transformation of *E. coli* via electroporation).

8.2. Duration

2–3 days

4.1 Ligate the annealed insert into the prepared vector as follows. Also prepare a no insert control reaction by leaving out the insert DNA in order to measure background colonies.

10× T4 DNA ligase buffer	2 μl
Prepared vector	50–100 ng
Annealed insert	100–300 ng
T4 DNA ligase	1 μl
Water	up to 20 μl

Incubate at room temperature for 1–2 h or overnight at 16 °C.

4.2 Following the ligation reaction, transform electro-competent or chemically competent *E. coli* cells of your choice following recommended instructions. Plate cells on selective media and incubate overnight.

4.3 Pick transformed colonies and screen for desired clone.

8.3. Tip

Do not use excessive amounts of ligation reaction when doing transformations. 1–2 μl of ligation reaction per 30 μl of competent cells should be sufficient.

See Fig. 9.5 for the flowcharts of Steps 3 and 4.

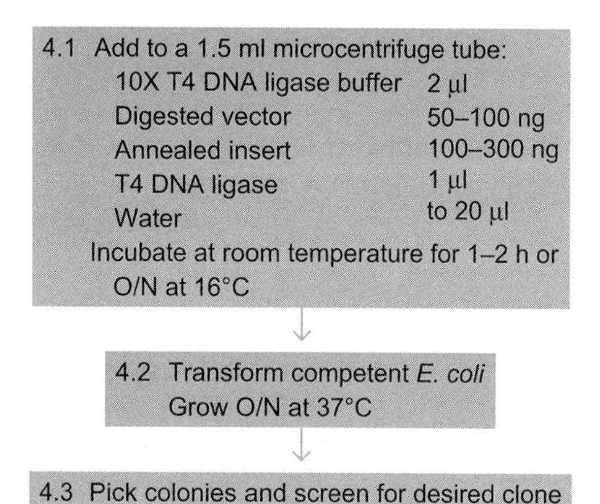

Figure 9.5 Flowchart of Steps 3 and 4.

REFERENCES

Referenced Protocols in Methods Navigator

Explanatory chapter: PCR -Primer design.
Agarose Gel Electrophoresis.
General PCR.
Explanatory Chapter: Nucleic Acid Concentration Determination.
Transformation of Chemically Competent *E. coli*.
Transformation of *E. coli* via electroporation.

CHAPTER TEN

Isolation of Plasmid DNA from Bacteria

Lefkothea-Vasiliki Andreou[1]
Ear Institute, University College London, London, United Kingdom
[1]Corresponding author: e-mail address: l.andreou@ucl.ac.uk

Contents

1. Theory 136
2. Equipment 136
3. Materials 137
 3.1 Solutions & buffers 138
4. Protocol 138
 4.1 Duration 138
 4.2 Preparation 138
5. Step 1 Harvesting the Bacteria 139
 5.1 Overview 139
 5.2 Duration 139
6. Step 2 Cell Lysis and Isolation of Plasmid DNA 139
 6.1 Overview 139
 6.2 Duration 140
 6.3 Caution 140
 6.4 Tip 141
 6.5 Tip 141
 6.6 Tip 141
References 142

Abstract

The purpose of this protocol is the isolation of plasmid DNA from bacteria. The boiling method for isolating plasmids by Holmes and Quigley (1981) is presented here. This method is rapid and simple and it allows for a large number of samples to be processed simultaneously (up to 40 samples). Thus, it is appropriate for the preparation of bacterial plasmids in order to screen a large number of colonies or small cultures for the presence of recombinant DNA inserts. The protocol can be effectively scaled up for the preparation of plasmids from liter cultures.

Methods in Enzymology, Volume 529
ISSN 0076-6879
http://dx.doi.org/10.1016/B978-0-12-418687-3.00010-0

1. THEORY

Plasmid DNA is used for a number of downstream applications such as transfection, sequencing, screening clones, restriction digestion, cloning, and PCR. A number of methods have been developed for the purification of plasmid DNA from bacteria. The most common method that is used is based on the alkaline lysis method (Birnboim and Doly, 1979). Nowadays, a variety of convenient and fast plasmid DNA purification kits are commercially available ensuring advanced quality and purity (see Explanatory Chapter: How Plasmid Preparation Kits Work). The size of the bacterial culture used (and effectively the plasmid DNA yield) defines the plasmid preparation as miniprep, midiprep, maxiprep etc. The small-scale mini preparation of plasmid DNA is commonly used to screen bacterial clones for the presence of recombinant DNA inserts. Here, we present a fast and simple protocol by Holmes and Quigley (1981) that allows the screening of a large number of colonies or small cultures for recombinant DNA inserts. Bacterial cells are collected and briefly boiled and the insoluble genomic DNA and cellular debris is removed by centrifugation. Plasmids are recovered by isopropanol precipitation and treated with RNase. The protocol works for plasmids ranging in size from 3.9 to 13.4 kb and yields enough plasmid DNA for several restriction digests (see Fig. 10.1). Multiple samples can be processed simultaneously (e.g., 20–40 samples).

2. EQUIPMENT

Shaking incubator
Microcentrifuge or refrigerated microcentrifuge
Boiling water bath (100 °C)
–or–
Heating block (e.g., Eppendorf thermomixer)
Refrigerator (4 °C)
Deep freezer (−80 °C)
Freezer (−20 °C)
Pipet-aid
5-ml disposable serological pipettes
Micropipettors

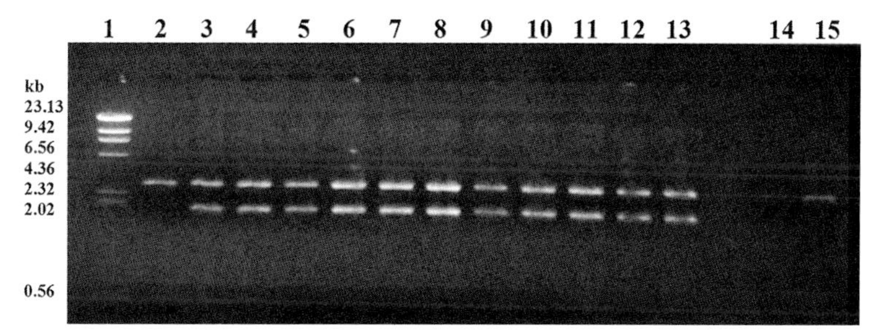

Figure 10.1 Restriction enzyme digestion of ligated plasmid DNA was electrophoresed on a 0.8% agarose gel and stained with ethidium bromide. Plasmids were prepared by the boiling protocol from microscale cultures of DH5α cells that were transformed with plasmid pUCotsA that contains a trehalose gene insert from *Brevibacterium lactofermentum* (ATCC 21799) in plasmid pUC18. Transformation was carried out as described in Kushner et al. (1978). Restriction digestion of the ligated plasmid with the endonucleases *Kpn*I and *Xba*I yields a 2.7 kb and a 1.8 kb band corresponding to the plasmid vector and the insert, respectively. Lane 1: Lambda DNA marker (λDNA/*Hin-d*III); lane 2: plasmid vector pUC18; lanes 3–13: plasmids isolated from white colonies; lanes 14 and 15: plasmids isolated from blue colonies (Andreou, 2005).

Pipettor tips
13-ml sterile polypropylene snap-cap tubes
1.5-ml microcentrifuge tubes
Toothpicks

3. MATERIALS

LB broth
Antibiotic (e.g., ampicillin, tetracycline)
Sucrose
Triton X-100
EDTA
Tris base
Lysozyme (Sigma)
Sodium acetate (NaOAc)
NaOH
Isopropanol
DNase-free RNase (Sigma)

3.1. Solutions & buffers

Step 2 STET Buffer

Component	Final concentration	Stock	Amount
Sucrose	8%		8 g
Triton X-100	0.5%	20%	2.5 ml
EDTA	50 mM	500 mM	10 ml
Tris–HCl, pH 8.0	10 mM	1 M	1 ml

Add purified water to 100 ml

Lysozyme (10 mg ml^{-1})

Dissolve 10 mg of lysozyme in 1 ml of 10 mM Tris.Cl (pH 8.0)

TE

Component	Final concentration	Stock	Amount
Tris–HCl, pH 8.0	10 mM	1 M	1 ml
EDTA	0.1 mM	500 mM	20 μl

Add purified water to 100 ml

4. PROTOCOL

4.1. Duration

Preparation	About 2 days
Protocol	About 2 h

4.2. Preparation

Transform DH5α with the desired plasmid DNA, either a purified plasmid or a ligation reaction to be screened for the presence of inserts, and plate the bacteria on the appropriate selective plates. Pick individual colonies of the transformed *Escherichia coli* strain and inoculate 3 ml of LB broth supplemented with 25 μg ml^{-1} ampicillin or another appropriate antibiotic in 13-ml sterile polypropylene snap-cap tubes (see also Growth Media for *E. coli*). Grow the bacterial cultures overnight (18–20 h) at 37 °C in a shaker for uniform growth to a density of about $A_{650}=1.2$.

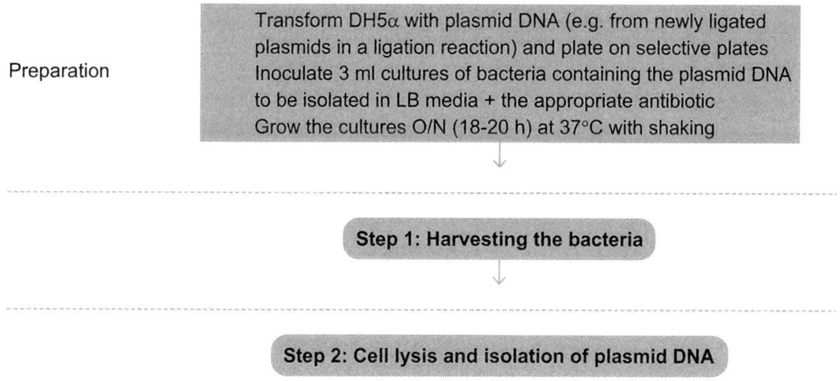

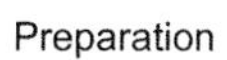

Preparation

Transform DH5α with plasmid DNA (e.g. from newly ligated plasmids in a ligation reaction) and plate on selective plates
Inoculate 3 ml cultures of bacteria containing the plasmid DNA to be isolated in LB media + the appropriate antibiotic
Grow the cultures O/N (18-20 h) at 37°C with shaking

Step 1: Harvesting the bacteria

Step 2: Cell lysis and isolation of plasmid DNA

Figure 10.2 Flowchart of the complete protocol, including preparation.

See Fig. 10.2 for the flowchart of the complete protocol, including preparation.

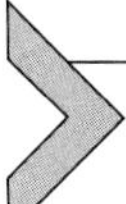

5. STEP 1 HARVESTING THE BACTERIA

5.1. Overview

The bacterial cells are collected by centrifugation.

5.2. Duration

20 min

1.1 Transfer 1.5 ml of the bacterial culture to a 1.5-ml microcentrifuge tube and store the rest of the culture in the refrigerator (4 °C).
1.2 Pellet the bacteria at 8000 rpm for 5 min and remove the media.
1.3 Leave the bacterial pellet to air-dry.

See Fig. 10.3 for the flowchart of Step 1.

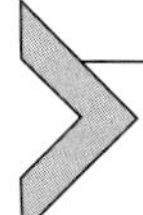

6. STEP 2 CELL LYSIS AND ISOLATION OF PLASMID DNA

6.1. Overview

The bacterial cells are boiled briefly at 100 °C in the presence of agents that weaken the cell wall and help prevent DNA degradation by nucleases. The partially denatured genomic DNA and denatured proteins are pelleted down by centrifugation. Plasmids are recovered by isopropanol precipitation and RNase degrades RNA into smaller components.

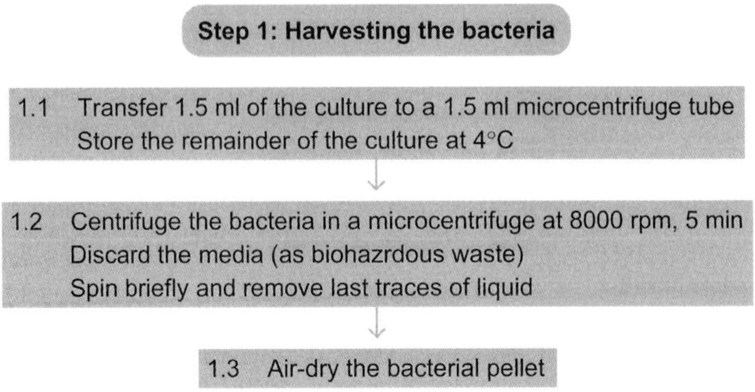

Step 1: Harvesting the bacteria

1.1 Transfer 1.5 ml of the culture to a 1.5 ml microcentrifuge tube
Store the remainder of the culture at 4°C

↓

1.2 Centrifuge the bacteria in a microcentrifuge at 8000 rpm, 5 min
Discard the media (as biohazrdous waste)
Spin briefly and remove last traces of liquid

↓

1.3 Air-dry the bacterial pellet

Figure 10.3 Flowchart of Step 1.

6.2. Duration

1.5 h

2.1 Resuspend the bacterial pellet in 0.35 ml of STET Buffer.

2.2 Add 25 μl of freshly prepared lysozyme (10 mg ml^{-1}) and mix by shaking gently for 3 s.

2.3 Place the tube in a boiling water bath (100 °C) for 40 s.

2.4 Centrifuge immediately at 14 000 rpm for 10 min at room temperature (or alternatively at 4 °C). Remove the pellet from the tube using a toothpick.

2.5 Add 40 μl of 3 M sodium acetate, pH 5.2, and 420 μl isopropanol and mix by vortexing briefly.

2.6 Place the sample in a −80 °C freezer for 30 min.

2.7 Centrifuge the sample at 14 000 rpm for 15 min at 4 °C. Remove the supernatant and air-dry the pellet.

2.8 Resuspend the pellet in 50 μl of TE that contains DNase-free RNase (50 μg ml^{-1}) and incubate the sample for 10 min at 37 °C.

2.9 The plasmid DNA is now ready to use in downstream applications, for example, restriction enzyme digestion or PCR and agarose gel electrophoresis (see General PCR, Agarose Gel Electrophoresis). Store the miniprep plasmid DNA at −20 °C.

6.3. Caution

Isopropanol is highly flammable. It should be kept away from heat and open flames. Poisoning can occur from inhalation, ingestion, or skin absorption and it may act as an irritant. Use in well-ventilated areas and wear safety glasses and protective gloves.

6.4. Tip

Air-dry the pellet instead of using a vacuum in order to avoid over-drying the pellet as this will make the DNA hard to dissolve.

6.5. Tip

The DNase-free RNase treatment removes the RNA that may cover small DNA fragments in agarose gels.

6.6. Tip

Resuspend the pellet in TE for long-term storage or in Tris–HCl, pH 8.5 to avoid EDTA inhibition of downstream reactions.

See Fig. 10.4 for the flowchart of Step 2.

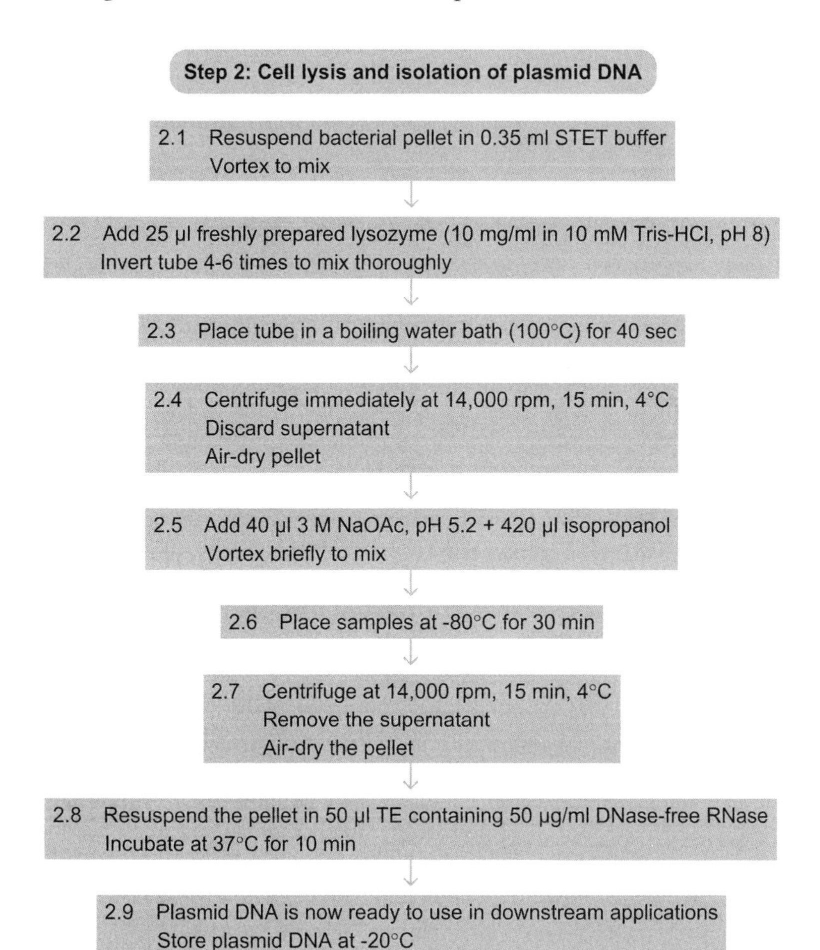

Figure 10.4 Flowchart of Step 2.

REFERENCES

Referenced Literature

Andreou, L. V. (2005). Cloning and heterologous expression of trehalose biosyntheticgenes of Corynebacteria. *Unpublished MSc thesis*. Ioannina, Greece: University of Ioannina.

Birnboim, H. C., & Doly, J. (1979). A rapid alkaline extraction procedure for screening recombinant plasmid DNA. *Nucleic Acids Research*, 7(6), 1513–1523.

Holmes, D. S., & Quigley, M. (1981). A rapid boiling method for the preparation of bacterial plasmids. *Analytical Biochemistry*, *114*, 193–197.

Kushner, S. R., Sheperd, J., Edwards, G., & Maples, V. F. (1978). *uvrD*, *uvrE* and *recL* represent a single gene. In P. C. Hanawalt, E. C. Friedberg, & C. F. Fox (Eds.), *DNA Repair Mechanisms* (pp. 251–254). New York: Academic Press.

Referenced Protocols in Methods Navigator

Explanatory Chapter: How Plasmid Preparation Kits Work.

Growth Media for *E. coli*.

General PCR.

Agarose Gel Electrophoresis.

CHAPTER ELEVEN

Preparation of Genomic DNA from Bacteria

Lefkothea-Vasiliki Andreou[1]
Ear Institute, University College London, London, United Kingdom
[1]Corresponding author: e-mail address: l.andreou@ucl.ac.uk

Contents

1. Theory 144
2. Equipment 144
3. Materials 144
   3.1 Solutions & buffers 145
4. Protocol 145
   4.1 Preparation 145
   4.2 Duration 145
5. Step 1 Cell Lysis 146
   5.1 Overview 146
   5.2 Duration 146
6. Step 2 Organic Extraction and Ethanol Precipitation of DNA 147
   6.1 Overview 147
   6.2 Duration 147
   6.3 Tip 148
   6.4 Tip 148
   6.5 Tip 148
   6.6 Tip 148
   6.7 Tip 148
   6.8 Caution 148
7. Step 3 DNA Quantity and Quality Assessment 149
   7.1 Overview 149
   7.2 Duration 149
   7.3 Tip 150
   7.4 Tip 150
   7.5 Caution 150
   7.6 Caution 151
References 151

Abstract

The purpose of this protocol is the isolation of bulk cellular DNA from bacteria (alternatively see Preparation of genomic DNA from *Saccharomyces cerevisiae* or Isolation of Genomic DNA from Mammalian Cells protocols).

Methods in Enzymology, Volume 529
ISSN 0076-6879
http://dx.doi.org/10.1016/B978-0-12-418687-3.00011-2

1. THEORY

High-quality, purified DNA is crucial for a number of downstream applications such as PCR, cloning, and DNA library construction. The isolation of DNA is traditionally achieved using organic extraction of the soluble DNA while the insoluble cell debris is removed. The DNA is then purified from soluble proteins and RNA by ethanol precipitation. Here, chromosomal DNA from bacteria is obtained by the method of Lovett and Keggins (Lovett and Keggins, 1979), with the exception of achieving the lysis of the protoplasts with 1% sodium dodecyl sulfate instead of Sarkosyl. The method can be applied for the isolation of genomic DNA from Gram-negative and Gram-positive bacteria and is particularly effective for rod-shaped Gram-positive bacteria such as *Bacillus subtilis*. A variety of convenient and fast DNA purification methods are also commercially available nowadays.

2. EQUIPMENT

Shaking incubator
Centrifuge
UV spectrophotometer
UV transparent cuvettes
Large-bore pipette
15-ml polypropylene centrifuge tubes

3. MATERIALS

Bacterial strains (source of DNA)
Lysogeny Broth (LB; or other appropriate bacterial liquid growth medium)
Tris base
Hydrochloric acid (HCl)
EDTA
Sodium chloride (NaCl)
Lysozyme
RNase
Proteinase K
Sodium dodecyl sulfate (SDS)

Tris-saturated Phenol (Roche)
Ethanol
Sterile deionized water

3.1. Solutions & buffers

Step 1 TES

Component	Final concentration	Stock	Amount
Tris–HCl, pH 7.5	20 mM	50 mM	40 ml
EDTA	5 mM	250 mM	2 ml
NaCl	100 mM	1 M	10 ml

Add water to 100 ml

RNase: Pancreatic, 2 mg ml^{-1} in water, heat-shocked at 80 °C for 15 min

Proteinase K: 10 mg ml^{-1} in TES, predigested at 37 °C for 90 min

10% SDS: Dissolve 10 g SDS in 100 ml water (final volume)

Step 2 Ethanol: 95%, ice-cold TE

Component	Final concentration	Stock	Amount
Tris–HCl, pH 8.0	10 mM	50 mM	2 ml
EDTA	0.1 mM	250 mM	4 μl

Add water to 10 ml

4. PROTOCOL

4.1. Preparation

Inoculate 10 ml of liquid broth with bacteria and let it grow overnight (18–20 h) at 37 °C with shaking (see also Growth Media for *E. coli*).

4.2. Duration

Preparation	18–20 h
Protocol	5 h

See Fig. 11.1 for the flowchart of the complete protocol, including preparation.

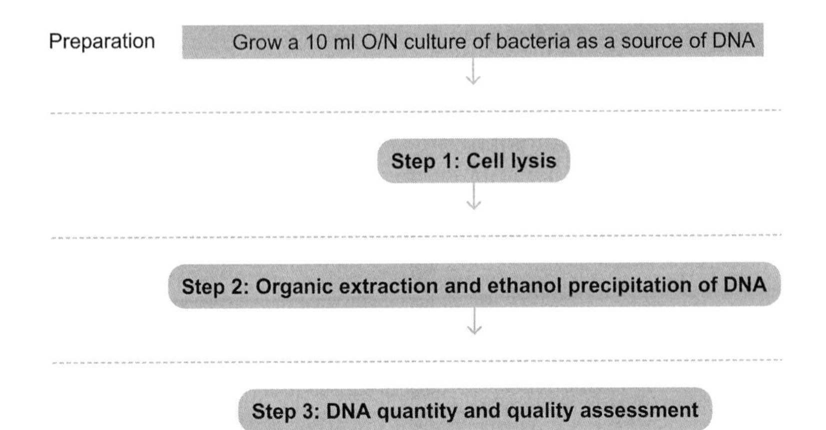

Figure 11.1 Flowchart of the complete protocol, including preparation.

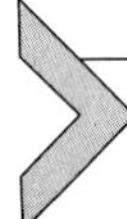

5. STEP 1 CELL LYSIS

5.1. Overview

The cells are collected and washed. A lysozyme buffer is used to damage cell walls and the detergent SDS further aids in disrupting cell membranes and lysing the cells. DNA, protein, and other cell components are released. RNase degrades RNA into smaller components and Proteinase K digests protein contaminants.

5.2. Duration

2.5 h

1.1 Centrifuge the 10 ml cell culture at 8000 rpm at 4 °C for 10 min. Discard supernatant.

1.2 Resuspend the cells in 2.5 ml TES buffer.

1.3 Centrifuge the cell suspension at 8000 rpm at 4 °C for 10 min. Discard supernatant.

1.4 Wash cells a second time with 2.5 ml TES.

1.5 Resuspend the cells in 2.5 ml TES buffer. Treat the cell suspension with 2 mg lysozyme. Swirl the suspension gently and incubate it at 37 °C for 30 min.

1.6 Add the RNase solution to 100 μg ml^{-1}, and incubate the suspension at 37 °C for 30 min.

1.7 Dilute the lysate to twice its original volume with the addition of 2.5 ml TES buffer.

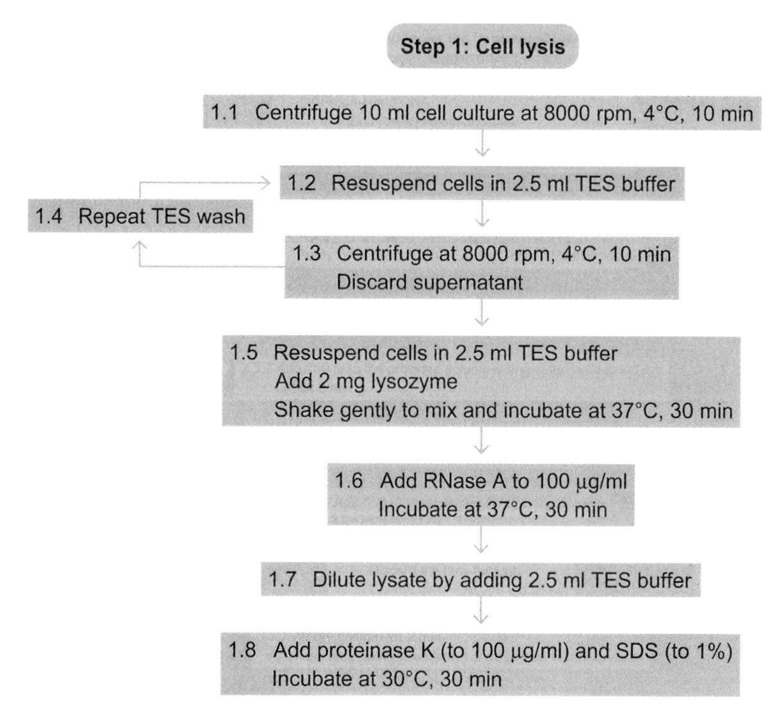

Figure 11.2 Flowchart of Step 1.

1.8 Treat the lysate with proteinase K (to 100 μg ml^{-1}) and SDS (to 1%) at 30 °C for 30 min.

See Fig. 11.2 for the flowchart of Step 1.

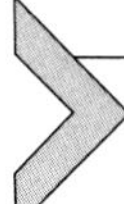

6. STEP 2 ORGANIC EXTRACTION AND ETHANOL PRECIPITATION OF DNA

6.1. Overview

DNA is separated from other cellular contaminants by organic extraction and is then precipitated from the solution by ethanol.

6.2. Duration

1.5 h

2.1 Shake the lysate gently with an equal volume (5 ml) of TES-saturated phenol at room temperature for 10 min.

2.2 Centrifuge the sample for 10 min at room temperature to separate the phases.

2.3 Transfer the aqueous phase with a large-bore pipette to a new 15-ml centrifuge tube.
2.4 Add an equal volume of TES-saturated phenol, shake the sample gently for 1–2 min and centrifuge it at room temperature for 5 min.
2.5 Repeat the phenol extraction (Steps 2.3–2.4) twice.
2.6 Transfer the final aqueous phase (~2.5 ml) to two volumes of cold 95% ethanol and hold it at 4 °C for at least 30 min.
2.7 Centrifuge the sample at 12 000 rpm at 4 °C for 20 min and leave the precipitate to air-dry for 15 min.
2.8 Resuspend the precipitate in 0.5–1 ml of TE (can take up to 12–18 h).
2.9 Store the dissolved DNA sample at 4 °C.

6.3. Tip

After the phenol extraction the sample can be extracted with one volume of chloroform or chloroform:isoamyl alcohol to remove any trace of phenol.

6.4. Tip

Transfer the upper aqueous phase, which contains DNA, without disturbing the interface in order to avoid the transfer of debris from the interface.

6.5. Tip

Once ethanol is added to the sample, the chromosomal DNA becomes visible as white fibers.

6.6. Tip

In order to desalt the DNA sample, resuspend the precipitate in 2 ml of TES and dialyse against 1000 volumes of TES at 4 °C for 12–18 h. The sample could be washed with ethanol 70% to remove salts, but dialysis is recommended.

6.7. Tip

TE is suitable for long-term storage of DNA but the EDTA it contains can interfere with certain downstream applications. If this is the case, precipitate DNA and resuspend it in sterile water.

6.8. Caution

Phenol can cause chemical burns that may not be noticed until the following day since it numbs the skin as it burns. Wear gloves, goggles and a lab coat. Also, keep tubes capped tightly. Work in a fume hood.

See Fig. 11.3 for the flowchart of Step 2.

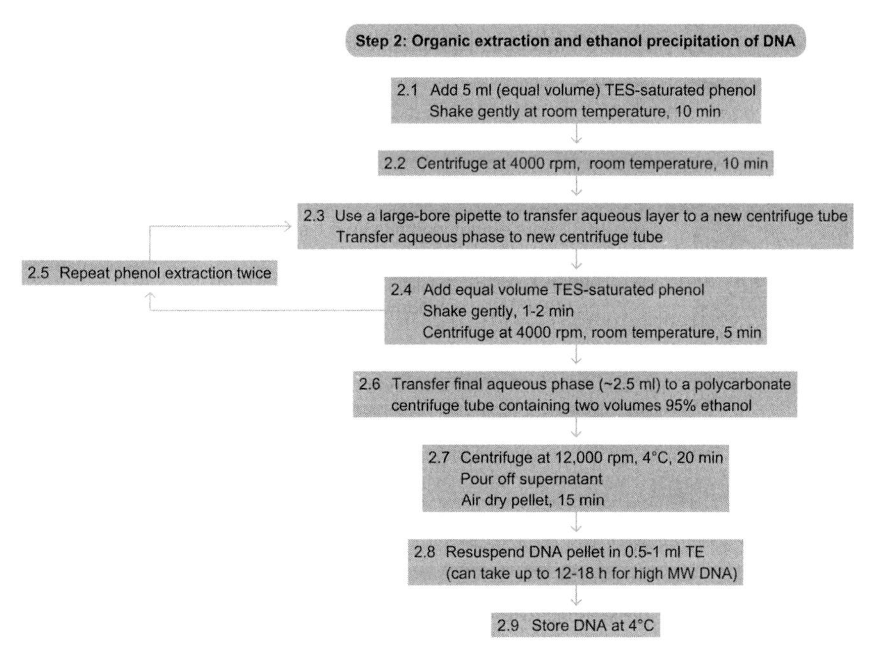

Figure 11.3 Flowchart of Step 2.

7. STEP 3 DNA QUANTITY AND QUALITY ASSESSMENT

7.1. Overview

Here you assess the quantity of DNA present in the sample by measuring the absorbance (A) at wave lengths of 260 and 280 nm and/or agarose gel electrophoresis (see also Explanatory Chapter: Nucleic Acid Concentration Determination and Agarose Gel Electrophoresis). *Gel electrophoresis is less accurate in quantitation but offers the advantage of allowing a qualitative analysis to take place.*

7.2. Duration

About 1 h

3.1 Scan a dilution of the DNA-containing solution in a UV spectrophotometer over the range of 240–320 nm to estimate DNA purity and DNA concentration.

3.2 Visualize the DNA present in the sample on a 0.8% agarose gel containing a DNA-intercalating dye such as ethidium bromide. Use a UV transilluminator to see the DNA band and to take a picture (Fig. 11.4).

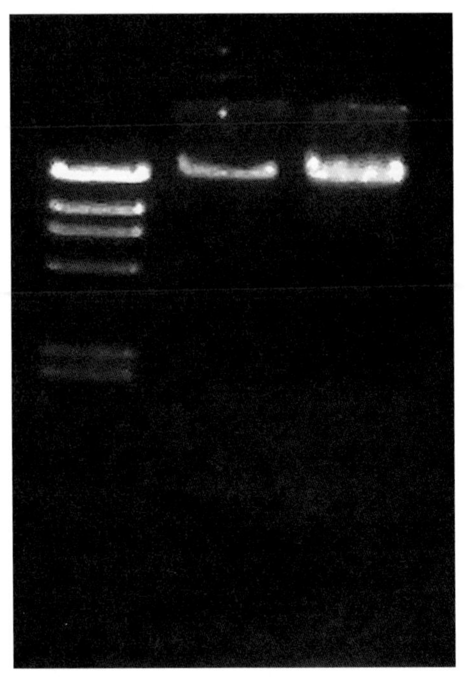

Figure 11.4 Agarose gel electrophoresis of microbial genomic DNA. Genomic DNA was electrophoresed on a 0.8% agarose gel and stained with ethidium bromide. Lane 1: Lambda DNA marker (λDNA/*Hind*III); lane 2: Genomic DNA isolated from *Corynebacterium glutamicum* ATCC 21253; lane 3: Genomic DNA isolated from *Brevibacterium lactofermentum* ATCC 21799 (Andreou, 2005).

7.3. Tip

An A_{260}/A_{280} ratio of around 1.8 (between 1.7 and 2.0) suggests a high-quality DNA sample. Smaller ratios indicate protein or phenol contamination (ratios >0.5). Once you have established that your sample contains pure DNA you can determine the DNA concentration. An A_{260} of 1.0 suggests a concentration of 50 μg ml^{-1} of dsDNA.

7.4. Tip

The presence of a single band near the wells is indicative of whole, uncut genomic DNA. A slight smear is acceptable but is indicative of partial degradation.

7.5. Caution

Ethidium bromide is a known mutagen and a carcinogen so it should be handled with extreme care. Wear gloves and a lab coat. Do not contaminate the surrounding lab bench area and equipment with EtBr. Dispose of EtBr waste in designated containers.

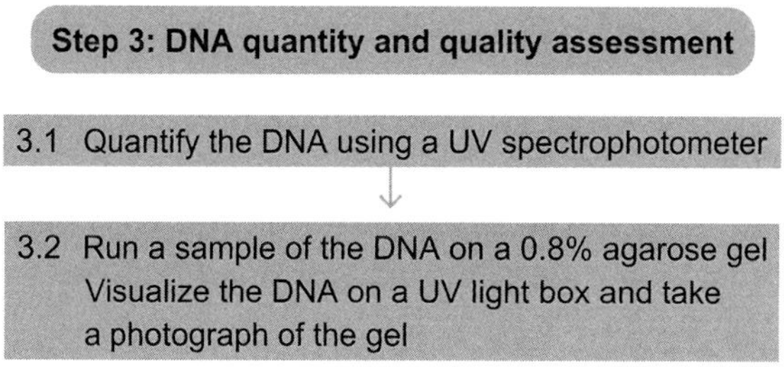

Step 3: DNA quantity and quality assessment

3.1 Quantify the DNA using a UV spectrophotometer

↓

3.2 Run a sample of the DNA on a 0.8% agarose gel
Visualize the DNA on a UV light box and take
a photograph of the gel

Figure 11.5 Flowchart of Step 3.

7.6. Caution

UV presents a safety hazard. Protect your eyes and skin from exposure to UV light.

See Fig. 11.5 for the flowchart of Step 3.

REFERENCES

Referenced Literature

Andreou, L. V. (2005). Cloning and heterologous expression of trehalose biosynthetic genes of Corynebacteria. *Unpublished MSc thesis*. Ioannina, Greece: University of Ioannina.

Lovett, P. S., & Keggins, K. M. (1979). *Bacillus subtilis* as a host for molecular cloning. In R. Wu (Ed.), *Methods in Enzymology*: *vol. 68*. (pp. 342–357). New York: Academic Press.

Referenced Protocols in Methods Navigator

Preparation of genomic DNA from *Saccharomyces cerevisiae*.
Isolation of Genomic DNA from Mammalian Cells.
Growth Media for *E. coli*.
Explanatory Chapter: Nucleic Acid Concentration Determination.
Agarose Gel Electrophoresis.

CHAPTER TWELVE

Preparation of Genomic DNA from *Saccharomyces cerevisiae*

Jessica S. Dymond[1]
The High Throughput Biology Center and Department of Molecular Biology and Genetics, Johns Hopkins University School of Medicine, Baltimore, MD, USA
[1]Corresponding author: e-mail address: jsiege16@jhmi.edu

Contents

1. Theory 154
2. Equipment 154
3. Materials 154
   3.1 Solutions & buffers 154
4. Protocol 155
   4.1 Preparation 155
   4.2 Duration 155
5. Step 1 Harvesting Cells from the Overnight Culture 155
   5.1 Overview 155
   5.2 Duration 155
   5.3 Tip 156
6. Step 2 Initial DNA Extraction 156
   6.1 Overview 156
   6.2 Duration 157
   6.3 Tip 157
   6.4 Tip 157
   6.5 Tip 157
   6.6 Tip 157
7. Step 3 Purification of the Crude DNA Preparation 157
   7.1 Overview 157
   7.2 Duration 158
   7.3 Tip 159
   7.4 Tip 159
References 160
Source References 160

Abstract

The ability to isolate genomic DNA rapidly and effectively for analysis by PCR, Southern blotting, or other methods is an essential skill. This protocol provides a fast and efficient method for obtaining genomic DNA from *S. cerevisiae*.

Methods in Enzymology, Volume 529
ISSN 0076-6879
http://dx.doi.org/10.1016/B978-0-12-418687-3.00012-4

1. THEORY

This method to extract genomic DNA from yeast is fast and reliably generates ample DNA for multiple applications. If high molecular weight genomic DNA is required, an alternative method employing enzymatic lysis, rather than mechanical lysis, should be employed (Boeke et al., 1985).

2. EQUIPMENT

Refrigerated centrifuge
Refrigerated microcentrifuge
Vortex mixer
Micropipettors
Micropipettor tips
15-ml polypropylene centrifuge tubes
1.5-ml microcentrifuge tubes
Glass beads, 425–600 μm, acid-washed

3. MATERIALS

YPD media
Tris base
EDTA
Hydrochloric acid (HCl)
Sodium chloride (NaCl)
Triton X-100
Sodium dodecyl sulfate (SDS)
Phenol/chloroform/isoamyl alcohol, 25:24:1 (v/v/v)
Ethanol
Ammonium acetate (NH_4OAc)
RNase A
Sterile H_2O

3.1. Solutions & buffers

Step 1 Lysis Buffer

Component	Final Concentration	Stock	Amount
Tris–HCl, pH 8.0	10 mM	1 M	10 ml
EDTA, pH 8.0	1 mM	0.5 M	2 ml

NaCl	100 mM	5 M	20 ml
Triton X-100	2% (v/v)	100%	20 ml
Sodium dodecyl sulfate (SDS)	1% (v/v)	10% (w/v)	100 ml

Add water to 1 l

TE Buffer

Component	Final concentration	Stock	Amount
Tris–HCl, pH 8.0	10 mM	1 M	10 ml
EDTA, pH 8.0	1 mM	0.5 M	2 ml

Add water to 1 l

4. PROTOCOL

4.1. Preparation

Inoculate a single yeast colony into 10 ml YPD and grow, shaking, at 30 °C overnight.

4.2. Duration

Preparation	1 day
Protocol	3–4 h

See Fig. 12.1 for the flowchart of the complete protocol.

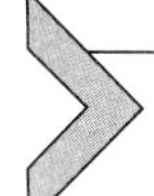

5. STEP 1 HARVESTING CELLS FROM THE OVERNIGHT CULTURE

5.1. Overview

Isolate yeast cells from the overnight culture and remove trace media.

5.2. Duration

15 min

1.1 Transfer the overnight culture into a 15-ml centrifuge tube.

1.2 Centrifuge at 3600 rpm at 4 °C for 5 min.

1.3 Resuspend the cell pellet in 1 ml H_2O. Transfer cells to a 1.5-ml microcentrifuge tube.

1.4 Centrifuge at 3600 rpm at 4 °C for 5 min. Pour off the supernatant.

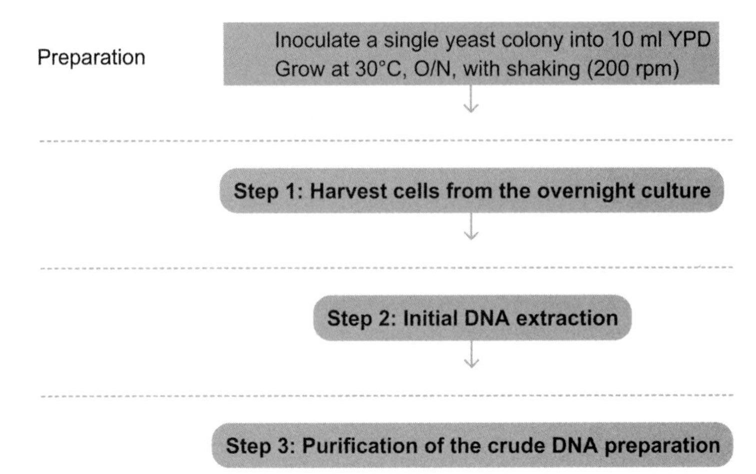

Figure 12.1 Flowchart of the complete protocol, including preparation.

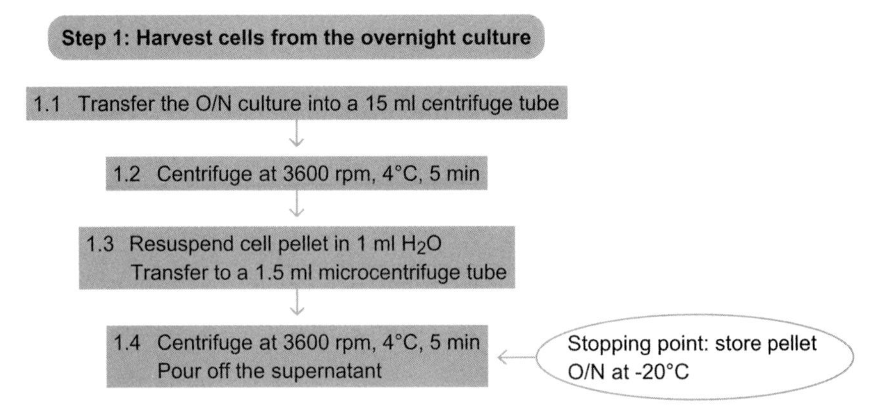

Figure 12.2 Flowchart of Step 1.

5.3. Tip

The protocol may be temporarily halted at this point; store the cell pellet at $-20\,^{\circ}C$ *overnight.*

See Fig. 12.2 for the flowchart of Step 1.

6. STEP 2 INITIAL DNA EXTRACTION

6.1. Overview

Cells are lysed and a crude DNA preparation is extracted.

6.2. Duration

35–45 min

2.1 Resuspend the cell pellet in 200 μl Lysis Buffer.
2.2 Add 200 μl glass beads and 400 μl phenol/chloroform/isoamyl alcohol (25:24:1).
2.3 Vortex for 2 min (single tube vortexer) or 10 min (multitube vortexer).
2.4 Add 400 μl TE Buffer and mix by vortexing briefly.
2.5 Centrifuge 10 min at maximum speed at room temperature.
2.6 Transfer 400 μl of the aqueous layer to a new microcentrifuge tube.
2.7 Add 1 ml ice-cold ethanol. Mix by inverting.
2.8 Centrifuge for 5 min at maximum speed at room temperature.
2.9 Wash the pellet with 70% ethanol.
2.10 Dry the pellet at room temperature for ~5 min.
2.11 Resuspend the pellet in 500 μl TE Buffer.

6.3. Tip

Remove glass beads from the top of the tube. Incomplete closure of the microcentrifuge tube will cause phenol/chloroform contamination of equipment and dissolution of tube labels.

6.4. Tip

Cell lysis can be confirmed by examining the resulting cell slurry under a microscope. If lysis is incomplete, continue vortexing in short intervals until lysis is nearly complete.

6.5. Tip

This preparation may be used for robust PCR applications, although the DNA will not be stable for long periods of time. For extremely stable genomic DNA, complete Step 3.

6.6. Tip

The protocol may be temporarily halted here. Store the pellet in TE at 4 °C overnight.

See Fig. 12.3 for the flowchart of Step 2.

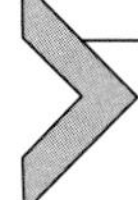

7. STEP 3 PURIFICATION OF THE CRUDE DNA PREPARATION

7.1. Overview

Contaminating RNA and cellular debris are removed.

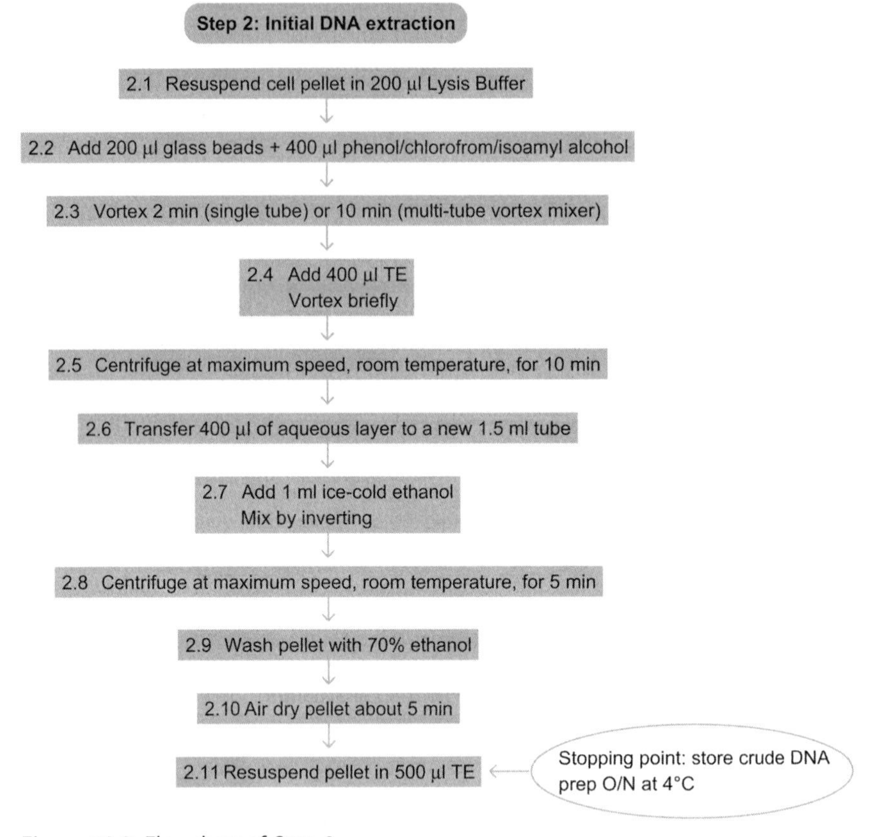

Figure 12.3 Flowchart of Step 2.

7.2. Duration

1–2 h

3.1 Add 15 µl 2 mg ml^{-1} RNase A.

3.2 Incubate at 37 °C for 30 min.

3.3 Add 500 µl phenol/chloroform/isoamyl alcohol (25:24:1). Vortex for 30 s (single tube vortexer) or 5 min (multitube vortexer) to mix.

3.4 Centrifuge for 5 min at maximum speed at room temperature.

3.5 Transfer 400 µl of the aqueous layer to a clean microcentrifuge tube.

3.6 Add 1 ml ice-cold ethanol, and 10 µl 4 M ammonium acetate. Invert to mix.

3.7 Centrifuge for 10 min at maximum speed at room temperature.

3.8 Wash the cell pellet with 70% ethanol.

3.9 Dry the pellet at room temperature for ~5 min.
3.10 Resuspend the pellet in 100 μl TE Buffer.

7.3. Tip

If a very high yield of genomic DNA is required, incubate at −20 °C for at least 1 h in Step 3.6.

7.4. Tip

If a very high yield of genomic DNA is required, perform the centrifugation for Step 3.7 at 4 °C.

See Fig. 12.4 for the flowchart of Step 3.

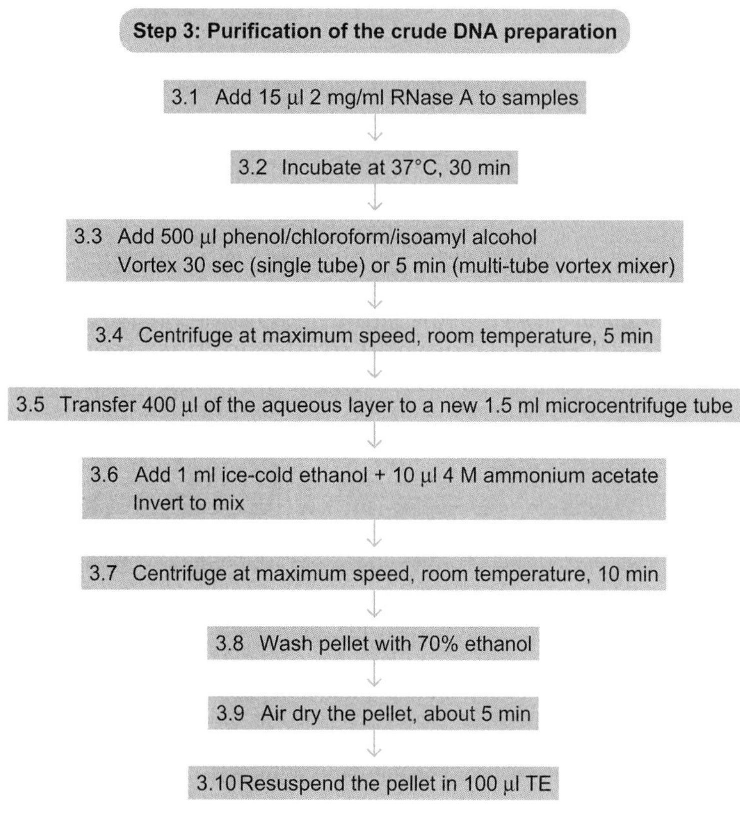

Figure 12.4 Flowchart of Step 3.

REFERENCES

Referenced Literature

Boeke, J. D., Garfinkel, D. J., Styles, C. A., & Fink, G. R. (1985). Ty elements transpose through an RNA intermediate. *Cell*, *40*, 491–500.

SOURCE REFERENCES

Hoffman, C. S. (1997). Preparation of yeast DNA. *Current Protocols in Molecular Biology*, *39*, 13.11.1–13.11.4.

CHAPTER THIRTEEN

Isolation of Genomic DNA from Mammalian Cells

Cheryl M. Koh[1]

Department of Pathology, The Johns Hopkins University School of Medicine, Baltimore, MD, USA
[1]Corresponding author: e-mail address: ckoh6@jhmi.edu

Contents

1. Theory 162
2. Equipment 162
3. Materials 162
   3.1 Solutions & buffers 163
4. Protocol 164
   4.1 Preparation 164
   4.2 Duration 164
   4.3 Caution 165
5. Step 1 Collection of Cells 165
   5.1 Overview 165
   5.2 Duration 165
6. Step 2 Cell Lysis 166
   6.1 Overview 166
   6.2 Duration 166
7. Step 3 Organic Extraction 167
   7.1 Overview 167
   7.2 Duration 167
8. Step 4 Ethanol Precipitation 167
   8.1 Overview 167
   8.2 Duration 168
   8.3 Tip 168
References 169

Abstract

The isolation of genomic DNA from mammalian cells is a routine molecular biology laboratory technique with numerous downstream applications. The isolated DNA can be used as a template for PCR, cloning, and genotyping and to generate genomic DNA libraries. It can also be used for sequencing to detect mutations and other alterations, and for DNA methylation analyses.

Methods in Enzymology, Volume 529
ISSN 0076-6879
http://dx.doi.org/10.1016/B978-0-12-418687-3.00013-6

1. THEORY

In general, isolation of genomic DNA from mammalian cells involves cell lysis, removal of proteins and other cellular contaminants, and organic extraction, followed by recovery of DNA. Typically, mammalian cells are lysed using a detergent-based buffer, which solubilizes lipids, thus disrupting the integrity of cell membranes. This releases cellular components into solution. Proteinase K is then added to facilitate the digestion and removal of proteins from the cell lysates. Next, organic extraction is carried out, in which a mixture of phenol, chloroform and isoamyl alcohol is added. DNA separates into the aqueous phase, while most other contaminants separate into the organic phase. An optional treatment with RNase A ensures that the sample is free from RNA contamination. DNA is then recovered by ethanol precipitation.

2. EQUIPMENT

Centrifuge
Incubator (50 °C)
Microcentrifuge
Vortex mixer
Magnetic stir plate
Micropipettors
Pipet-aid
15-ml Conical centrifuge tubes
1.5-ml Microcentrifuge tubes
Micropipettor tips
Beaker, 1 l
Magnetic stir bars
10-ml pipettes
pH indicator paper
Aluminum foil

3. MATERIALS

Sterile deionized water
Tris base
Hydrochloric acid (HCl)
Sodium chloride (NaCl)
EDTA disodium ($Na_2EDTA.2H_2O$)

Sodium dodecyl sulfate (SDS)
Proteinase K
Phenol
8-hydroxyquinoline
Chloroform
Isoamyl alcohol
Ammonium acetate (NH_4OAc)
Ethanol
Phosphate buffered saline (PBS)
RNase A (optional)

3.1. Solutions & buffers

Step 2 Lysis Buffer

Component	Final Concentration	Stock	Amount
NaCl	100 mM	1 M	1 ml
Tris–HCl, pH 8.0	10 mM	50 mM	2 ml
EDTA, pH 8.0	25 mM	250 mM	1 ml
SDS	0.5%	10%	0.5 ml
Proteinase K	0.1 mg ml^{-1}	1 mg ml^{-1}	1 ml

Add sterile water to 10 ml. Add proteinase K fresh before use

Step 3 Buffered Phenol

Component	Stock	Amount
Phenol		500 ml
8-hydroxyquinoline		0.5 g
Tris–HCl, pH 8.0	50 mM	Variable

See preparation step for instructions to make it up

Phenol Extraction Buffer

Component	Amount
Buffered phenol	25 ml
Chloroform	24 ml
Isoamyl alcohol	1 ml

Mix well. Wrap container in aluminium foil to protect from light

Step 4 7.5 M ammonium acetate

Dissolve 57.8 g ammonium acetate in 100 ml (final volume) of sterile deionized water.

70% Ethanol

Mix 70 ml 100% ethanol and 30 ml sterile deionized water.

TE Buffer

Component	Final concentration	Stock	Amount
Tris–HCl, pH 8	10 mM	50 mM	2 ml
EDTA	1 mM	250 mM	40 μl

Add sterile deionized water to 10 ml

4. PROTOCOL

4.1. Preparation

Make the buffered phenol. Add 0.5 g 8-hydroxyquinoline to a glass beaker containing a stir bar. Add 500 ml phenol and 500 ml 50 mM Tris–HCl, pH 8.0.

Cover with aluminium foil to protect light-sensitive reagents from oxidation.

Stir for 10 min at room temperature, allowing the phases to separate.

Decant most of the upper aqueous phase into an appropriate waste container. Carefully remove the remainder with a 10-ml pipette. Add another 500 ml 50 mM Tris–HCl, pH 8.0. Stir and decant aqueous phase as before. Check the pH of the lower phenol phase with pH paper. Repeat equilibrations with 50 mM Tris–HCl, pH 8.0 until the pH of the phenol phase reaches 8.0.

Add 250 ml of 50 mM Tris–HCl, pH 8.0. Store at 4 °C in either a brown glass bottle or a clear glass bottle wrapped in aluminium foil to protect from light.

Have cells ready to extract DNA.

4.2. Duration

Preparation	Variable
Protocol	About 2 days

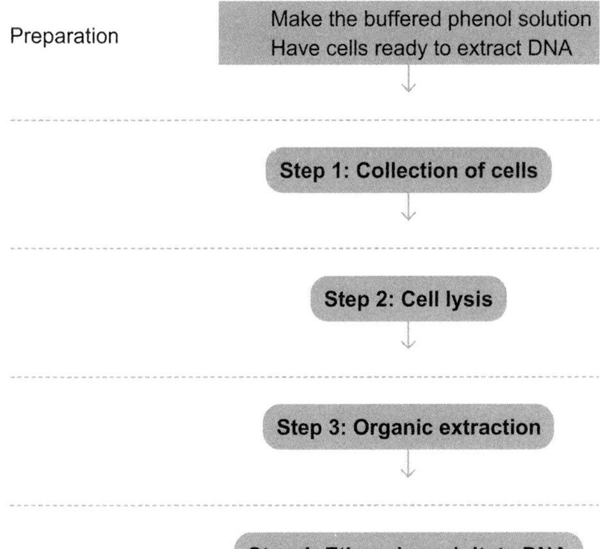

Figure 13.1 Flowchart of the complete protocol, including preparation.

4.3. Caution

Phenol is highly corrosive. It should be handled with care and should be opened only in a fume hood. Used phenol should be disposed appropriately according to chemical safety regulations, not down the sink.

See Fig. 13.1 for the flowchart of the complete protocol.

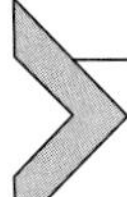

5. STEP 1 COLLECTION OF CELLS

5.1. Overview

Cells grown in culture are harvested and washed.

5.2. Duration

20 min

1.1 Collect cells in a 15-ml conical centrifuge tube. If starting with tissue culture cells in suspension, directly collect suspension in a conical tube. If starting with adherent cells in culture, trypsinize and collect cells in a conical tube.

1.2 Determine the number of cells collected.

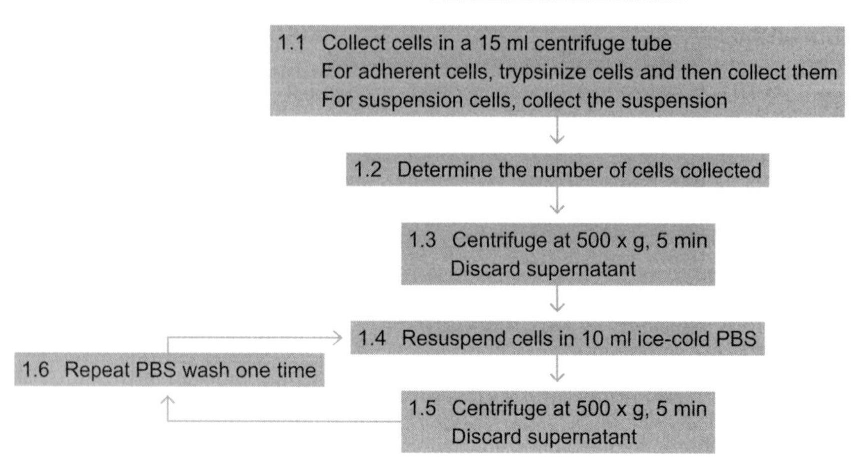

Figure 13.2 Flowchart of Step 1.

1.3 Centrifuge at 500 × g for 5 min. Discard supernatant.
1.4 Resuspend in 10 ml ice-cold PBS.
1.5 Centrifuge at 500 × g for 5 min. Discard supernatant.
1.6 Repeat PBS wash (Steps 1.4 and 1.5).
See Fig. 13.2 for the flowchart of Step 1.

6. STEP 2 CELL LYSIS

6.1. Overview

A detergent-based buffer is used to disrupt cell membranes and release DNA, protein, and other cell components. Proteinase K removes protein contaminants (see also Lysis of mammalian and Sf9 cells).

6.2. Duration

12–18 h

2.1 Resuspend cells in a suitable amount of Lysis Buffer. In general, use 1 ml of buffer per 10^8 cells.
2.2 Transfer to a 1.5-ml microcentrifuge tube. Vortex the samples.
2.3 Incubate at 50 °C for 12–18 h.
See Fig. 13.3 for the flowchart of Step 2.

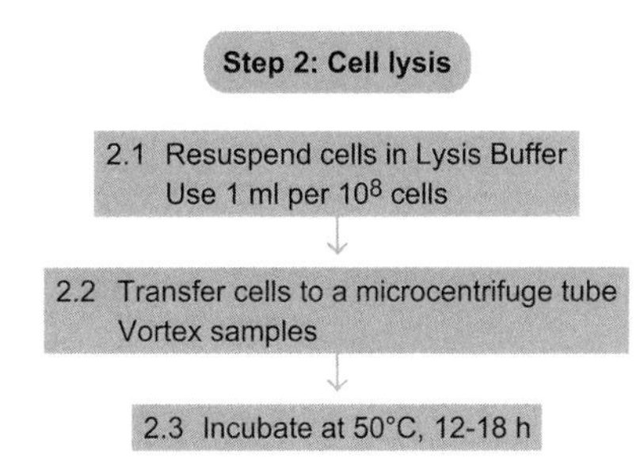

Figure 13.3 Flowchart of Step 2.

7. STEP 3 ORGANIC EXTRACTION

7.1. Overview

Organic extraction is used to separate DNA from other cellular contaminants.

7.2. Duration

10 min

3.1 Add an equal volume of phenol extraction buffer to the lysed cell suspension.

3.2 Vortex for 10 s.

3.3 Centrifuge at $2000 \times g$ for 5 min in a microcentrifuge at room temperature.

3.4 Transfer the upper aqueous phase, which contains DNA, to a new microcentrifuge tube. Determine the volume of the aqueous phase obtained.

3.5 Optional: Add RNase A to a final concentration of 20 μg ml^{-1}. Incubate sample at 37 °C for 20 min.

See Fig. 13.4 for the flowchart of Step 3.

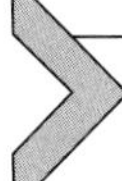

8. STEP 4 ETHANOL PRECIPITATION

8.1. Overview

Ethanol is used to precipitate DNA from solution.

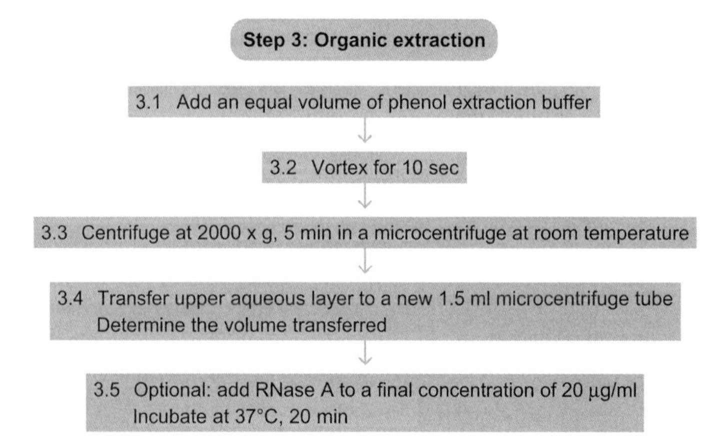

Figure 13.4 Flowchart of Step 3.

8.2. Duration

40 min

4.1 Add 0.5 volumes of 7.5 M ammonium acetate and 2 volumes of 100% ethanol. Vortex.

4.2 Centrifuge at 2000 × g for 10 min in a microcentrifuge at room temperature.

4.3 Remove supernatant carefully, without disrupting DNA pellet.

4.4 Add 1 ml of 70% ethanol and invert tube several times to wash the DNA pellet.

4.5 Centrifuge at 1700 × g for 5 min in a microcentrifuge at room temperature.

4.6 Remove supernatant. Uncap the tube and allow the pellet to air-dry for 10–15 min.

4.7 Resuspend DNA in TE buffer or sterile water.

8.3. Tip

Resuspension in TE buffer stabilizes the DNA for long-term storage. However, the EDTA in the TE buffer may interfere with certain downstream applications, in which case, resuspension in water is preferred.

See Fig. 13.5 for the flowchart of Step 4.

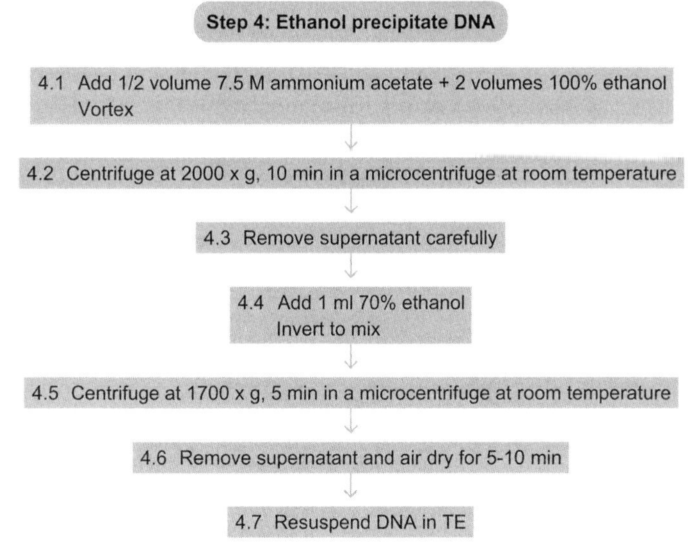

Figure 13.5 Flowchart of Step 4.

REFERENCES

Related Literature

Ausubel, F. M., et al. (1992). Preparation and analysis of DNA. *Short Protocols in Molecular Biology: vol. 2.1*. New York: Wiley.

Referenced Protocols in Methods Navigator

Lysis of mammalian and Sf9 cells

CHAPTER FOURTEEN

Sanger Dideoxy Sequencing of DNA

Sarah E. Walker, Jon Lorsch[1]

Department of Biophysics and Biophysical Chemistry, Johns Hopkins University School of Medicine, Baltimore, MD, USA

[1]Corresponding author: e-mail address: jlorsch@jhmi.edu

Contents

1. Theory 172
2. Equipment 172
3. Materials 173
   3.1 Solutions & buffers 174
4. Protocol 175
   4.1 Preparation 175
   4.2 Duration 175
   4.3 Caution 176
5. Step 1 Anneal Primer to DNA 176
   5.1 Overview 176
   5.2 Duration 176
   5.3 Tip 177
6. Step 2 Labeling 177
   6.1 Overview 177
   6.2 Duration 177
   6.3 Tip 178
   6.4 Tip 178
7. Step 3 Extension/Termination 178
   7.1 Overview 178
   7.2 Duration 178
8. Step 4 Resolution of Labeled Products by Urea-PAGE 179
   8.1 Overview 179
   8.2 Duration 179
   8.3 Tip 182
   8.4 Tip 182
   8.5 Tip 182
   8.6 Tip 182
   8.7 Tip 182
References 184

Methods in Enzymology, Volume 529
ISSN 0076-6879
http://dx.doi.org/10.1016/B978-0-12-418687-3.00014-8

Abstract

While the ease and reduced cost of automated DNA sequencing has largely obviated the need for manual dideoxy sequencing for routine purposes, specific applications require manual DNA sequencing. For instance, in studies of enzymes or proteins that bind or modify DNA, a DNA ladder is often used to map the site at which an enzyme is bound or a modification occurs. In these cases, the Sanger method for dideoxy sequencing provides a rapid and facile method for producing a labeled DNA ladder.

1. THEORY

Sequencing of DNA has been revolutionized by the advent of dideoxynucleotides and the Sequenase enzyme. Dideoxynucleotide triphosphates are readily incorporated into a growing DNA chain, but lack the 3′ hydroxyl group necessary to allow the chain to continue, and effectively terminate polymerization. Sequenase is a T7 DNA polymerase that has been chemically (Sequenase 1.0) or genetically (Sequenase 2.0) modified to remove its 3′–5′ exonuclease activity (Fuller et al., 1996). Sequencing DNA using Sequenase and dideoxynucleotides takes place in four steps. First a primer is annealed to single-stranded DNA. Single-stranded DNA can be synthesized by a bacteriophage, or as in the more frequently used method described below, by denaturing double-stranded plasmid DNA. Next, the annealed primer is extended to form various lengths of labeled DNA. Label can be incorporated in the form of a radiolabeled primer used in the annealing step, or, as described below, a radioactive nucleotide can be incorporated during the extension step. We choose to use [α-^{32}P]-deoxythymidine, as it is less costly than other radioactive nucleotides, but any [^{35}S]-, [α-^{33}P]-, or [α-^{32}P]-deoxynucleotide can be used. Finally, the extension reaction is divided between four different termination reactions, each containing either dideoxy-adenosine, -cytosine, -guanosine, or -thymidine. These four reactions are then resolved on a denaturing polyacrylamide gel and visualized by autoradiography.

2. EQUIPMENT

Microcentrifuge
Heating block (68 and 95 °C)
Incubator (42 °C)
Sequencing gel apparatus (e.g., BRL Model S2, available from Labrepco or Lab-X, or equivalent)

Power supply (capable of 65 W)
Sequencing gel plates (Labrepco for BRL Model S2 plates)
Sharktooth comb (CBS scientific, 0.2 mm × 62 well comb)
0.2 mm spacers with foam ears, slightly longer than plates
0.4 mm spacers (CBS scientific)
Large binder clips (or other gel clamps)
Gel dryer
Phosporimager (or film processor)
Phosphor screen (or autoradiography film)
Micropipettors
Micropipettor tips
1.5-ml microcentrifuge tubes
50-ml conical tubes
200 ml beaker
500 ml Sidearm vacuum flask
Buchner funnel
Microtiter plate (10 μl wells)
Glass fiber prefilter
Whatman 3MM chromatography paper
Syringe, 60 ml
Needle, 18 gauge

3. MATERIALS

DNA to be sequenced
Sequencing primer
dNTP stocks (dATP, dCTP, dGTP, dTTP, 10 mM)
[α-^{32}P]-dTTP (800 Ci mmol^{-1}, 10 mCi ml^{-1})
ddNTP stocks (ddATP, ddCTP, ddGTP, ddTTP, 10 mM)
N-tris(hydroxy–methyl)methyl-2-aminoethanesulfonic acid (TES)
Sodium chloride (NaCl)
Magnesium chloride ($MgCl_2$)
Hydrochloric acid (HCl)
Formamide
EDTA
Xylene cyanol FF
Bromphenol blue
Sodium hydroxide (NaOH)
Dithiothreitol (DTT)

Sequenase (USB Biochemicals)
40% Acrylamide:bisacrylamide solution (19:1)
Tris base
Boric acid (H_3BO_3)
Urea
Ammonium persulfate (APS)
Tetramethylethylenediamine (TEMED)
95% Ethanol
Siliconizing agent (e.g., Sigma-Cote or a nonsilicone smoothing agent)

3.1. Solutions & buffers

Step 1 TMN

Component	Final concentration	Stock	Amount
TES	560 mM	2.8 M	1 ml
NaCl	400 mM	2 M	1 ml
$MgCl_2$	160 mM	0.8 M	1 ml
HCl	1.21 M	12.1 M (37%)	1 ml
H_2O			1 ml

Step 2 Termination Mixes (ddA, ddC, ddG, ddT)

Component	Final concentration	Stock	Amount
dATP	80 μM	10 mM	8 μl
dCTP	80 μM	10 mM	8 μl
dGTP	80 μM	10 mM	8 μl
dTTP	80 μM	10 mM	8 μl
Particular ddNTP	8 μM	10 mM	0.8 μl
H_2O			967 μl

-dT mix

Component	Final concentration	Stock	Amount
dATP	1.5 μM	10 mM	0.75 μl
dCTP	1.5 μM	10 mM	0.75 μl
dGTP	1.5 μM	10 mM	0.75 μl
H_2O			5 ml

Step 3 STOP dye

Component	Final concentration	Stock	Amount
Formamide	95%		47.5 ml
EDTA, pH 8.0	20 mM	500 mM	2 ml
Xylene cyanol FF	0.05%		2.5 mg
Bromophenol blue	0.05%		2.5 mg

Add water to 50 ml

Step 4 10 × TBE

Component	Final concentration	Amount
Tris base	89 mM	108 g
Boric acid	89 mM	55 g
EDTA	2 mM	7.44 g

Add water to 1 l

10% APS

Dissolve 5 g ammonium persulfate in 50 ml H_2O in a 50-ml conical tube.

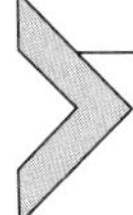

4. PROTOCOL

4.1. Preparation

Isolate plasmid DNA (see Isolation of plasmid DNA from bacteria) using either a standard miniprep kit or CsCl density gradient centrifugation for best results.

Order or synthesize an oligonucleotide primer (see Explanatory chapter: PCR -Primer design) designed to anneal ~30 nucleotides from where you wish to start reading the DNA sequence, as you would for PCR, but use only one primer for each sequence. Both plasmid and oligonucleotides should be dissolved in water or a low salt solution without EDTA.

4.2. Duration

Preparation	About 1 day
Protocol	5–9 h + overnight

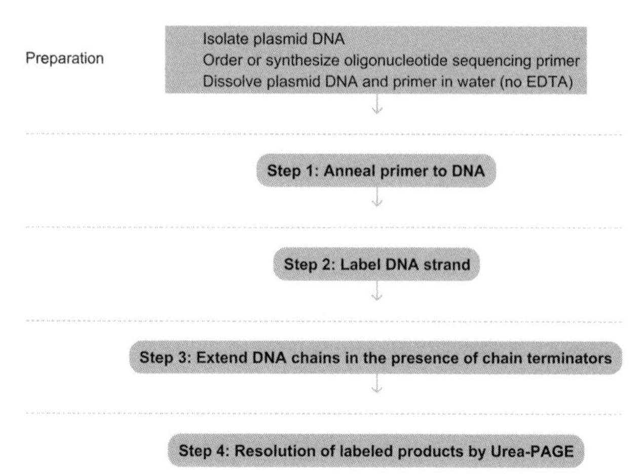
Preparation
Isolate plasmid DNA
Order or synthesize oligonucleotide sequencing primer
Dissolve plasmid DNA and primer in water (no EDTA)
Step 1: Anneal primer to DNA
Step 2: Label DNA strand
Step 3: Extend DNA chains in the presence of chain terminators
Step 4: Resolution of labeled products by Urea-PAGE

Figure 14.1 Flowchart of the complete protocol, including preparation.

4.3. Caution

Consult your institute Radiation Safety Officer for proper ordering, handling, and disposal of radioactive materials.

See Fig. 14.1 for the flowchart of the complete protocol.

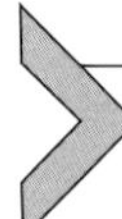

5. STEP 1 ANNEAL PRIMER TO DNA

5.1. Overview

Anneal a complementary primer ~30 nucleotides from where you wish to start reading the DNA sequence.

5.2. Duration

30 min

1.1 Mix the following in a 1.5-ml tube:
1–10 μg DNA
1 μl primer at 3 pmol μl^{-1}
1 μl 1 M NaOH
H_2O to bring volume to 10 μl

1.2 Incubate for 10 min at 68 °C.

1.3 Add 2 μl of TMN buffer as each tube is removed from the water bath or heat block.

1.4 Transfer tubes to a rack at room temperature for 10 min.

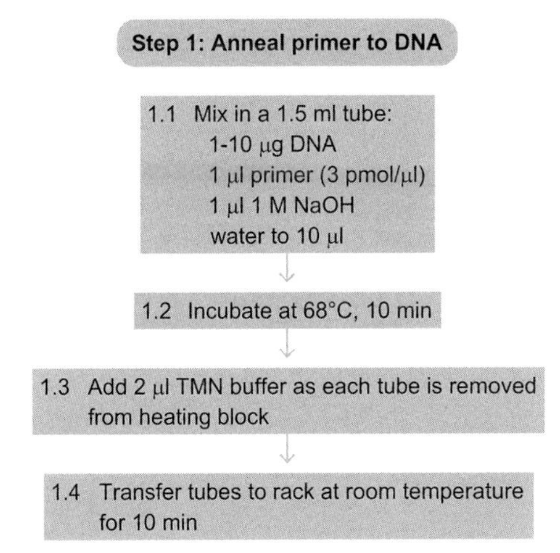

Figure 14.2 Flowchart of Step 1.

5.3. Tip

Make labeling mix (for Step 2) and aliquot termination mixes (for Step 3) during incubations (Steps 1.2 and 1.4) for best use of time.

See Fig. 14.2 for the flowchart of Step 1.

6. STEP 2 LABELING

6.1. Overview

Extend the annealed primer, incorporating ^{32}P-labeled TTP to label each strand.

6.2. Duration

5 min

2.1 Aliquot 2.5 μl of termination mixes to four wells of a microtiter dish for each reaction. Add ddA to one column, ddC to the next, ddG to the third, and ddT to the fourth column, with as many rows as you have reactions.

2.2 Make a labeling master mix for 10–12 reactions plus 2 extra for pipetting loss.

For each reaction include:

2 μl – dT mix
1 μl 100 mM DTT
0.5–1 μl [α-^{32}P]-TTP

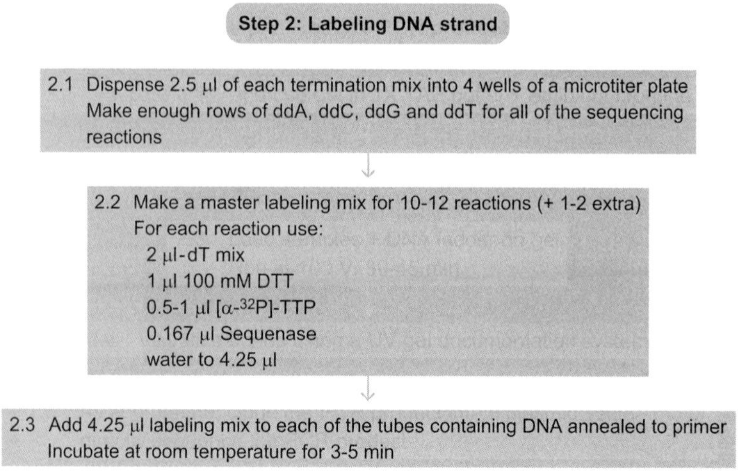

Step 2: Labeling DNA strand

2.1 Dispense 2.5 μl of each termination mix into 4 wells of a microtiter plate
Make enough rows of ddA, ddC, ddG and ddT for all of the sequencing reactions

2.2 Make a master labeling mix for 10-12 reactions (+ 1-2 extra)
For each reaction use:
2 μl-dT mix
1 μl 100 mM DTT
0.5-1 μl [α-^{32}P]-TTP
0.167 μl Sequenase
water to 4.25 μl

2.3 Add 4.25 μl labeling mix to each of the tubes containing DNA annealed to primer
Incubate at room temperature for 3-5 min

Figure 14.3 Flowchart of Step 2.

0.167 μl Sequenase
H_2O to make 4.25 μl

2.3 Add 4.25 μl labeling mix to each tube from Step 1 (containing DNA annealed with primer) and leave at room temperature for 3–5 min.

6.3. Tip

If a 10-μl microtiter dish is not available, PCR plates or tubes also work, and can be placed in a thermal cycler to incubate.

6.4. Tip

Use 0.5 μl [α-^{32}P]-TTP if less than 2-weeks old or 1 μl if older.

See Fig. 14.3 for the flowchart of Step 2.

7. STEP 3 EXTENSION/TERMINATION

7.1. Overview

Labeled DNAs will be extended and terminated by the incorporation of dideoxynucleotides.

7.2. Duration

30 min

3.1 Remove 3.5 μl each labeling reaction and add to each of the four termination mixes in the microtiter dish.

Step 3: Extend DNA chains in the presence of chain terminators
3.1 Remove 3.5 μl of each labeling reaction and add to each of the 4 termination mixes
3.2 Incubate microtiter dish in a 42°C incubator, 10 min
3.3 Add 4 μl STOP dye to each reaction

Figure 14.4 Flowchart of Step 3.

3.2 Incubate microtiter dish in a 42 °C incubator for 10 min.
3.3 Add 4 μl STOP dye to each reaction.
See Fig. 14.4 for the flowchart of Step 3.

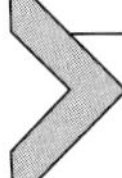

8. STEP 4 RESOLUTION OF LABELED PRODUCTS BY UREA-PAGE

8.1. Overview

Pour a sequencing gel and resolve the labeled DNA products by electrophoresis (see Analysis of RNA by analytical polyacrylamide gel electrophoresis).

8.2. Duration

4–8 h + overnight

4.1 Place sequencing plates on two stable racks. Wipe plates with 95% ethanol until dry.
4.2 Treat plates with a siliconizing or nonsilicone smoothing agent. Do not use excessive amounts (>1 ml per plate) as this can allow the gel to migrate upwards during electrophoresis.
4.3 Assemble plate sandwich as shown in Fig. 14.5. Clamp together with five large binder clips on each side on the center of the spacers.
4.4 Make gel solution:
42 g Urea
10 ml 10× TBE
15 ml 40% acrylamide:bisacrylamide solution (19:1)
H_2O to make 100 ml
Solution will get cold. Heat in a microwave for 10 s or on a hot plate at high temperature to bring solution just to room temperature. Do not allow the acrylamide solution to get warmer than room

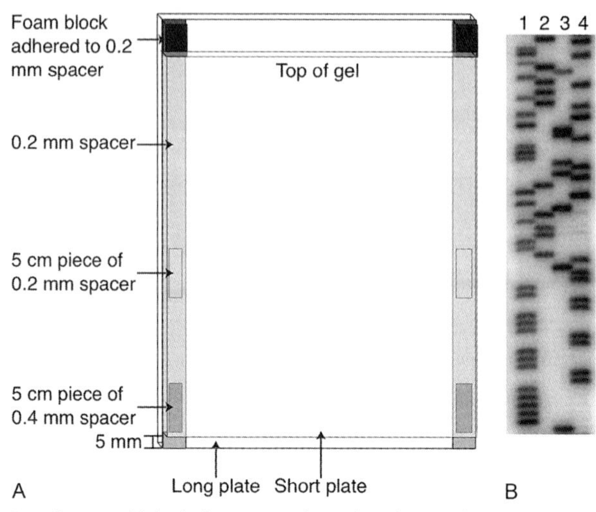

Figure 14.5 Resolution of labeled DNA products by electrophoresis through denaturing polyacrylamide gels. (a) Assembly of plates and spacers. A plate sandwich is assembled by first placing the spacers on the siliconized surface along each edge of the long plate. To create a wedge, a ~5–10 mm piece cut from a 0.4 mm spacer is placed at the bottom of the plate, and a 5–10 mm piece of a 0.2 mm spacer is placed ~10 cm from the bottom end of the plate. A 0.2 mm spacer with a foam block adhered to the top end is placed directly on top of these spacers to create a sandwich that is 0.2–0.6 mm thick. The short plate is then placed on top of the spacers so that the foam blocks are flush with the top edge of the short plate and the bottom edge of the short plate is ~5 mm from the bottom of the long plate, leaving a gap for pouring the gel. The sandwich is clamped together on the center of each spacer using binder clips. (b) Analysis of a sequencing gel. A gel was loaded with four reactions each containing (1) ddT, (2) ddG, (3) ddC, or (4) ddA and the DNA was sequenced from a reverse primer. The portion of the gel that is shown is read from top to bottom 5′-CTAATACGACTCACTATAGGTAAAGTGTCATAGCACCAACTGTTAATTAAATTAAATTAAAAAG-3′.

temperature as this can cause premature polymerization. Invert several times or stir to dissolve urea. Filter solution through a glass fiber filter in a Buchner funnel attached to a vacuum flask and degas briefly.

4.5 Transfer the filtered gel solution to a 250 ml beaker, and add 500 μl 10% APS and 50 μl of TEMED. Quickly stir solution with a 60 ml syringe, and draw solution into the syringe. Pour gel from bottom to top, by placing the syringe outlet at the center of the gel sandwich and slowly pushing the gel solution between the plates. Allow gel to move across the bottom section, and then tilt the plates slightly to facilitate gradual movement of the gel front into the plates until the syringe is almost empty. Lower the plates, refill the syringe, and continue tilting and pushing gel into plates until the gel front reaches the top of the short plate.

4.6 Insert the flat side of a sharktooth comb into the sandwich to make a large well ~0.5 cm into the gel (Combs from CBS scientific have a line marking how far to insert comb). Clamp the comb in place with three evenly spaced binder clips.

4.7 Rinse the syringe for future use. Allow the gel to polymerize for ~30 min to 1 h.

4.8 Remove comb and rinse well with a gentle stream of water to remove any strings of polyacrylamide that may be present.

4.9 Place the gel in sequencing apparatus without disturbing spacers. Fill the upper and lower reservoirs with 1× TBE and use a 60 ml syringe with a bent 18-gauge needle to rinse the well and underneath the gel to remove any bubbles.

4.10 Prerun gel at 65 W for 15 min to 1 h.

4.11 Place the microtiter dish (from Step 3.3) on a heat block at 95 °C for 3 min to denature DNA, and then place on ice. Wash well just before loading to remove any urea that has seeped out of gel, and insert sharktooth comb with teeth into the gel to ~1 mm deep.

4.12 Load 2 µl per lane in the wells between the teeth of the comb. After loading the first set of four sequencing lanes, load one lane with stop dye and then continue loading the remaining sets in adjacent lanes so that loading is asymmetrical for later analysis.

4.13 Run the gel at 65 W for ~1 h 45 min to 2 h to run bromphenol blue dye near the bottom of the gel in order to resolve the first ~100–150 nucleotides. Run the gel longer to see sequence further from the primer, keeping in mind that Sequenase extends only ~600–800 nucleotides.

4.14 Drain buffer from the upper chamber, and lay gel sandwich on a covered area of bench.

4.15 Remove spacers and use a metal spatula or wedge separator to carefully pry plates apart. If the gel is stuck to both plates, leave it until it cools completely and then attempt to pry apart.

4.16 Place Whatman 3MM paper on the gel and press gloved hands firmly around entire surface to stick the gel to Whatman paper. Carefully remove gel/paper from plate, starting from top. Cover the gel with plastic wrap and trim any areas that were not loaded to speed up the drying process.

4.17 Place the gel on a second piece of Whatman 3MM paper on the gel dryer and dry at 85 °C for 1 h, or longer, under vacuum.

4.18 Expose to X-ray film or phosphor screen for autoradiography overnight, and process (Fig. 14.5).

8.3. Tip

Handle the sequencing plates very carefully as they are much more easily broken than most electrophoresis plates. The slightest bump into the side of a sink or a benchtop will chip the plates, which can render the plates unusable.

8.4. Tip

Reactions may be stored at −20 °C for several days to several weeks, but the quality of sequence may deteriorate over time. It is recommended that samples be run on a gel as soon as possible after sequencing.

8.5. Tip

To maximize time, the sequencing gel can be poured before setting up annealing reactions (Step 1), and set up and prerun during extension/termination (Step 3).

8.6. Tip

To get more information from one set of sequencing reactions, simply load the reactions multiple times. Load a set of samples and run the gel for 1–2 h, pause running and load the samples again in empty lanes, and run the bromphenol blue from these samples to the bottom of the gel. The first set of samples loaded will have sequence further from the primer, while the samples run last will have sequence starting from the primer.

8.7. Tip

Drying times may vary depending on the pump. Allow the gel to cool before removing vacuum to prevent cracking if you are unsure that gel is dry.

See Fig. 14.6 for the flowchart of Step 4.

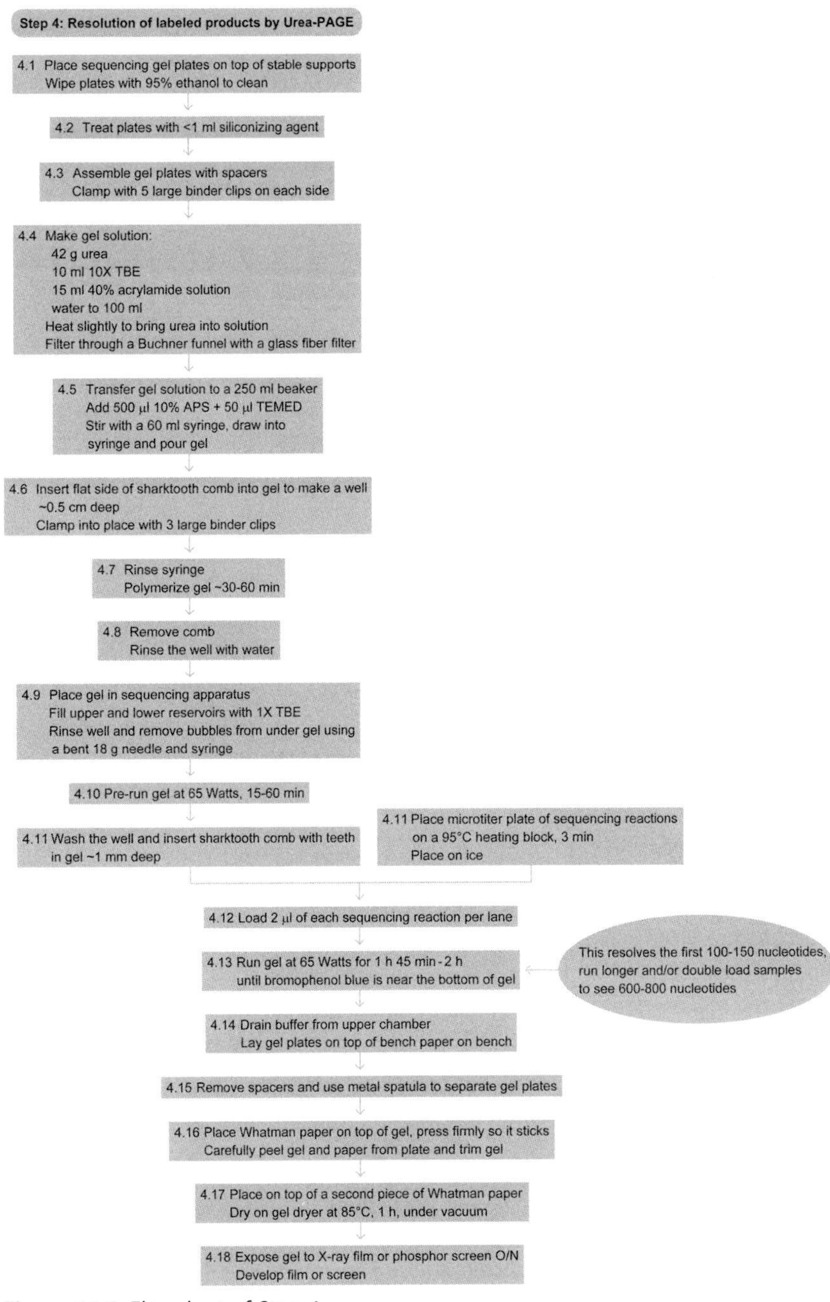

Figure 14.6 Flowchart of Step 4.

REFERENCES

Referenced Literature

Fuller, C. W., McArdle, B. F., Griffin, A. M., & Griffin, H. G. (1996). DNA sequencing using sequenase version 2.0 T7 DNA polymerase. *Methods in Molecular Biology*, *58*, 373–387.

Referenced Protocols in Methods Navigator

Isolation of plasmid DNA from bacteria
Explanatory chapter: PCR -Primer design
Analysis of RNA by analytical polyacrylamide gel electrophoresis

CHAPTER FIFTEEN

Preparation of Fragment Libraries for Next-Generation Sequencing on the Applied Biosystems SOLiD Platform

Srinivasan Yegnasubramanian[1]
Sidney Kimmel Comprehensive Cancer Center, Johns Hopkins University School of Medicine, Baltimore, MD, USA
[1]Corresponding author: e-mail address: syegnasu@jhmi.edu

Contents

1. Theory 186
2. Equipment 187
3. Materials 188
   3.1 Solutions & buffers 189
4. Protocol 190
   4.1 Preparation 190
   4.2 Duration 190
   4.3 Caution 191
5. Step 1 Shear the DNA to Generate Random Fragments 191
   5.1 Overview 191
   5.2 Duration 191
   5.3 Caution 192
6. Step 2 End-Repair the Fragmented DNA 193
   6.1 Overview 193
   6.2 Duration 193
   6.3 Tip 194
   6.4 Tip 194
   6.5 Tip 194
7. Step 3 Adaptor Ligation 194
   7.1 Overview 194
   7.2 Duration 194
   7.3 Tip 196
8. Step 4 Size Selection of Library 196
   8.1 Overview 196
   8.2 Duration 196
   8.3 Tip 197

Methods in Enzymology, Volume 529
ISSN 0076-6879
http://dx.doi.org/10.1016/B978-0-12-418687-3.00015-X

Abstract

The primary purpose of this protocol is to prepare genomic DNA libraries that can then be analyzed by massively parallel next-generation sequencing on the Applied Biosystems SOLiD platform. This protocol can be adapted to next-generation sequencing workflows to ultimately generate up to 1 billion 50 bp sequence tags from the ends of each of the DNA molecules in the library in a single next-generation sequencing run.

1. THEORY

The term next-generation sequencing refers to technologies that have enabled the massively parallel analysis of DNA sequence facilitated through the convergence of advancements in molecular biology, biochemistry and nucleic acid chemistry, electrical and mechanical engineering, and computational biology. Commercialized next-generation sequencing technologies are capable of generating massively parallel sequence data for tens to hundreds of millions of DNA templates simultaneously and generate more than four gigabases of sequence in a single day. These technologies have largely started to replace or significantly augment high-throughput Sanger sequencing for large-scale genomic projects, a trend that is likely to expand as the technologies continue to evolve and advance (Metzker, 2010). These technologies have created significant enthusiasm for the advent of a new era of individualized medicine.

The majority of the existing next-generation sequencing technologies employ massively parallel 'short-read' sequencing in which a short sequence (between 25 and 100 bp, depending on the technology and user options) at one or both ends of each of many hundreds of millions of DNA molecules are sequenced simultaneously. These reads are then aligned back to a reference genome assembly to allow measurement of genomic variation such as mutations and single-nucleotide polymorphisms compared to that reference genome assembly. In contrast, classical Sanger sequencing produces long reads of a single DNA molecule at a time, and high-throughput instruments allowing parallel sequencing of hundreds to thousands of sequences in parallel are in routine use. The advantage of the next-generation sequencing platforms is that the cost and throughput of sequencing is significantly lower (estimated to be 100-fold or more cheaper depending on the platform used). The major disadvantage of next-generation sequencing is the inability to generate long reads at this sequence throughput. However, this disadvantage is quickly diminishing due to increasing read-lengths and the use of mate-paired libraries that can circumvent much of the need for long sequence reads.

Because these technologies produce sequence from the ends of each molecule, it is common to fragment genomic DNA to produce DNA libraries with random ends to facilitate whole genome sequence coverage. Such a library, commonly referred to as a fragment library, is the most common type of library sequenced with Next-Generation Sequencing technologies. For each of the existing Next-Generation Sequencing technologies, the protocols used to generate fragment libraries compatible with massively parallel sequencing on that platform are different, but the basic principles are the same. In each case, genomic DNA is randomly fragmented by shearing, sonication, or nebulization. These fragments are modified with adaptor sequences that contain universal amplification and/or sequencing primer sequences. The resulting library of DNA molecules represents the fragment library. In this chapter we describe in detail the procedures used to generate fragment libraries for one of the commercially available next-generation sequencing platforms – the SOLiD platform marketed by Applied Biosystems. In the accompanying explanatory chapter (see Explanatory Chapter: Next Generation Sequencing), we review concepts and procedures for this platform as well as the other commercially available ones, covering the principles behind preparing the fragment libraries for massively parallel sequencing and the sequencing chemistries.

2. EQUIPMENT

Covaris S2 Sonication System
Microcentrifuge (e.g., Eppendorf model 5417R)
96-well plate PCR thermal cycler (e.g., GeneAmp® PCR System 9700)
NanoDrop® ND-1000 Spectrophotometer
E-Gel iBase and E-Gel Safe Imager (Invitrogen, G6466)
Vortex mixer
PicoFuge® (e.g., Stratagene)
Micropipettors (2, 20, 200, and 1000 μl)

Aerosol barrier pipet tips
Covaris microTUBEs
0.5-ml Eppendorf LoBind centrifuge tubes
1.5-ml Eppendorf LoBind centrifuge tubes
PCR strip tubes
Agilent bioanalyzer 2100 instrument (optional)

3. MATERIALS

Note: Several of the components below should be purchased from New England Biolabs (NEB) and/or from Life Technologies/Applied Biosystems using the indicated catalog numbers. All oligonucleotide sequences for adaptors and primers can be ordered from commercial vendors such as Integrated DNA Technologies (IDT) or Life Technologies.

SOLiD Library P1 adaptor plus strand (5′-CCACTACGCCTCCGCTTTCCTCTCTATGGGCAGTCGGTGAT-3′)
SOLiD Library P1 adaptor minus strand (5′-ATCACCGACTGCCCATAGAGAGGAAAGCGGAGGCGTAGTGGTT-3′)
SOLiD Library P2 adaptor plus strand (5′-AGAGAATGAGGAACCCGGGGCAGTT-3′)
SOLiD Library P2 adaptor minus strand (5′-CTGCCCCGGGTTCCTCATTCTCT-3′)
SOLiD Library PCR primer 1 (5′-CCACTACGCCTCCGCTTTCCTCTCTATG-3′)
SOLiD Library PCR primer 2 (5′-CTGCCCCGGGTTCCTCATTCT-3′)
NEBNext End Repair Reaction Buffer Mix (10×) (NEB, E6062A or E6062AA, or as part of E6060S/L, or see recipe below)
NEBNext End Repair Enzyme Mix (NEB, E6061A or E6061AA, or as part of E6060S/L)
SOLiD Library Column Purification Kit (Applied Biosystems, 4443744)
NEBNext Quick Ligation Reaction Buffer (NEB, E6064A or E6064AA, or as part of E6060S/L, or see recipe below)
T4 DNA Ligase (NEB, E6063A or E6063AA or as part of E6060S/L)
LongAmp Taq 2× Master Mix (NEB, E6065A or E6065AA or as part of E6060S/L)
E-Gels compatible with E-Gel iBase system
Tris base
Hydrochloric acid (HCl)
EDTA

Nuclease-free water
Isopropyl alcohol
Ethylene glycol
50-bp DNA ladder (e.g., Invitrogen 10416-014)

3.1. Solutions & buffers

Step 1 1× Low TE Buffer

Component	Final concentration	Stock	Amount
Tris–HCl (pH 8.0)	10 mM	1 M	10 ml
EDTA (pH 8.0)	0.1 mM	0.5 M	200 μl

Add water to 1 l

Step 2 10× End Repair Reaction Buffer Mix

Component	Final concentration	Stock	Amount
Tris–HCl (pH 7.5)	500 mM	1 M	2.5 ml
$MgCl_2$	100 mM	1 M	500 μl
Dithiothreitol	100 mM	1 M	500 μl
ATP	10 mM	100 mM	500 μl
dATP	4 mM	100 mM	200 μl
dCTP	4 mM	100 mM	200 μl
dGTP	4 mM	100 mM	200 μl
dTTP	4 mM	100 mM	200 μl

Add water to 5 ml
Note: This buffer can be purchased from New England Biolabs as part of the NEBNext Master Mix 3 (cat# E6060S/L) or as an individual buffer (cat# E6062A/AA)

Step 3 and also used in Preparation of Adaptor Oligos 5× NEBNext Quick Ligation Reaction Buffer

Component	Final concentration	Stock	Amount
Tris–HCl (pH 7.6)	330 mM	1 M	1.65 ml
$MgCl_2$	50 mM	1 M	250 μl
Dithiothreitol	5 mM	1 M	25 μl
ATP	5 mM	100 mM	250 μl
Polyethylene glycol (PEG 6000)	30%	100%	1.5 ml

Add water to 5 ml

Note: This buffer can be purchased from New England Biolabs as part of the NEBNext Master Mix 3 (cat# E6060S/L) or as an individual buffer (cat# E6062A/AA)

4. PROTOCOL

4.1. Preparation

Isolate relatively high-molecular-weight (>5 kb) genomic DNA to use as starting material. One to five microgram of genomic DNA is needed for this protocol (see Preparation of Genomic DNA from Bacteria, Preparation of genomic DNA from *Saccharomyces cerevisiae* or Isolation of Genomic DNA from Mammalian Cells).

Prepare double-stranded P1 and P2 adaptors at 50 μM stock concentration.

Mix equal volumes of 125 μM P1 adaptor plus and minus strand oligos. Add 1 part of 5× NEBNext Quick Ligation Reaction Buffer to 4 parts of the plus and minus strand oligo mixture. This will give a final concentration of 50 μM of each oligo and 1× Quick Ligation Buffer. Hybridize oligos by running the following program in a thermal cycler:

Temperature (°C)	Time (min)
95	5
72	5
60	5
50	3
40	3
30	3
20	3
10	3
4	Hold

Store the hybridized oligos at −20 °C until ready for use.

4.2. Duration

Preparation	1–5 h
Protocol	1–2 days

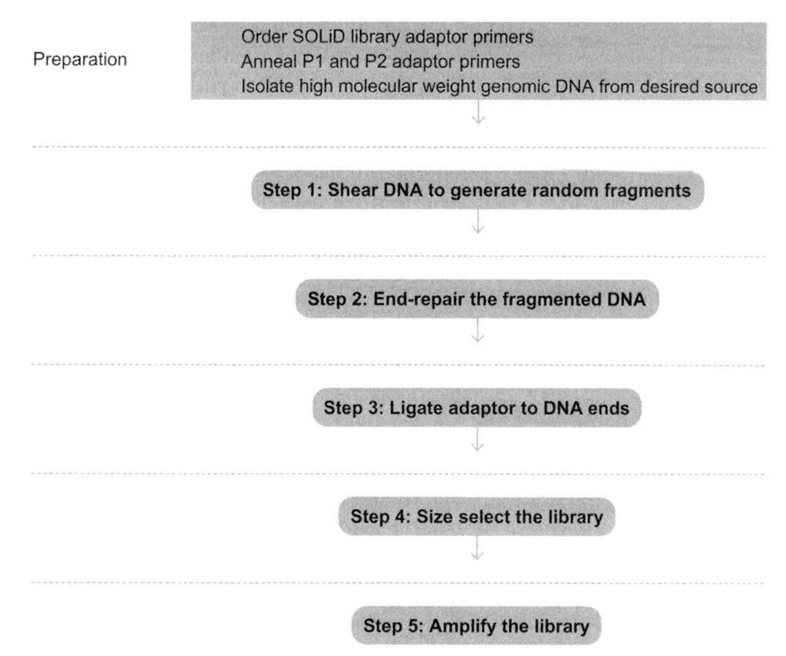

Figure 15.1 Flowchart of the complete protocol, including preparation.

4.3. Caution

Consult your institute's BioSafety Officer for proper handling of potentially infectious DNA samples and of potentially harmful reagents.

See Fig. 15.1 for the flowchart of the complete protocol.

5. STEP 1 SHEAR THE DNA TO GENERATE RANDOM FRAGMENTS

5.1. Overview

Fragment the input genomic DNA to make small DNA molecules with a modal size of ~100–200 bp with random ends. This is done by sonication in a Covaris S2 System.

5.2. Duration

30 min

1.1 Dilute the desired amount of DNA to 100 μl in 1× Low TE Buffer in a 1.5-ml LoBind centrifuge tube.

1.2 Slowly transfer the DNA to a microTUBE loaded into the Covaris S2 through the presplit septa, being careful not to introduce any bubbles into the bottom of the tube.

1.3 Shear the DNA using the following Covaris S2 System conditions:
Number of cycles: 6
Bath temperature: 5 °C
Bath temperature limit: 30 °C
Mode: frequency sweeping
Water quality testing function: off
Duty cycle: 10%
Intensity: 5
Cycles/burst: 100
Time: 60 s

1.4 Insert a pipette tip through the presplit septa, and slowly transfer the sheared DNA from the microTUBE to a new 1.5-ml LoBind centrifuge tube.

5.3. Caution

Follow Covaris S2 System instructions for proper setup, operation, and maintenance of the instrument.

See Fig. 15.2 for the flowchart of Step 1.

Step 1: Shear DNA to generate random fragments
1.1 Dilute the desired amount of DNA to 100 ml in 1X Low TE Buffer in a 1.5 ml LoBind centrifuge tube
1.2 Slowly transfer the DNA into a microTUBE loaded into the Covaris S2 through the pre-split septa
Do not introduce any bubbles
1.3 Shear the DNA using the following Covaris S2 System conditions:
Number of Cycles: 6
Bath Temperature: 5°C
Bath Temperature Limit: 30°C
Mode: Frequency Sweeping
Water Quality Testing Function: Off
Duty Cycle: 10%
Intensity: 5
Cycles/Burst 100
Time: 60 seconds
1.4 Transfer the sheared DNA into a clean 1.5 ml LoBind centrifuge tube

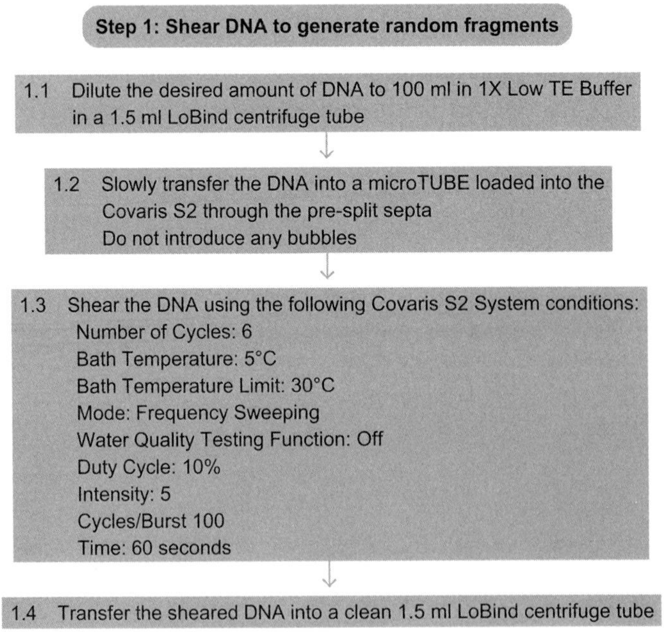

Figure 15.2 Flowchart of Step 1.

6. STEP 2 END-REPAIR THE FRAGMENTED DNA

6.1. Overview

DNA fragments generated by sonication in Step 1 will be end polished to produce a library of DNA fragments with blunt, 5′-phosphorylated ends that are ready for ligation. The end polishing is accomplished by using the T4 DNA polymerase, which can fill in 5′ overhangs via its polymerase activity and recess 3′ overhangs via its 3′→5′ exonuclease activity. The phosphorylation of 5′ ends is accomplished by T4 polynucleotide kinase.

6.2. Duration

~1 h

2.1 Mix the following components in a sterile LoBind centrifuge tube:

Fragmented DNA	~100 μl (recovered from sonicator)
10× End Repair Reaction Buffer Mix	20 μl
End Repair Enzyme Mix	10 μl
Sterile water	to 200 μl

2.2 Incubate at 20 °C for 30 min.

2.3 Purify DNA using SOLiD Library Column Purification Kit:

- **2.3.1** Add 4 volumes (in this case add 800 μl since the reaction is 200 μl) of Binding Buffer (B2-S) with 55% isopropanol to the reaction from above.
- **2.3.2** Add 700 μl of this mixture to the provided column sitting in the collection tube and let column stand for 2 min.
- **2.3.3** Centrifuge the column at >10 000 × g for 1 min and discard the flow-through.
- **2.3.4** Repeat Steps 2.3.2 and 2.3.3 until all of the sample has been loaded onto the column.
- **2.3.5** Add 650 μl of Wash Buffer (W1) to the column.
- **2.3.6** Centrifuge the column at >10 000 × g for 2 min and discard flow through. Repeat centrifugation again without adding any additional W1 buffer to remove the residual Wash Buffer.
- **2.3.7** Air-dry the column for 2 min and transfer the column from the collection tube to a clean 1.5-ml LoBind centrifuge tube.
- **2.3.8** Add 50 μl of Elution Buffer (E1) to the column and let stand for 2 min.

2.3.9 Centrifuge the column at >10 000 × g for 1 min.

2.3.10 Add the eluate from Step 2.3.9 back to the column, and let stand for 2 min. Centrifuge the column at >10 000 × g for 1 min.

2.3.11 Calculate the final concentration using the NanoDrop ND-1000 spectrophotometer according to manufacturer's instructions (alternatively see Explanatory Chapter: Nucleic Acid Concentration Determination).

6.3. Tip

Read the kit manual before first use to make sure all buffers are properly prepared and to get familiarized with the kit components and procedures.

6.4. Tip

The best yields are achieved when 5 μg or less of genomic DNA were initially used. If more than this amount was used in the above steps, it is best to use multiple columns.

6.5. Tip

This is a stopping point and purified DNA can be stored at 4 °C, or can be taken directly for Adaptor Ligation.

See Fig. 15.3 for the flowchart of Step 2.

7. STEP 3 ADAPTOR LIGATION

7.1. Overview

The primary goal of this step is to ligate adaptors to the DNA library. These adaptors are necessary for subsequent amplification and sequencing of the DNA library on the SOLiD instrument. Ligation of double-stranded DNA adaptors is accomplished by use of T4 DNA ligase. The double-stranded adaptors do not have 5′ phosphates and contain a 5′ overhang on one end to prevent ligation in the incorrect orientation. The lack of 5′ phosphate leads to generation of a nick on one strand of the ligated DNA. This will be nick translated and removed during an upcoming amplification step.

7.2. Duration

30–60 min

3.1 Calculate the amount of 50 μM stock P1 and P2 adaptors needed for ligation. The goal is to have a 30:1 ratio of each adaptor to DNA molecules in the library. Assuming a modal size of ~165 bp using the

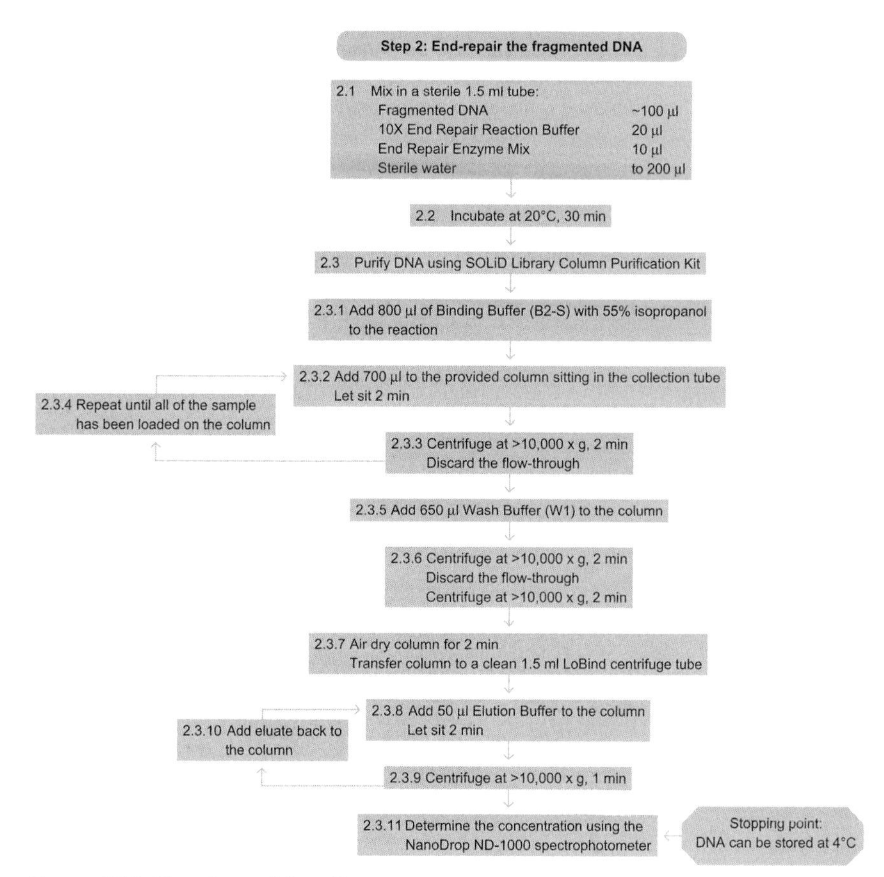

Figure 15.3 Flowchart of Step 2.

sonication conditions listed above and 660 pg per base pair of DNA, one would have 9.2 pmol μg^{-1} of DNA. Therefore, for a 30:1 ratio, we would need 5.5 µl of the 50 µM P1 and P2 adaptor stocks per 1 µg of DNA library.

3.2 Mix the following components in a sterile LoBind centrifuge tube:

End repaired DNA from Step 2	Variable
5× NEBNext Quick Ligation Reaction Buffer	40 µl
P1 DNA adaptor (50 µM)	Variable
P2 DNA adaptor (50 µM)	Variable
T4 DNA ligase	10 µl
Sterile water	to 200 µl

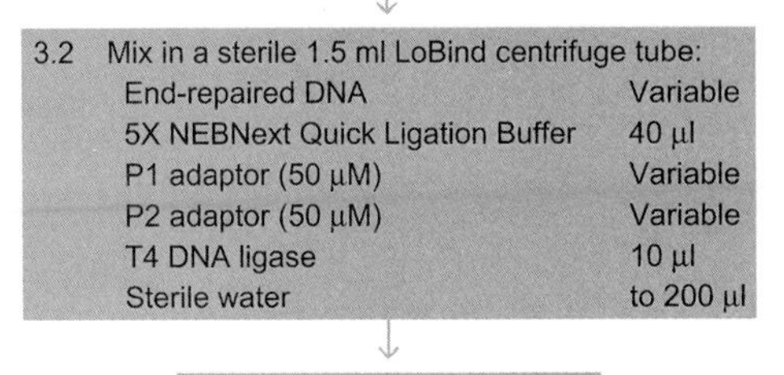

Figure 15.4 Flowchart of Step 3.

3.3 Incubate at 20 °C for 15 min.

3.4 Purify DNA as described in Step 2.3, except that the DNA should be eluted in 65 μl instead of 50 μl of Elution Buffer (E1).

7.3. Tip

This is a stopping point and purified DNA can be stored at 4 °C, or can be taken directly for size selection.

See Fig. 15.4 for the flowchart of Step 3.

8. STEP 4 SIZE SELECTION OF LIBRARY

8.1. Overview

In this step, the adaptor-ligated library will be size-selected to ~200–250 bp size range. By doing this, we remove unligated adaptors and also select the optimal size-range for subsequent emulsion PCR and sequencing. In the procedure described below, we will use the Invitrogen E-Gel system for convenient loading, running, and elution of the size selected DNA.

8.2. Duration

1 h

4.1 Remove the combs from the E-Gel and load onto the E-Gel iBase system linked to the E-gel Safe Imager Real-Time Transilluminator according to the manufacturer's instructions.

4.2 Load 20 μl of the ligated, purified DNA from Step 3 into each of three wells in the top row for each sample. When loading, skip the center well, the wells to the right and left of the center well, and the two outer most wells.

4.3 Load 10 μl of the 50-bp ladder (0.1 μg μl^{-1}) to the center top well. Add 7 μl of water to fill the well.

4.4 Fill empty wells in the top row with 20 μl of nuclease-free water.

4.5 Fill the middle row of wells in the gel with 20 μl of nuclease-free water. Add 20 μl of nuclease free water to the center middle row of wells.

4.6 Run the gel with iBase program SizeSelect 2% (preprogrammed in iBase) for a total of precisely 11 min and 40 s. Monitor the gel in real-time using the real-time transilluminator. If needed, the wells in the middle row should be filled with nuclease-free water. When the 200 bp band from the ladder is at the bottom of the middle row center well but still within the well, if the run has not stopped already, then stop the run.

4.7 Collect the solutions from the wells, which should now contain DNA in the range of 200–250 bp, and pool them according to samples.

4.8 Wash each of the middle row collection wells with 25 μl of nuclease-free water and collect the wash solution and add to the retrieved material collected in Step 4.7.

8.3. Tip

Alternatively, the user could run the library on a standard 1% agarose gel along with a 50-bp DNA ladder (see Agarose Gel Electrophoresis). After running, the portion of the DNA library corresponding to 200–250 bp could be excised using a sterile blade. This excised gel fragment can be purified with a standard DNA gel extraction procedure such as those available in commercialized kits (e.g., Qiagen Gel Extraction kit). However, the procedure described above will allow more standardized generation of fragment libraries.

See Fig. 15.5 for the flowchart of Step 4.

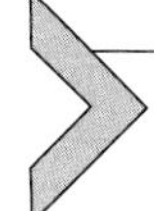

9. STEP 5 AMPLIFICATION OF THE LIBRARY

9.1. Overview

In this step, the size-selected fragment library is amplified to a small extent to prepare for downstream emulsion PCR and sequencing steps. The initial phase of amplification involves a step for nick translation of the nick

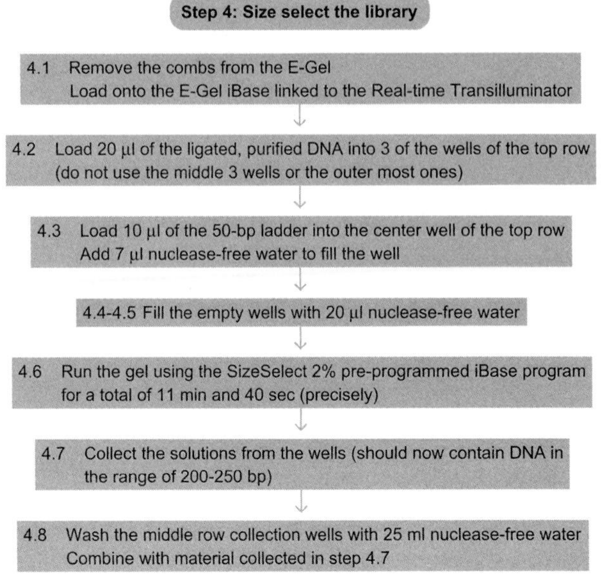

Figure 15.5 Flowchart of Step 4.

remaining at the 5′ ligation junction of the adaptors. This nick will get translated to the end of the adaptor to remove the nick completely. The next steps use standard PCR with primers specific to the P1 and P2 adaptors to amplify the library. The polymerase used for amplification can also carry out the nick translation step, and therefore only one enzyme is needed.

9.2. Duration

~1–2 h

5.1 Mix the following components in a sterile LoBind centrifuge tube:

DNA from size selection step	Variable
P1 adaptor primer (50 µM stock)	10 µl
P2 adaptor primer (50 µM stock)	10 µl
2× LongAmp Taq Master Mix	250 µl
Sterile water	to 500 µl

5.2 Alliquot 125 μl into each of four PCR tubes.

5.3 Place in a thermal cycler and carry out the following cycling program:

Cycle process	Temperature	Time	Cycles
Nick translation	72 °C	20 min	1
Initial denaturation	95 °C	5 min	1
Denaturation	95 °C	15 s	2–10[a]
Annealing	62 °C	15 s	
Extension	70 °C	1 min	
Final extension	70 °C	5 min	1
Hold	4 °C	∞	1

[a]Adjust the number of cycles based on starting amount of DNA as follows:
100 ng to 1 μg: 6–8 cycles
1–2 μg: 4–6 cycles
2–5 μg: 2–3 cycles

5.4 Clean up amplified library according to protocols described in Step 2.3.

5.5 Quantitate and quality control the generated library using an Agilent Bioanalyzer 2100 (optional), according to the manufacturer's recommendations. Samples should then be ready for sequencing workflows including emulsion PCR, bead enrichment, and sequencing. Refer to Applied Biosystems web site or consult a SOLiD Next-Generation Sequencing service facility for protocols on these subsequent steps.

9.3. Tip

These cycling conditions represent starting guidelines. It may be necessary to optimize the number of cycles to avoid overamplification while obtaining enough library material to proceed to downstream emulsion PCR steps. Amplified products should be tested for quality by analysis with an Agilent Bioanalyzer or agarose gel to ensure that the predominant products (>75%) are within the size selected range. Otherwise, it is likely that the product was overamplified producing PCR artifacts that will compromise sequence real estate.

See Fig. 15.6 for the flowchart of Step 5.

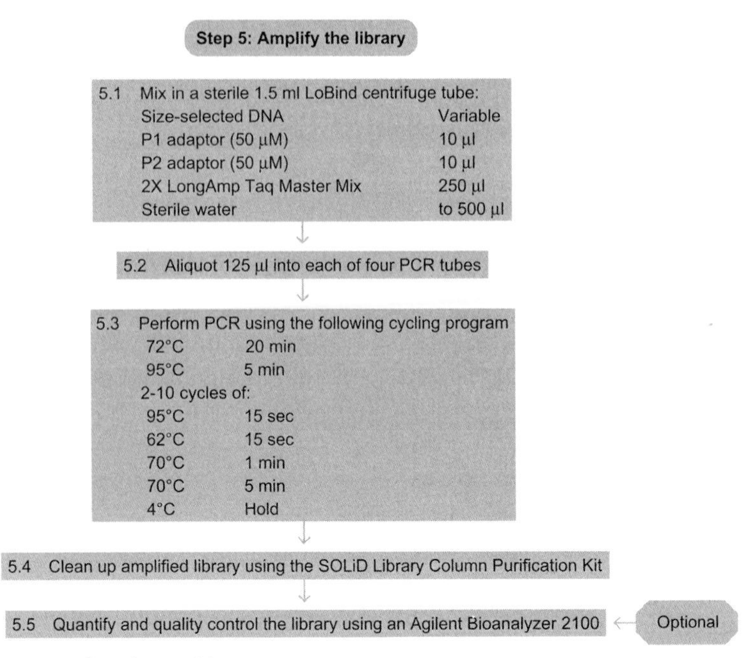

Figure 15.6 Flowchart of Step 5.

REFERENCES

Referenced Literature

Metzker, M. L. (2010). Sequencing technologies – The next generation. *Nature Reviews Genetics*, *11*, 31–46.

Related Literature

These protocols were adapted from the Applied Biosystems protocols in the Applied Biosystems SOLiD™ 4 System Library Preparation Guide (www.appliedbiosystems.com), as well as the protocols supplied with the NEBNext DNA Sample Prep Master Mix Set 3 (NEB E6060S/L) kit.

Referenced Protocols in Methods Navigator

Explanatory Chapter: Next Generation Sequencing
Preparation of Genomic DNA from Bacteria
Preparation of genomic DNA from *Saccharomyces cerevisiae*
Isolation of Genomic DNA from Mammalian Cells
Explanatory Chapter: Nucleic Acid Concentration Determination
Agarose Gel Electrophoresis

Explanatory Chapter: Next Generation Sequencing

Srinivasan Yegnasubramanian[1]
Sidney Kimmel Comprehensive Cancer Center, Johns Hopkins University School of Medicine, Baltimore, MD, USA
[1]Corresponding author: e-mail address: syegnasu@jhmi.edu

Contents

Abstract

Technological breakthroughs in sequencing technologies have driven the advancement of molecular biology and molecular genetics research. The advent of high-throughput Sanger sequencing (for information on the method, see Sanger Dideoxy Sequencing of DNA) in the mid- to late-1990s made possible the accelerated completion of the human genome project, which has since revolutionized the pace of discovery in biomedical research. Similarly, the advent of next generation sequencing is poised to revolutionize biomedical research and usher a new era of individualized, rational medicine.

The term next generation sequencing refers to technologies that have enabled the massively parallel analysis of DNA sequence facilitated through the convergence of advancements in molecular biology, nucleic acid chemistry and biochemistry, computational biology, and electrical and mechanical engineering. The current next generation sequencing technologies are capable of sequencing tens to hundreds of millions of DNA templates simultaneously and generate >4 gigabases of sequence in a single day. These technologies have largely started to replace high-throughput Sanger sequencing for large-scale genomic projects, and have created significant enthusiasm for the advent of a new era of individualized medicine.

Methods in Enzymology, Volume 529
ISSN 0076-6879
http://dx.doi.org/10.1016/B978-0-12-418687-3.00016-1

1. THEORY

1.1. Overview of commercialized next generation sequencing platforms

Given the promise of and the demand for next generation sequencing technologies, there has been intense competition for the development of NGS platforms. 454 life technologies, later acquired by Roche, was the first to commercially release an NGS platform. Solexa, now part of Illumina, released the next platform, with Applied Biosystems marketing the third commercialized platform, which was acquired from Agencourt. Helicos was the first company to release a single-molecule sequencing NGS platform, and more recently several new companies have entered the arena, including Complete Genomics, Pacific Biosciences, and Ion Torrents, with others likely to follow in the near future.

The major steps involved in next generation sequencing technologies that are generically applicable to all of the current technologies are library choice/construction, preparation of libraries for sequencing, and massively parallel sequencing. We first discuss some of the terminology used in Next Generation Sequencing experiments and then discuss each of these steps below and highlight the broad similarities and differences between platforms.

1.2. Terminology in next generation sequencing

- Read: refers to a single contiguous stretch of sequence returned from the instrument.
- Fragment read: a read generated from a fragment library; these reads are generated from a single end of a small DNA fragment that is typically in the order of 100–500 bps depending on the sequencing platform.
- Fragment paired-end reads: two reads generated from each end of a DNA fragment from a fragment library.
- Mate-paired read: two reads generated from each end of a large DNA fragment with a predefined size-range.
- Coverage: the average number of times each base pair in the target genome was covered by reads. For example, 30× coverage implies that each base pair in the reference genome was covered by 30 reads on average.

1.3. NGS library choice and construction

Two major types of libraries can be employed depending on the application: fragment libraries, and mate-paired libraries. For fragment libraries, genomic DNA from a sample is randomly fragmented to a small modal size, typically just 1–5 times the size of the sequencing platform's read length. Sequencing adaptors are then attached to these library molecules to allow sequencing from a single end of each DNA fragment in the library. The protocols used to generate such fragment libraries for the Applied Biosystems SOLiD platform are described in detail in the accompanying protocols chapter (see Preparation of fragment libraries for Next Generation Sequencing on the Applied Biosystems SOLiD platform). More recently, it has become possible to sequence from both ends of such library DNA fragments using a process referred to as fragment paired-end sequencing. Fragment libraries are ideal for analysis of single-nucleotide substitutions/variations. Each DNA fragment in the library produces a single read and multiple overlapping fragments are sequenced for each position in the genome. A coverage of >30× is usually needed to confidently distinguish true variation from sequencing errors and for robustly distinguishing homozygous and heterozygous SNPs. Additionally, fragment libraries can also provide information on genomic copy number. This can be done by taking all of the fragment library reads within fixed genomic bins and carrying out analyses to assess whether the number of reads observed is different from the number expected by random chance (e.g., Xie and Tammi, 2009), representing an extension of digital karyotyping analyses (Wang et al., 2002). Fragment libraries can also be target-enriched with microarray- or solution-based hybrid capture strategies for targeted resequencing (Albert et al., 2007; Gnirke et al., 2009). In these analyses, first, a fragment library is prepared. Next, the library is subjected to target sequence enrichment by hybridization to target-complementary oligonucleotides, called 'baits.' The oligonucleotide 'baits' can be immobilized on the surface of a microarray. Agilent and Nimblegen, among other companies, offer this as a standard or custom design product. More recently, the 'bait' oligonucleotides are synthesized in situ on microarrays, then released by cleavage from the microarray, amplified, and modified with biotin and immobilized on magnetic beads to allow solution-based capture of targets (Gnirke et al., 2009). Agilent markets this as their SureSelect solution-capture-based target enrichment process, and kits have been released for use with the SOLiD and Illumina NGS platforms.

Such approaches have allowed targeted resequencing of any portion of the genome, such as all exons in the human genome (Maher, 2009).

A mate-paired library is constructed by first randomly shearing or fragmenting genomic DNA to a modal size that is typically >1000 bps, which significantly exceeds the read lengths produced by most of the currently commercialized platforms. This library is then size-separated on a gel, and the part of the library corresponding to a specific size range, for example, 2–3 kbp, is excised and purified. These fragments are then circularized via ligation of an adapter under conditions that promote circularization of library molecules with the adaptor separating the two ends, as opposed to ligation of different library molecules together. This geometry allows generation of a library consisting of DNA fragments comprised of subfragments from the two ends of the original size-selected DNA library. The two mate-paired subfragments are then sequenced to reveal the sequences at the two ends of each 2–3 kbp library template. Because we know a priori the possible distances between the two sequences comprising the mate-paired read, after alignment to the reference genome, we can assess whether there was likely to be an amplification, deletion, or translocation between the mate-paired sequences. Similarly, the orientation of the sequences can be used to detect potential inversions. Therefore, mate-paired libraries not only provide information on single nucleotide substitutions, but also on genomic structural variation, as has been demonstrated in several recent reports (Korbel et al., 2007; McKernan et al., 2009).

With the advent of more recent NGS platforms, other library types are also possible. Pacific Biosciences, for instance, has developed ultra-long read lengths of >1000 base pairs. They have deployed these highly processive reads to generate repeated serial reads of both strands of double-strand DNA after circularization of a fragment library with hairpin adaptors ligated to each end of the fragments. The resulting 'SMRT Bell' libraries allow high-fidelity sequencing where the accuracy increases with the number of times the polymerase traverses the circularized SMRT Bell fragments (http://www.pacificbiosciences.com/). This company is also developing strobe-sequencing, where the progress of the processive polymerase in copying long template DNA is followed in an on-off periodic fashion as a way to generate several mate-tags of sequence from a long DNA template, with all tags oriented in the same direction. Complete Genomics has introduced highly complex library generation strategies involving serial cutting and circularization to fabricate DNA nanoballs for unchained ligation-based sequencing (Drmanac et al., 2010). This strategy has been used for resequencing of whole human genomes (Roach et al., 2010). Other library configurations and geometries are likely to be introduced as the diversity of NGS platforms increases.

1.4. Preparation of libraries for sequencing on different NGS platforms

The steps involved in preparing libraries for sequencing on a specific NGS platform are usually tailor-made. For the Roche 454 and Applied Biosystems SOLiD systems, this involves emulsion PCR (Dressman et al., 2003) to amplify individual template DNA molecules clonally on the surface of a bead. In emulsion PCR, individual DNA templates are sequestered along with PCR reagents, such as nucleotide triphosphates, primers, and Taq polymerase, and a primer-coated bead within an aqueous droplet surrounded by a hydrophobic shell within an oil-in-water emulsion. Subjecting these droplets to PCR allows clonal amplification of each template DNA molecule on the surface of the bead. In the case of Roche 454, the beads are then deposited in picoliter wells of a plate. These beads serve as the substrate for sequencing on the instrument (Margulies et al., 2005). In the case of Applied Biosystems, the clonally amplified DNA molecules on the surface of the bead are end-modified and covalently and randomly attached to the surface of a glass slide (http://www.appliedbiosystems.com). This glass slide is then loaded for sequencing on the instrument. Recent improvements in the automation of the emulsion PCR process have streamlined these otherwise cumbersome steps. For the Illumina/Solexa Genome Analyzer and HiSeq platforms, DNA libraries are clonally bridge amplified to generate clonal clusters of each DNA template in situ on the surface of lanes in a flow cell (http://www.illumina.com). These flow cells are then subjected to massively parallel sequencing as described in the following section. For Helicos, library generation is simpler and does not require any clonal amplification steps. In their true single-molecule sequencing (tSMS) platform, library fragments are tailed with poly-adenosine using the terminal deoxynucleotidyl transferase (TdT) enzyme and hybridized onto oligo-dT primer-conjugated flow cells, which are then subjected to sequencing via extension from the oligo-dT primers (Harris et al., 2008).

1.5. Massively parallel sequencing of libraries on different NGS platforms

Each of the currently commercialized NGS platforms uses distinct chemistries to allow massively parallel sequencing of many millions to billions of template DNA molecules. The differences in chemistries confer various strengths and weaknesses to each platform. Because these technologies are rapidly evolving, we focus our discussion on the broad characteristics of the chemistries that are likely to remain stable for the currently commercialized platforms and only touch briefly on up-and-coming platforms that have not yet seen widespread adoption.

The Roche 454 system (http://www.454.com/) uses a sequence-by-synthesis strategy in which DNA templates on the surface of a bead are copied by a DNA polymerase which is forced to add a single-nucleotide species one at a time by cycling the flow of each nucleotide in turn and repeating these cycles for several iterations (Margulies et al., 2005). The pyrophosphates released by the polymerase are converted to light by a pyrophosphatase-based pyrosequencing process in which the amount of light emitted can be used to calculate the number of a specific nucleotide added at each cycle. One somewhat persistent problem with this method is that mononucleotide repeat tracks (e.g., a run of 12 adenines in a row) can lead to errors. This method allows sequencing read lengths of 400 or more base pairs in current implementations. However, the overall throughput is limited by the number of picoliter wells on a plate that can be sequenced, and this platform currently has the lowest sequence capacity per time or per dollar compared to other commercialized platforms.

Illumina (http://www.illumina.com/) and Helicos (http://www.helicosbio.com/) also use a sequence-by-synthesis strategy, but avoid errors associated with mononucleotide runs by using fluorescently labeled reversible chain terminator nucleotides allowing controlled addition of only a single nucleotide at a time, even in stretches of mononucleotide repeats. Because these platforms halt at the addition of every single nucleotide, the coupling efficiencies become limiting and read lengths are typically less than 100 bp. In the case of Helicos, which uses tSMS technology, there appears to be a persistent issue of 'dark bases' in which the nucleotide incorporation is not associated with fluorescence generation. This will probably be an issue with other emerging tSMS platforms as well.

The Applied Biosystems (http://solid.appliedbiosystems.com) (now Life Technologies) SOLiD platform uses a sequence-by-ligation approach in which a DNA ligase, instead of a DNA polymerase, is used to assess sequence via sequential ligation of fluorescently labeled oligonucleotide probes that can interrogate each combination of two adjacent bases (16 combinations possible). However, there are only four different fluorescent dyes, and each one must interrogate one of four possible dinucleotide combinations.

Because of this, an individual ligation reaction does not uniquely identify the corresponding dinucleotide combination. Each base in the sequence is interrogated twice in this degenerate fashion and the combined data across an entire read can be deconvoluted to decipher the final sequence. The first step of the sequencing reaction is to anneal a sequencing primer to the P1 adaptor on the library template (see accompanying fragment library preparation chapter for details) and then to add a mixture of the 16 possible labeled probes. The appropriate di-base probe binds to the first and second base of the template and is ligated to the sequencing primer only if there is a perfect match. The fluorophore associated with this probe is then registered and the probe is enzymatically processed to allow sequential ligation of another probe to query the sixth and seventh bases. This process is carried out a total of ten times for the first primer. After the last ligation step, the reaction is 'reset' by denaturing and washing away the newly synthesized DNA strand from the template DNA that is covalently linked to the bead (see emulsion PCR description above). A new sequencing primer designed to hybridize to a sequence that is offset by one base from the first primer is then annealed so that the first ligation reaction stemming from this sequencing primer interrogates the last base of the adaptor sequence (position 0) and the first base of the template. This primer also goes through a total of ten ligation steps. There are a total of five different sequencing primers that each undergo ten ligation steps. This results in each base being interrogated twice and a sequencing length of 50 base pairs per read.

1.6. The near- and long-term horizon

Each of the platforms described above is routinely making advancements in sequencing throughput in terms of both time and cost per gigabase pairs of sequence output. In the meanwhile, other platforms such as Complete Genomics (http://www.completegenomics.com/), Pacific Biosciences (http://www.pacificbiosciences.com/), Ion Torrents (http://www.iontorrent.com/), and possibly several other players are preparing to enter the market. As a result of this intense competition, cost and time per gigabase pair of sequence produced is rapidly declining. Another consequence of this is that currently commercialized platforms may lose or fail to gain market share and be in danger of folding.

Nonetheless, with the rapid declines in cost, it will be possible to carry out large-scale genomic projects to elucidate novel biology at an unprecedented scale. Additionally, it may become routine to sequence entire human genomes in the context of health and disease and apply such technologies to entire populations and not just individuals. This information can serve as a source of individualized biomarkers that can provide individualized guidance for therapeutic decision-making. The key will be to develop, in parallel, the computational, biostatistical, and bioinformatics solutions to harness the power of these increasingly cost-effective technologies and deploy them not only to understand novel biology, but also to improve the practice and delivery of health care.

REFERENCES

Referenced Literature

Albert, T. J., Molla, M. N., Muzny, D. M., et al. (2007). Direct selection of human genomic loci by microarray hybridization. *Nature Methods*, *4*, 903–905.

Dressman, D., Yan, H., Traverso, G., Kinzler, K. W., & Vogelstein, B. (2003). Transforming single DNA molecules into fluorescent magnetic particles for detection and enumeration of genetic variations. *Proceedings of the National Academy of Sciences of the United States of America*, *100*, 8817–8822.

Drmanac, R., Sparks, A. B., Callow, M. J., et al. (2010). Human genome sequencing using unchained base reads on self-assembling DNA nanoarrays. *Science*, *327*, 78–81.

Gnirke, A., Melnikov, A., Maguire, J., et al. (2009). Solution hybrid selection with ultra-long oligonucleotides for massively parallel targeted sequencing. *Nature Biotechnology*, *27*, 182–189.

Harris, T. D., Buzby, P. R., Babcock, H., et al. (2008). Single-molecule DNA sequencing of a viral genome. *Science*, *320*, 106–109.

Korbel, J. O., Urban, A. E., Affourtit, J. P., et al. (2007). Paired-end mapping reveals extensive structural variation in the human genome. *Science*, *318*, 420–426.

Maher, B. (2009). Exome sequencing takes centre stage in cancer profiling. *Nature*, *459*, 146–147.

Margulies, M., Egholm, M., Altman, W. E., et al. (2005). Genome sequencing in microfabricated high-density picolitre reactors. *Nature*, *437*, 376–380.

McKernan, K. J., Peckham, H. E., Costa, G. L., et al. (2009). Sequence and structural variation in a human genome uncovered by short-read, massively parallel ligation sequencing using two-base encoding. *Genome Research*, *19*, 1527–1541.

Roach, J. C., Glusman, G., Smit, A. F., et al. (2010). Analysis of genetic inheritance in a family quartet by whole-genome sequencing. *Science*, *328*, 636–639.

Wang, T. L., Maierhofer, C., Speicher, M. R., et al. (2002). Digital karyotyping. *Proceedings of the National Academy of Sciences of the United States of America*, *99*, 16156–16161.

Xie, C., & Tammi, M. T. (2009). CNV-seq, a new method to detect copy number variation using high-throughput sequencing. *BMC Bioinformatics*, *10*, 80.

Referenced Protocols in Methods Navigator

Sanger Dideoxy Sequencing of DNA

Preparation of fragment libraries for Next Generation Sequencing on the Applied Biosystems SOLiD platform

CHAPTER SEVENTEEN

Generating Mammalian Stable Cell Lines by Electroporation

Patti A. Longo, Jennifer M. Kavran, Min-Sung Kim, Daniel J. Leahy[1]

Johns Hopkins University School of Medicine, Baltimore, MD, USA
[1]Corresponding author: e-mail address: dleahy@jhmi.edu

Contents

1. Theory 210
2. Equipment 211
3. Materials 212
   3.1 Solutions & Buffers 213
4. Protocol 214
   4.1 Preparation 214
   4.2 Duration 215
5. Step 1 Dilute Plasmid DNA 215
   5.1 Overview 215
   5.2 Duration 216
   5.3 Tip 216
   5.4 Tip 216
6. Step 2 Prepare Cells for Electroporation 216
   6.1 Overview 216
   6.2 Duration 216
   6.3 Tip 218
   6.4 Tip 218
7. Step 3 Electroporate the Cells 219
   7.1 Overview 219
   7.2 Duration 219
   7.3 Tip 219
   7.4 Tip 219
8. Step 4 Plating Electroporated Cells 220
   8.1 Overview 220
   8.2 Duration 220
   8.3 Tip 220
   8.4 Tip 220
   8.5 Tip 220
9. Step 5 Picking Single Colonies of Cells 221
   9.1 Overview 221
   9.2 Duration 221
   9.3 Tip 222

Methods in Enzymology, Volume 529
ISSN 0076-6879
http://dx.doi.org/10.1016/B978-0-12-418687-3.00017-3

9.4 Tip 222
9.5 Tip 222
9.6 Tip 223
10. Step 6 Methotrexate Amplification 223
10.1 Overview 223
10.2 Duration 223
10.3 Tip 224
10.4 Tip 224
10.5 Tip 224
10.6 Tip 224
References 225

Abstract

Expression of functional, recombinant mammalian proteins often requires expression in mammalian cells (see Single Cell Cloning of a Stable Mammalian Cell Line). If the expressed protein needs to be made frequently, it can be best to generate a stable cell line instead of performing repeated transient transfections into mammalian cells. Here, we describe a method to generate stable cell lines via electroporation followed by selection steps.

This protocol will be limited to the CHO *dhfr–* Urlaub et al. (1983) and LEC1 cell lines, which in our experience perform the best with this method.

1. THEORY

Electroporation is a popular technique to introduce foreign DNA into host cells. The DNA enters the cells following a quick electric pulse that generates temporary openings in the cell membrane. After electroporation, selection steps must be followed to allow for cells containing only the target DNA to grow.

While outside the scope of this protocol, we have listed some other cell lines and appropriate selection markers in Table 17.1. Stable cell lines also can be generated by other transfection methods (e.g., using lipid-like transfection reagents or calcium phosphate transfection. See also Rapid creation of stable mammalian cell lines for regulated expression of proteins using the Gateway® Recombination Cloning Technology and Flp-In T-REx® lines) followed by appropriate selection steps.

Two different selection methods are described here for the first round of selection immediately following electroporation. The first selection method relies on co-transfection with the pcDNA3.1 vector, which carries the *neo* gene that confers resistance to geneticin, an aminoglycoside antibiotic (Southern and Berg, 1982). It is generally not necessary for the gene of

Table 17.1 Summary of selection method

Cell type	Transfection method	Selection markers	Round 1 selection	Days until die-off begins	Round 2 selection
CHO *dhfr–*	Electroporation	pSV2-*dhfr*	Alpha MEM, 5% D-FBS	12–14	Alpha MEM, 5% D-FBS, MTX (0.1–0.4 μM)
LEC1	Electroporation	pcDNA3.1, pSV2-*dhfr*	Alpha MEM, 5% FBS, 0.5 mg ml^{-1} geneticin	15–17	Alpha MEM, 5% D-FBS, MTX (0.025–0.1 μM)
HEK *293 GnTi-*	PEI	pcDNA3.1	DMEM:F12, 5% FBS, 2 mg ml^{-1} geneticin	7–10	N/A
CHO-S	PEI	pcDNA3.1	DMEM:F12, 5% FBS, 0.5 mg ml^{-1} geneticin	7–10	N/A

interest and the selectable marker to be on the same plasmid. Positively transfected cells are selected by growth in the presence of geneticin. The second selection method relies on dihydrofolate reductase (DHFR) activity (Wigler et al., 1980) and its inhibitor methotrexate (MTX). DHFR selection works best with the cells lacking DHFR activity, such as CHO *dhfr–*. But cells having wild-type DHFR activity can also be selected with high MTX concentrations (Table 17.1). The DHFR/MTX selection step can be repeated with higher concentrations of MTX until a desirable level of protein expression is achieved.

Stable cell lines often lose their protein expression with time as a result of a heterogeneity in the transfected population of cells. A more homogeneous population of cells can be obtained by limiting dilution cloning or picking individual colonies of drug-resistant cells.

2. EQUIPMENT

Laminar flow hood
CO_2 incubator
Centrifuge

Electroporator (BIO-RAD Gene Pulser II)
Sterile 2 mm gap cuvettes
Water bath (37 °C)
Inverted microscope
Hemacytometer
0.22-μm sterile filters
T75 tissue culture flasks
Sterile 50-ml polypropylene conical tubes
Sterile 1.5-ml polypropylene tubes
Sterile 150-mm tissue culture dishes
Sterile 100-mm tissue culture dishes 24-well tissue culture plates
Sterile pregreased glass cloning cylinders (Sigma C1059)
Sterile pipette tips
Sterile disposable pipettes

3. MATERIALS

Plasmid DNA expressing gene of interest
Cell line (CHO *dhfr*−, LEC1) (ATCC)
Fetal bovine serum (FBS, Invitrogen)
Dialysed FBS (D-FBS, Invitrogen)
Geneticin® (Invitrogen)
Methotrexate (MTX, Sigma)
Hanks Balanced Salt Solution (HBSS w/o Ca, Mg; Invitrogen 14170)
HT supplement, 100× (Invitrogen)
TrypLE™ Express (Invitrogen)
MEM α (containing Earl's Salts and L-glutamine, but no ribonucleosides, deoxyribonucleosides, $NaCO_3$; Invitrogen 12000)
DMEM/F12 (with L-glutamine, but no HEPES, $NaHCO_3$; Invitrogen 12500)
Freestyle™ 293 medium (Invitrogen 12338-026)
Hybridoma SFM (Invitrogen 12045)
pcDNA 3.1 (Invitrogen)
pSV2-dhfr (ATCC)

Note *Some of the stock solutions come with the pH indicator phenol red. This supplement does not affect the application and might be useful if the researcher wishes to visualize any pH changes that can occur in the solutions over time. In the case of non-CO_2 incubators (e.g., when scaling-up the production of adherent cells in roller bottles), HEPES-buffered media can be used to keep the pH stable.*

Note *Catalog numbers are from the US website of Invitrogen and may differ on other local websites.*

3.1. Solutions & Buffers

Step 2 Lec1 Growth Medium: Alpha MEM + 5% FBS

Add 50 ml FBS to 1 l of Alpha MEM

CHO *dhfr–* Growth Medium: Alpha MEM + 5% FBS + HT solution

Add 50 ml FBS and 10 ml of 100× HT solution to 1 l of Alpha MEM

Steps 4–6 50 mg ml^{-1} Geneticin® (active)

Add enough active Geneticin to 50 ml of water to make a 50 mg ml^{-1} stock solution. Each lot of Geneticin will have a different active concentration. For example, if a 5 g bottle of Geneticin has an active concentration of 750 μg mg^{-1} of powder, it contains an active weight of 3.75 g Geneticin. Therefore dissolve the 5 g of powder in 75 ml of water to obtain an active concentration of 50 mg ml^{-1}. Mix to dissolve, pass through a 0.2 mm filter to sterilize, and dispense aliquots into sterile tubes. Store short term at 4 °C or long term at −20 °C

Lec1 Selection Medium

Add 50 ml FBS and 10 ml of 50 mg ml^{-1} Geneticin to 1 l of Alpha MEM

CHO *dhfr–* Selection Medium

Add 50 ml D-FBS to 1 l of Alpha MEM

Step 6 30 mM MTX

Component	Final Concentration	Stock	Amount
Methotrexate	30 mM		0.136 g
Alpha MEM			9.5 ml
NaOH	50 mM	1 M	0.5 ml

Mix well until MTX is dissolved and pass through a 0.2 μm filter to sterilize. Store in sterile aliquots at −20 °C

1 mM MTX

Add 33.3 μl of 30 mM MTX to 1 ml Alpha MEM

MTX selection concentrations

Component	Final concentration	Stock	Amount
MTX	25 nM	1 mM	12.5 μl
MTX	50 nM	1 mM	25 μl
MTX	75 nM	1 mM	37.5 μl
MTX	100 nM	1 mM	50 μl
MTX	200 nM	1 mM	100 μl
MTX	300 nM	1 mM	150 μl
MTX	400 nM	1 mM	200 μl
MTX	500 nM	1 mM	250 μl
MTX	1 μM	1 mM	500 μl

Add the indicated amount of MTX to 500 ml Alpha MEM containing 5% FBS

4. PROTOCOL

4.1. Preparation

Before transfection, sterile high-quality DNA must be prepared. The vector containing the appropriate expression promoter (see Molecular Cloning) and the gene of interest should be transformed into a recA– strain of *E. coli* (see Transformation of Chemically Competent *E. coli* or Transformation of *E. coli* via electroporation) and then the plasmid DNA isolated (see Isolation of plasmid DNA from bacteria). Commercially available, endotoxin-free kits for large-scale plasmid DNA isolation produce sufficiently high-quality DNA. High-quality DNA is characterized as having an OD260/280 ratio between 1.88 and 1.92, an OD260/230 ratio of 2.1–2.2, and a concentration above 0.5 mg ml^{-1} (see Explanatory Chapter: Nucleic Acid Concentration Determination).

All steps are carried out using sterile technique in a laminar flow hood. Solutions should be sterile filtered through 0.22-μm filters. All plastic and glassware, if not purchased as sterile, should be autoclaved twice. Cell growth media are warmed to 37 °C prior to contact with cells. Before electroporating, cells should have undergone at least five passages from thawing. Grow LEC1 or CHO *dhfr–* cells to 80% confluency in a T75 flask. One T75 flask at 80% confluence yields enough cells for 1–2 electroporations. No residual trypsin can be present during electroporation.

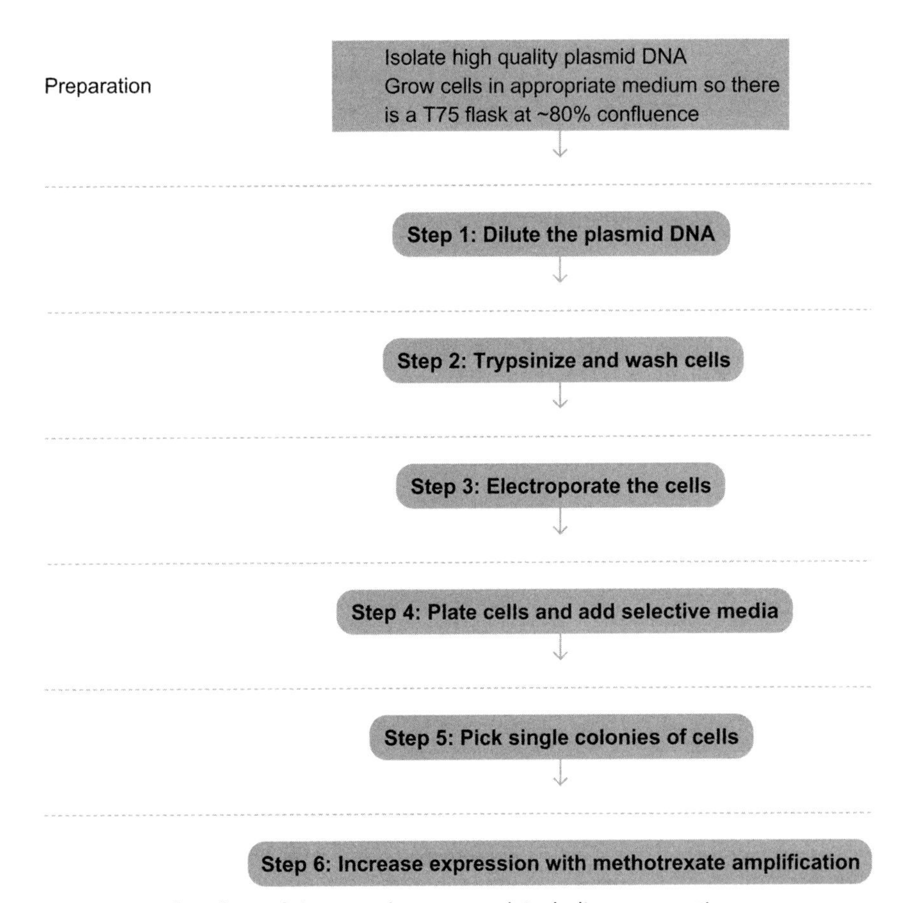

Figure 17.1 Flowchart of the complete protocol, including preparation.

4.2. Duration

Preparation	1 week
Protocol	~1–3 months

See Fig. 17.1 for the complete protocol, including preparation.

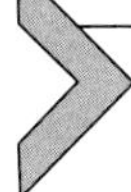

5. STEP 1 DILUTE PLASMID DNA

5.1. Overview

The proper amount of each high-quality plasmid DNA is diluted into the appropriate media for electroporation.

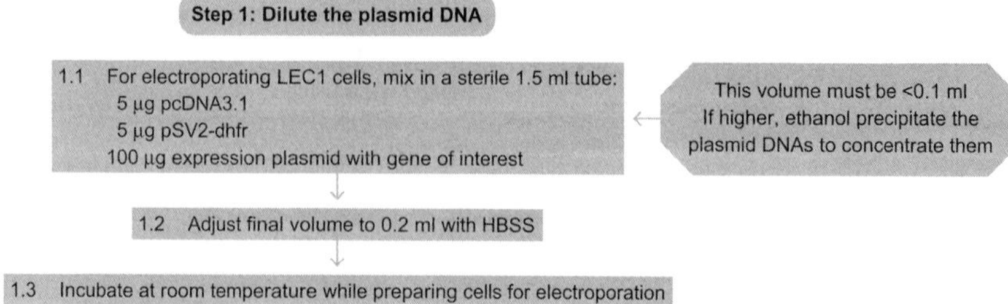

Figure 17.2 Flowchart of Step 1.

5.2. Duration

5 min

1.1 In a sterile tube, combine 5 µg pcDNA3.1, 5 µg pSV2-*dhfr*, and 100 µg of the expression plasmid containing the gene of interest.

1.2 Adjust the final volume to 0.2 ml with HBSS.

1.3 Incubate at room temperature while preparing cells.

5.3. Tip

These plasmids contain the selection markers for LEC1 cells. See Table 17.1 for the selection markers for other cell lines.

5.4. Tip

The total volume of the plasmid DNA mixture must be <0.1 ml, (i.e., half of the final volume). If the volume in Step 1.1 is larger than 0.1 ml, the electroporation efficiency will decrease. If this occurs, increase the starting concentrations of each DNA stock.

See Fig. 17.2 for the flowchart of Step 1.

6. STEP 2 PREPARE CELLS FOR ELECTROPORATION

6.1. Overview

Cells are trypsinized from a T75 flask, counted, and then washed into the appropriate media for electroporation.

6.2. Duration

30 min

2.1 Aspirate the media from the flask.

2.2 Add 4 ml TrypLE™ Express.

2.3 Lay flask flat to ensure that the entire surface is covered with the liquid and incubate for 1 min at room temperature.

2.4 Remove TrypLE™ Express solution.

2.5 View cells under microscope and wait until cells have released from the bottom of flask, appearing round in the microscope.

2.6 Immediately add 20 ml of the appropriate growth media to the flask (Table 17.2).

2.7 Pipette gently up and down against the cell growth surface of the flask to resuspend cells.

2.8 Transfer cell suspension to a 50-ml conical centrifuge tube.

2.9 Aliquot a small amount of suspension and count the number of cells using a hemacytometer.

2.10 Centrifuge the cell suspension at 200 × g for 5 min at room temperature.

2.11 Aspirate the supernatant.

2.12 Add 20 ml HBSS to wash the cells. Gently pipette up and down to resuspend the cell pellet.

2.13 Centrifuge the cell suspension at 200 × g for 5 min at room temperature. Aspirate the supernatant.

2.14 Add HBBS so that the final cell concentration is 20×10^6 cells ml^{-1}.

2.15 Gently resuspend the cell pellet with serological pipette.

Table 17.2 Cell lines and growth conditions

Cell line	Ideal growth condition	Optimal media	Cell source
HEK293S *Gnti-*	Suspension	Freestyle293 'completed'	ATCC CRL3022
HEK293S *Gnti-*	Adherent	DMEM/F12, 5% FBS	ATCC CRL3022
HEK293T/17	Adherent	DMEM/F12, 5% FBS	ATCC CRL11268
CHO-S	Suspension	Hybridoma, 1% FBS	INVITROGEN R800-07
CHO-S	Adherent	DMEM/F12, 5% FBS	INVITROGEN R800-07
CHO *dhfr–*	Adherent	Alpha MEM, 5% FBS, HT	ATCC CRL-9096
LEC1	Adherent	Alpha MEM, 5% FBS	ATCC CRL-1735
COS7	Adherent	DMEM/F12, 5% FBS	ATCC CRL1651

6.3. Tip

Adding the growth media at Step 2.6 serves to inactivate the enzyme in TrypLE™ Express.

6.4. Tip

All traces of trypsin must be removed prior to electroporation or the efficiency will be negatively affected. Do not skip the wash steps.

See Fig. 17.3 for the flowchart of Step 2.

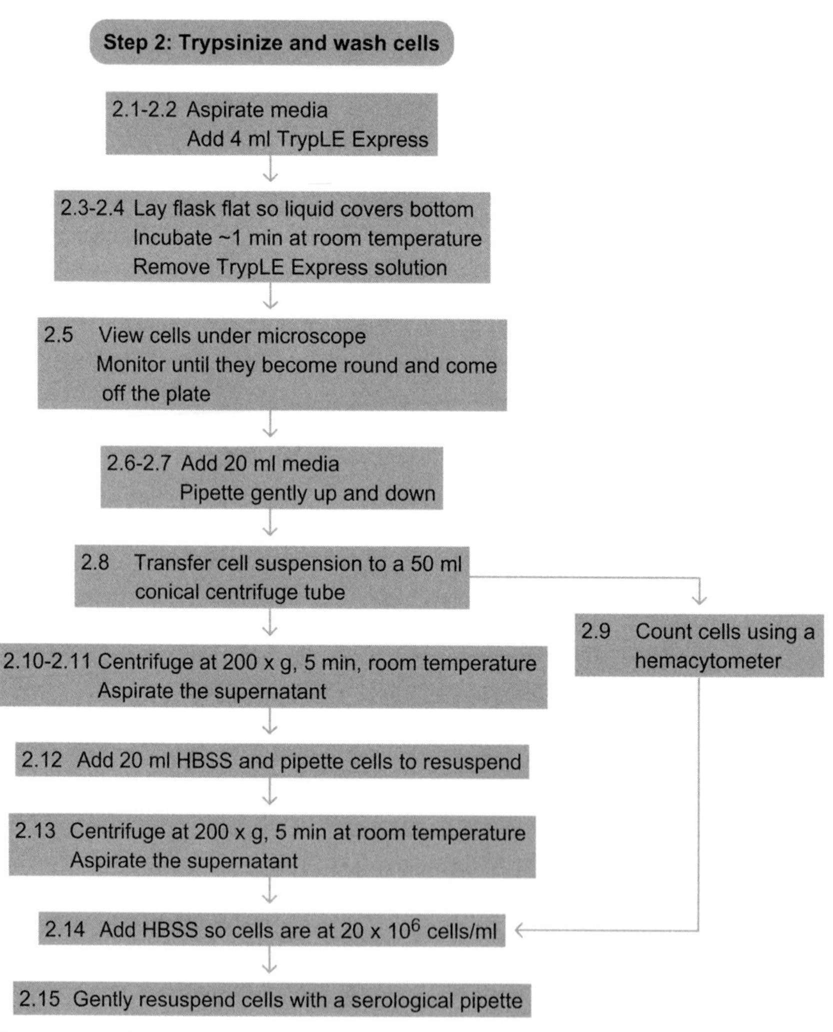

Figure 17.3 Flowchart of Step 2.

7. STEP 3 ELECTROPORATE THE CELLS

7.1. Overview

Cells are transfected using an electroporation device.

7.2. Duration

1–2 min

3.1 Add 0.2 ml of the cell suspension (from Step 2.15) to the diluted DNA (from Step 1.3).

3.2 Mix gently and transfer the mixture to a 2-mm sterile electroporation cuvette.

3.3 Cap the cuvette and transfer to the electroporator.

3.4 Electroporate cells according to the manufacturer's instructions.

7.3. Tip

If multiple samples are being processed, place the cell/DNA mixture in the cuvette on ice until all samples are ready to electroporate.

7.4. Tip

We find the optimal settings for CHO dhfr– cells and LEC1 cells to be at high capacitance, 0.174 kV, and 400 μF. These variables may need to be optimized for each cell line.

See Fig. 17.4 for the flowchart of Step 3.

Step 3: Electroporate the cells

3.1 Add 0.2 ml cell suspension to the diluted DNA

↓

3.2 Mix gently
Transfer to a 2 mm electroporation cuvette

↓

3.3 Cap the cuvette
Transfer to the electroporator

↓

3.4 Electroporate the cells
For CHO *dhfr*-or LEC1 cells use settings of
high capacitance, 0.174 kV and 400 μF

Figure 17.4 Flowchart of Step 3.

8. STEP 4 PLATING ELECTROPORATED CELLS

8.1. Overview

Electroporated cells are plated into first selection medium and returned to the incubator. Untransfected cells typically start to die off between day 12 and day 17 in culture.

8.2. Duration

20 min

12–17 days for cells to die off and resistant cells to grow.

4.1 Transfer the cells to a fresh sterile 1.5-ml microcentrifuge tube.

4.2 Adjust the total cell volume to 1 ml with HBSS and mix gently.

4.3 Transfer 0.1 ml of the cell suspension to a 150-mm dish containing 25 ml of appropriate round 1 selection medium (Table 17.1).

4.4 Place the dish in a CO_2 incubator at 37 °C.

4.5 Seven days after electroporation, aspirate media from the cells and refeed with 25 ml of the appropriate round 1 selection medium.

4.6 Aspirate and add fresh selection media every 3–4 days until there is an obvious die-off of cells as judged by the disappearance of the monolayer of cells. Viable cells will remain in selected areas, giving the appearance of individual colonies.

8.3. Tip

If all cells die or none of the cells dies, then either the quality of cells or DNA is nonideal. Thaw and culture new cells and/or prepare a higher quality batch of DNA.

8.4. Tip

Die-off is usually observed between 14 and 17 days after electroporation but that varies for different cell lines. See Table 1 for approximate times for several cell lines.

8.5. Tip

Multiple 150-mm dishes can be prepared by plating different amounts of the cell/DNA mix. We usually plate 0.10, 0.15, 0.25, and 0.5 ml of the cell/DNA suspension, respectively, into four separate dishes containing 25 ml each of the appropriate first round selection media.

See Fig. 17.5 for the flowchart of Step 4.

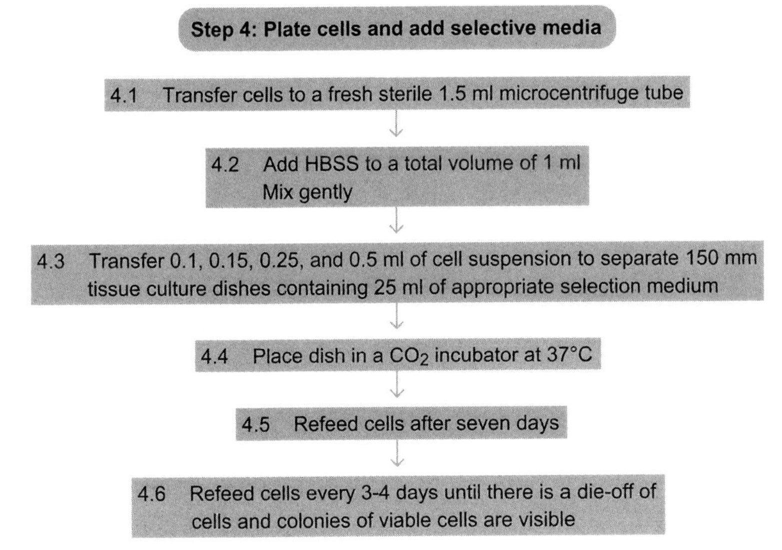

Figure 17.5 Flowchart of Step 4.

9. STEP 5 PICKING SINGLE COLONIES OF CELLS

9.1. Overview

After most cells die off, single colonies of transfected cells are isolated.

9.2. Duration

20 min active, plus time for cells to grow

5.1 With media still in the dish, carefully hold it above eye level and note visible colonies. They appear as round opaque areas against a clear background.

5.2 Mark a circle around each colony on the bottom of the dish with a black marker. Do not circle colonies that are too close together, or the cloning rings will not be able to fit around a single colony. Choose at least 24 colonies per transfection plate.

5.3 View the cells under the microscope to confirm that the areas you have circled are colonies of viable cells.

5.4 Carefully tilt the dish and remove all media with a pipette.

5.5 Using forceps, grease a cloning ring and place it around a colony, so as to make a well around each colony.

5.6 Pipette 0.2 ml TrypLE™ Express into each cloning ring.

5.7 Cover the plate and place on microscope stage.

5.8 Watch for cells to round up and pull away from the plate. They will become round and highly refractile. This usually takes 30–60 s.

5.9 Return plate to hood. Using a pipette, gently remove TrypLE™ Express from all 24 cloning rings.

5.10 Add 0.2 ml appropriate cell selection medium to six rings. The remaining rings remain empty.

5.11 Beginning with the first cylinder containing selection medium, gently pipette up and down to dislodge the cells from the surface.

5.12 Transfer the cell suspension to one well of a 24-well dish containing 1 ml of the appropriate selection media.

5.13 Transfer the cells from the next five rings containing selection media.

5.14 Add 0.2 ml media to another six cloning rings and transfer cells. Continue until the cells have been removed from all 24 cloning rings and transferred to a well of the 24-well plate.

5.15 Put the 24-well plate in the CO_2 incubator. Replace the appropriate selection media every 3 days. Assay for expression of your protein when the cells are 100% confluent.

5.16 Choose at least three of the highest expressing colonies. Expand cells for freezing stocks and for growth in a T75 flask to be used in a second round of selection with methotrexate.

9.3. Tip

Try to transfer the cells from all 24 cloning rings within 20 min to avoid cell death due to long exposure to TrypLE™ Express or due to drying out. Alternatively, you can begin by filling all 24 cloning rings with appropriate selection media. Proceed to remove media from 6 wells and treat them with TrypLE™ Express. Resuspend each well in selection media and transfer each into a separate well of a 24-well plate containing 1 ml appropriate selection media. Repeat for the next group of 6 wells until all 24 wells are collected.

9.4. Tip

Dispense 1 ml of the appropriate selection media in each well of a 24-well plate prior to placing the cloning rings on the transfected plate.

9.5. Tip

At this stage, if the level of protein expression is satisfactory, no further selection is needed and the stable cell line can be used for protein production.

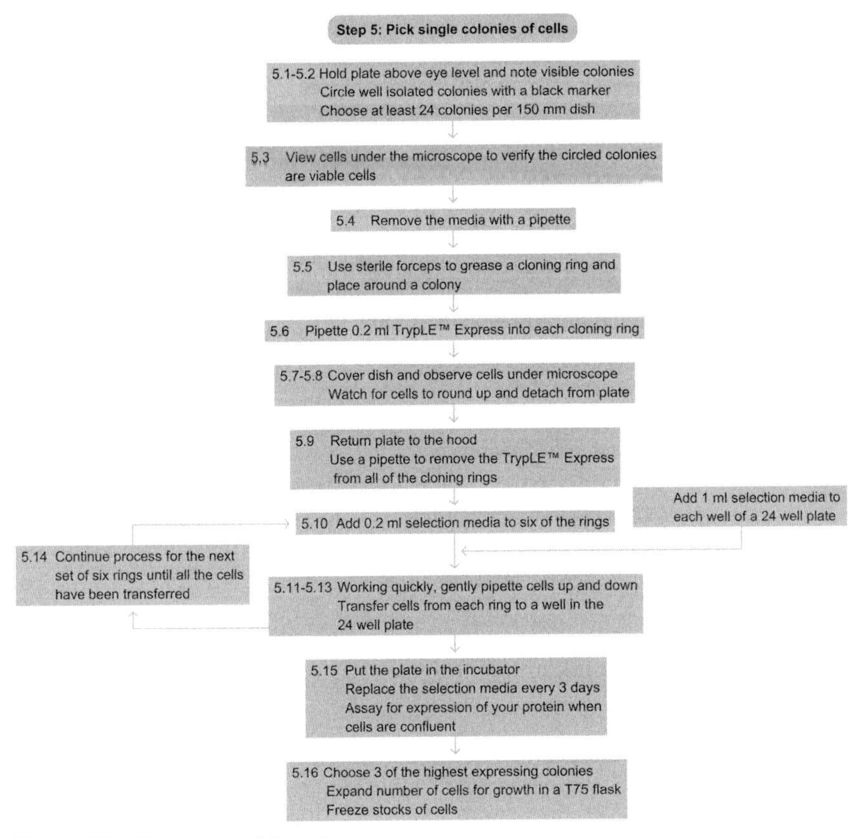

Figure 17.6 Flowchart of Step 5.

9.6. Tip

Once the cells in the 24-well plate are growing, the 150 mm master plate on which the cloning rings were placed can be discarded.

See Fig. 17.6 for the flowchart of Step 5.

10. STEP 6 METHOTREXATE AMPLIFICATION

10.1. Overview

For cells transfected with the pSV2-*dhfr* vector, target protein expression can be increased with further rounds of methotrexate selection.

10.2. Duration

Variable (weeks)

6.1 Grow the three highest expressing colonies in a T75 flask until they are 80% confluent.

6.2 Trypsinize and count the cells.

6.3 Resuspend the cells at a concentration of 1×10^6 cells ml^{-1} in second round selection media without MTX (Table 17.1).

6.4 Prepare four 100-mm plates by adding 10 ml of second round selection media containing MTX spanning the concentration range of MTX suggested in Table 17.1.

6.5 Add 0.3 ml of cell suspension to each plate.

6.6 Gently swirl the plate to evenly distribute the cells.

6.7 Return dishes to the incubator and grow for 3 days.

6.8 Remove media from plate and replace with 10 ml of fresh second round selection media with MTX. Change the media every 3 days until cells die off. Typical times are listed in Table 17.1.

6.9 After most of the cells have died and colonies are visible, pick colonies as before.

6.10 Assay for the best expressing clones as appropriate for your protein of interest.

6.11 Expand and freeze the three highest expressing colonies.

10.3. Tip

Each cell type and expressed protein will have different optimal MTX concentrations. We typically plate cells at four concentrations of MTX and pick colonies from each concentration.

10.4. Tip

If no amplification is seen at the first round of MTX selection, MTX selection can be stopped.

10.5. Tip

Several vials (at least three) of positive clones should be frozen at each round of MTX selection.

10.6. Tip

If the first round of MTX selection results in colonies with higher levels of protein expression, two more rounds of MTX selection can be performed. Each round should be performed at four concentrations of MTX but using MTX concentrations that are two- to tenfold higher than those used in the previous round of selection.

See Fig. 17.7 for the flowchart of Step 6.

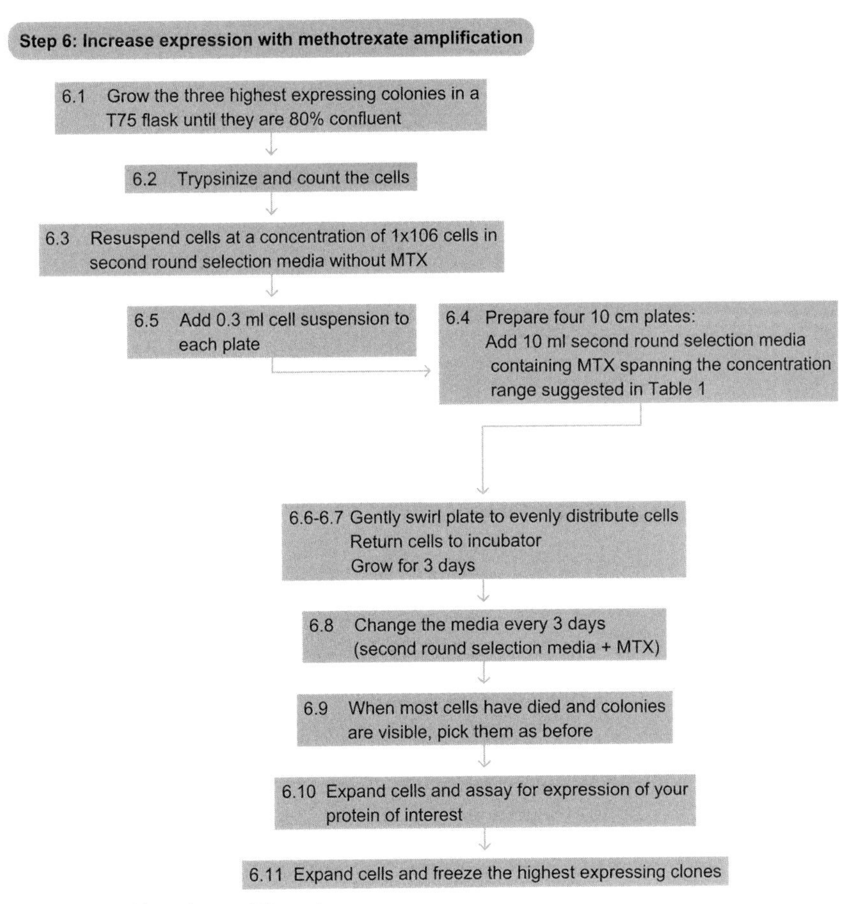

Figure 17.7 Flowchart of Step 6.

REFERENCES

Referenced Literature

Urlaub, G., et al. (1983). Deletion of the diploid dihydrofolate reductase locus from cultured mammalian cells. *Cell*, *33*(2), 405–412.

Southern, P. J., & Berg, P. (1982). Transformation of mammalian cells to antibiotic resistance with a bacterial gene under control of the SV40 early region promoter. *Journal of Molecular and Applied Genetics*, *1*(4), 327–341.

Wigler, M., et al. (1980). Transformation of mammalian cells with an amplifiable dominant-acting gene. *Proceedings of the National Academy of Sciences of the United States of America*, *77*(6), 3567–3570.

Referenced Protocols in Methods Navigator

Single Cell Cloning of a Stable Mammalian Cell Line

Rapid creation of stable mammalian cell lines for regulated expression of proteins using the Gateway® Recombination Cloning Technology and Flp-In T-REx® lines

Molecular Cloning
Transformation of Chemically Competent *E. coli*
Transformation of *E. coli* via electroporation
Isolation of plasmid DNA from bacteria
Explanatory Chapter: Nucleic Acid Concentration Determination

CHAPTER EIGHTEEN

Transient Mammalian Cell Transfection with Polyethylenimine (PEI)

Patti A. Longo, Jennifer M. Kavran, Min-Sung Kim, Daniel J. Leahy[1]
Johns Hopkins University School of Medicine, Baltimore, MD, USA
[1]Corresponding author: e-mail address: dleahy@jhmi.edu

Contents

1. Theory 228
2. Equipment 229
3. Materials 229
   3.1 Solutions & buffers 230
4. Protocol 231
   4.1 Preparation 231
   4.2 Duration 232
5. Step 1 Small-Scale Transient Transfection 232
   5.1 Overview 232
   5.2 Duration 232
6. Step 1.1 Seed Adherent Cells for Transfection 232
   6.1 Overview 232
   6.2 Duration 233
7. Step 1.2 Transiently Transfect Cells 234
   7.1 Overview 234
   7.2 Duration 234
   7.3 Tip 234
   7.4 Tip 234
   7.5 Tip 234
   7.6 Tip 234
8. Step 1.3 Harvest Cells and Analyze Protein Expression 235
   8.1 Overview 235
   8.2 Duration 235
   8.3 Tip 235
   8.4 Tip 235
   8.5 Tip 236
9. Step 2 Large-Scale Transient Transfection of Suspension Cells 236
   9.1 Overview 236
   9.2 Duration 236

Methods in Enzymology, Volume 529
ISSN 0076-6879
http://dx.doi.org/10.1016/B978-0-12-418687-3.00018-5

10. Step 2.1 Prepare the Cells To Be Transfected 236
10.1 Overview 236
10.2 Duration 236
11. Step 2.2 Transfect Cells 237
11.1 Overview 237
11.2 Duration 237
12. Step 2.3 Harvest Cells and Process Protein as Needed 238
12.1 Overview 238
12.2 Duration 239
12.3 Tip 239
12.4 Tip 239
12.5 Tip 239
References 240

Abstract

Standard protein expression systems, such as *E. coli*, often fail to produce folded, monodisperse, or functional eukaryotic proteins (see Small-scale Expression of Proteins in *E. coli*). The expression of these proteins is greatly benefited by using a eukaryotic system, such as mammalian cells, that contains the appropriate folding and posttranslational machinery. Here, we describe methods for both small- and large-scale transient expression in mammalian cells using polyethylenimine (PEI). We find this procedure to be more cost-effective and quicker than the more traditional route of generating stable cell lines. First, optimal transfection conditions are determined on a small-scale, using adherent cells. These conditions are then translated for use in large-scale suspension cultures. For further details on generating stable cell lines please (see Rapid creation of stable mammalian cell lines for regulated expression of proteins using the Gateway® Recombination Cloning Technology and Flp-In T-REx® lines or Generating mammalian stable cell lines by electroporation).

1. THEORY

DNA can be introduced into a host cell by transfection with polyethylenimine (PEI), a stable cationic polymer (Boussif et al., 1995). PEI condenses DNA into positively charged particles that bind to anionic cell surfaces. Consequently, the DNA:PEI complex is endocytosed by the cells and the DNA released into the cytoplasm (Sonawane et al., 2003). Our laboratory uses PEI over other cell transfection reagents because of its low cost.

This protocol is appropriate for two suspension cell lines, CHO-S and HEK *293 GnTi-*. Many cell lines can be transfected successfully with PEI but in our experience these two cell lines express the highest level of protein compared to other cells.

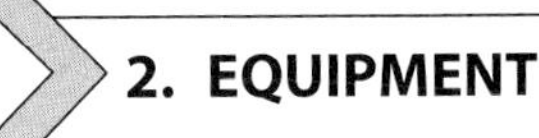

2. EQUIPMENT

Laminar flow hood
CO_2 incubator
Platform shaker
Centrifuge
Water bath (37 °C)
Inverted microscope
Hemacytometer
Sterile 0.22 μm filters
Sterile 250-ml polypropylene centrifuge tubes
Sterile 50-ml polypropylene conical tubes
Sterile 1.5-ml polypropylene tubes
Sterile 6-well tissue culture plates micropipettors
Sterile micropipettor tips
Sterile disposable serological pipettes
Sterile square polypropylene bottles

3. MATERIALS

Plasmid DNA directing your protein of interest
Fetal bovine serum (FBS, Invitrogen)
Polyethylenimine 'Max' (linear, MW 25 000) (Polysciences, Inc.)
L-Glutamine 100× (Invitrogen)
Sodium hydroxide (NaOH)
MEM α (containing Earl's Salts and l-glutamine, but no ribonucleosides, deoxyribonucleosides, $NaCO_3$; Invitrogen 12000)
DMEM/F12 (with L-glutamine, but no HEPES, $NaHCO_3$; Invitrogen 12500)
Freestyle™ 293 medium (Invitrogen 12338-026)
FreeStyle™ CHO-S (Invitrogen R800-07)
Hybridoma SFM (Invitrogen 12045)
Opti-MEM® (Invitrogen)
HEK293S GnTI- (ATCC# CRL-3022)
HEK293T/17 (ATCC# 11268)

Note *Some of the stock solutions come with the pH indicator phenol red. This supplement does not affect the application and might be useful if the researcher wishes to visualize any pH changes that can occur in the solutions over time. In the case of non-CO_2 incubators (e.g., when scaling-up the production of adherent cells in roller bottles), HEPES-buffered media can be used to keep the pH stable.*

Note *Catalog numbers are from the US website of Invitrogen and may differ on other local websites.*

3.1. Solutions & buffers

PEI 'Max'

Dissolve 1 g PEI 'Max' in 900 ml distilled water. Adjust the pH to 7.0 with 1 N NaOH. Add distilled water to 1 l

Note: Stable at least 9 months at 4 °C

Make smaller volumes depending on how much is needed. PEI 'Max' cannot be frozen!

FreeStyle™ 293 'Completed'

Component	Stock	Amount
FreeStyle™ 293 medium		1 l
FBS	100%	10 ml
L-Glutamine	200 mM	10 ml

DMEM:F12, 5% FBS

Add 50 ml FBS to 1 l of DMEM:F12

Alpha MEM, 5% FBS

Add 50 ml FBS to 1 l of Alpha MEM

Hybridoma SFM, 1% FBS

Add 10 ml FBS to 1 l of Hybridoma SFM

4. PROTOCOL

4.1. Preparation

Before transfection, sterile high-quality DNA must be prepared. The vector containing the appropriate expression promoter (see Molecular Cloning) and the gene of interest should be transformed into a recA- strain of *E. coli* (see Transformation of Chemically Competent *E. coli* or Transformation of *E. coli* via electroporation) and then the plasmid DNA isolated (see Isolation of plasmid DNA from bacteria). Commercially available, endotoxin-free kits for large-scale plasmid DNA isolation produce sufficiently high-quality DNA. High-quality DNA is characterized as having an OD260/280 ratio between 1.88 and 1.92, an OD260/230 ratio of 2.1–2.2, and a concentration above 0.5 mg ml^{-1} (see Explanatory Chapter: Nucleic Acid Concentration Determination).

Cells must be greater than five passages from liquid nitrogen, adapted to media, free of mycoplasma contamination, and single cells if in suspension culture. All steps are carried out using sterile technique in a laminar flow hood. Solutions should be sterile-filtered through 0.22-μm filters. All plastic and glassware, if not purchased as sterile, should be double autoclaved. Cell growth media should be warmed to 37 °C prior to contact with cells. Different growth media are needed for each cell line and growth condition. These media are listed in Table 18.1. Each time the protocol says to use 'media,' use the appropriate media as outlined in Table 1 for the specific cell line and growth conditions. Cell type used depends on specific needs of protein of interest. In our laboratory, optimized protein expression conditions determined for adherent cells translate well into large-scale suspension conditions for the same cell line.

Table 18.1 Cells: Growth characteristics and medium

Cell line	Growth type	Ideal medium	Cell source
HEK293S GnTI-	Suspension	Freestyle™ 293 'completed'	ATCC# CRL-3022
HEK293S GnTI-	Adherent	DMEM:F12, 5% FBS	ATCC# CRL-3022
HEK293T/17	Adherent	DMEM:F12, 5% FBS	ATCC# CRL-11268
CHO-S	Suspension	Hybridoma SFM, 1% FBS	Invitrogen R800-07
CHO-S	Adherent	DMEM:F12, 5% FBS	Invitrogen R800-07

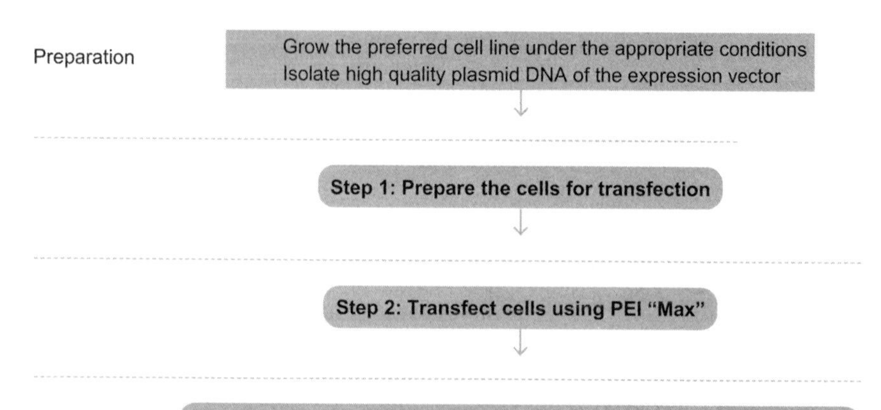

Figure 18.1 Flowchart of the complete protocol, including preparation.

4.2. Duration

Preparation	1 week
Protocol	1–2 weeks

See Fig. 18.1 for the flowchart of the complete protocol.

5. STEP 1 SMALL-SCALE TRANSIENT TRANSFECTION

5.1. Overview

This step will prescreen a variety of transfection conditions including media, cell type, ratio of PEI to DNA, and expression time to maximize protein expression before scaling up to a large-scale transfection.

5.2. Duration

5 days

Note	*Transfection must be done in the absence of antibiotics.*

6. STEP 1.1 SEED ADHERENT CELLS FOR TRANSFECTION

6.1. Overview

The appropriate amount of cells are transferred to a 6-well dish and allowed to become adherent.

6.2. Duration

1.5 days

1.1.1 Wash the cells with PBS, trypsinize, add 10 ml DMEM:F12 + 5% FBS, and gently pipet the cells several times to ensure an even suspension before counting. Transfer the cells to a 15-ml sterile centrifuge tube. Count the cells using a hemacytometer.

1.1.2 For each transfection condition to be tested, aliquot 3×10^5 cells into a sterile centrifuge tube.

1.1.3 Spin the cells at $200 \times g$ at room temperature for 5 min. Aspirate supernatant.

1.1.4 Add DMEM:F12 + 5% FBS to the cell pellet to a final concentration of 3×10^5 cells ml^{-1}.

1.1.5 Resuspend the cells with a serological pipette.

1.1.6 Add 2 ml of DMEM:F12 + 5% FBS to each well of a 6-well plate.

1.1.7 Transfer 1 ml of the cell suspension to each well.

1.1.8 Place dish in the 37 °C incubator, 5% CO_2.

1.1.9 After 24 h, remove the media from each well.

1.1.10 Add 2.7 ml of fresh DMEM:F12 + 5% FBS to each well.

1.1.11 Return the dish to incubator.

See Fig. 18.2 for the flowchart of Step 1.1.

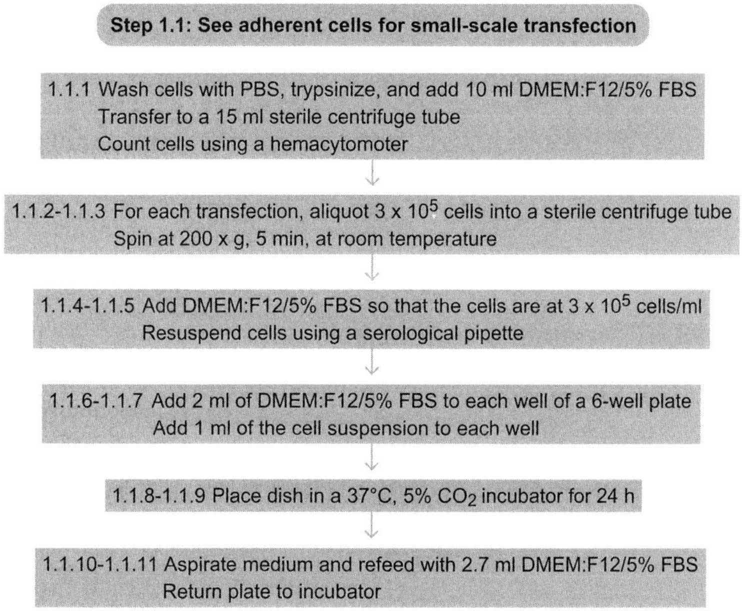

Step 1.1: See adherent cells for small-scale transfection

1.1.1 Wash cells with PBS, trypsinize, and add 10 ml DMEM:F12/5% FBS
Transfer to a 15 ml sterile centrifuge tube
Count cells using a hemacytomoter

↓

1.1.2-1.1.3 For each transfection, aliquot 3 x 10^5 cells into a sterile centrifuge tube
Spin at 200 x g, 5 min, at room temperature

↓

1.1.4-1.1.5 Add DMEM:F12/5% FBS so that the cells are at 3 x 10^5 cells/ml
Resuspend cells using a serological pipette

↓

1.1.6-1.1.7 Add 2 ml of DMEM:F12/5% FBS to each well of a 6-well plate
Add 1 ml of the cell suspension to each well

↓

1.1.8-1.1.9 Place dish in a 37°C, 5% CO_2 incubator for 24 h

↓

1.1.10-1.1.11 Aspirate medium and refeed with 2.7 ml DMEM:F12/5% FBS
Return plate to incubator

Figure 18.2 Flowchart of Step 1.1.

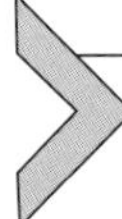

7. STEP 1.2 TRANSIENTLY TRANSFECT CELLS

7.1. Overview

Cells are transfected by adding DNA and PEI 'Max' to the cells. The PEI-DNA mixture is prepared and added to the cells on the same day as changing the media.

7.2. Duration

45 min active time; 4 days total

1.2.1 Dilute 9 μg of PEI 'Max' into a total volume of 150 μl of Opti-MEM. The amount of PEI can be varied.
1.2.2 Dilute 3 μg of DNA into a total volume of 150 μl of Opti-MEM.
1.2.3 Add the diluted PEI 'Max' to the diluted DNA.
1.2.4 Incubate the mixture at room temperature for 30 min.
1.2.5 Carefully add the PEI-DNA mixture to a well of adherent cells. Take care to gently pipette the solution down the side of the well and not on top of the cells, so as not to disrupt the adherent cells.
1.2.6 Return the dish to the 5% CO_2 incubator.

7.3. Tip

Opti-MEM can be replaced with Hybridoma Media without serum.

7.4. Tip

The protocol outlined here uses a 3:1 ratio of PEI to DNA (w/w). We have found this ratio to be optimal for most genes we have expressed. However, this ratio should be screened for each gene tested. We routinely screen ratios between 1:1 and 5:1.

7.5. Tip

In general, use 1 μg of DNA per 1 ml of culture to be transfected. PEI and DNA should each be diluted into 1/20 of the total culture volume before being combined.

7.6. Tip

Small-scale transfections can be performed with suspension-adapted cells. The protocol for small scale is essentially the same. For suspension culture, we use square plastic bottles designed to hold 125 ml; however, we add only 5–12 ml of cell medium for optimal aeration and agitation.

See Fig. 18.3 for the flowchart of Step 1.2.

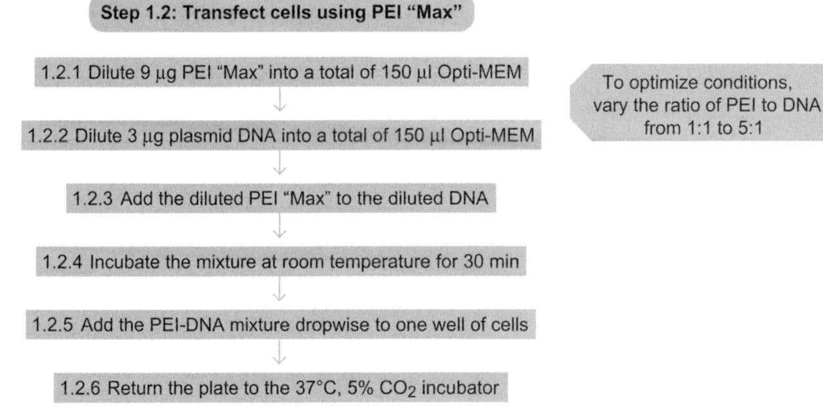

Figure 18.3 Flowchart of Step 1.2.

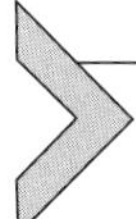

8. STEP 1.3 HARVEST CELLS AND ANALYZE PROTEIN EXPRESSION

8.1. Overview

Harvest and lyse cells (see Lysis of mammalian and Sf9 cells). Analyze protein expression by Western blotting (see Western Blotting using Chemiluminescent Substrates).

8.2. Duration

1–2 days

1.3.1 96 h after transfection, collect samples to be analyzed. For secreted proteins collect and save the media. For membrane or intracellular proteins, remove media. Wash the cells with PBS and lyse the cells.

1.3.2 Analyze protein expression by Western blotting or ELISA as appropriate.

8.3. Tip

HEK293 GnTI⁻ cells can be resuspended by pipetting gently up and down with a serological pipette. CHO-S cells adhere more tightly to the dish and need to be manually resuspended with a cell scraper.

8.4. Tip

Some protocols call for trypsin digestion to detach the cells from the dish. This can be avoided by manual scraping of the cells. Trypsin could degrade the expressed protein if it is a membrane protein.

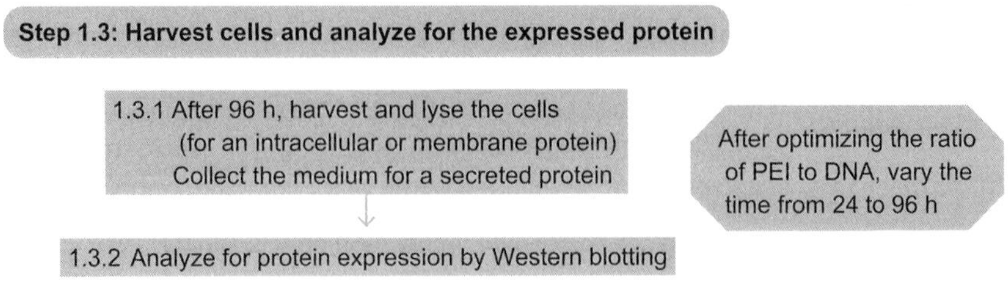

Step 1.3: Harvest cells and analyze for the expressed protein

1.3.1 After 96 h, harvest and lyse the cells (for an intracellular or membrane protein) Collect the medium for a secreted protein

↓

1.3.2 Analyze for protein expression by Western blotting

After optimizing the ratio of PEI to DNA, vary the time from 24 to 96 h

Figure 18.4 Flowchart of Step 1.3.

8.5. Tip

Once a cell type, media, and optimal ratio of PEI to DNA are established, this protocol can be repeated and samples taken between 24 and 96 h posttransfection to optimize the length of expression.

See Fig. 18.4 for the flowchart of Step 1.3.

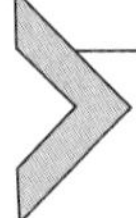

9. STEP 2 LARGE-SCALE TRANSIENT TRANSFECTION OF SUSPENSION CELLS

9.1. Overview

Preparative scale expression of protein in suspension culture. For this protocol, the parameters optimized in Step 1 are expanded to larger volume cultures. You will need 400 ml of cells at a density of 2–3 $\times$ 10^6 cells ml^{-1}.

9.2. Duration

4–8 days

10. STEP 2.1 PREPARE THE CELLS TO BE TRANSFECTED

10.1. Overview

Harvest and count suspension cells to ensure that they are at the proper density. Centrifuge the cells and resuspend them in a total of 360 ml of fresh suspension growth medium.

10.2. Duration

30 min

2.1.1 Grow 400 ml of cells in the appropriate suspension growth medium to a density between 2 and 3 $\times$ 10^6 cells ml^{-1}.

2.1.2 Transfer the cell suspension to sterile centrifuge bottles.

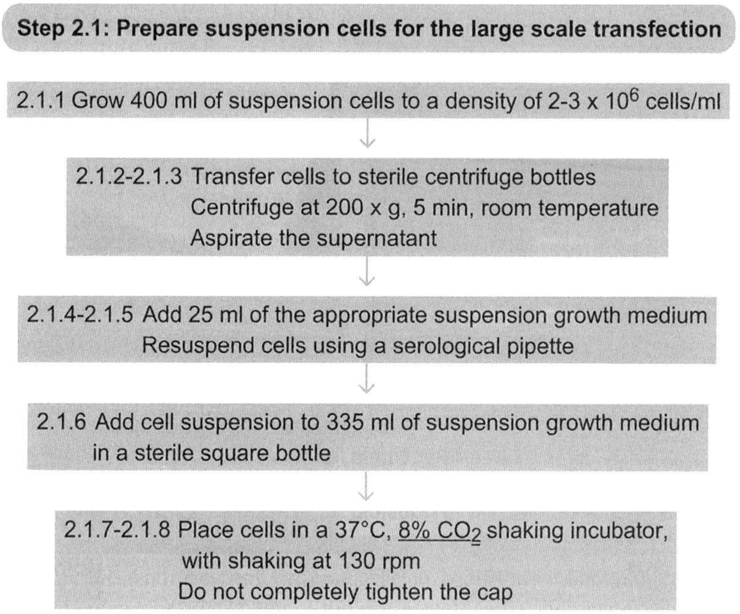

Step 2.1: Prepare suspension cells for the large scale transfection

2.1.1 Grow 400 ml of suspension cells to a density of 2-3 x 10^6 cells/ml

↓

2.1.2-2.1.3 Transfer cells to sterile centrifuge bottles
Centrifuge at 200 x g, 5 min, room temperature
Aspirate the supernatant

↓

2.1.4-2.1.5 Add 25 ml of the appropriate suspension growth medium
Resuspend cells using a serological pipette

↓

2.1.6 Add cell suspension to 335 ml of suspension growth medium in a sterile square bottle

↓

2.1.7-2.1.8 Place cells in a 37°C, 8% CO_2 shaking incubator, with shaking at 130 rpm
Do not completely tighten the cap

Figure 18.5 Flowchart of Step 2.1.

2.1.3 Spin the cells at 200 × g at room temperature for 5 min. Aspirate the supernatant.
2.1.4 Add 25 ml of the appropriate fresh suspension growth medium.
2.1.5 Gently resuspend the cells with a serological pipette.
2.1.6 Add the cells to 335 ml of fresh suspension growth medium in a sterile square bottle.
2.1.7 Do not tighten the bottle cap all the way.
2.1.8 Place the cells into a 37 °C incubator shaker set at 8% CO_2, with shaking at 130 rpm.

See Fig. 18.5 for the flowchart of Step 2.1.

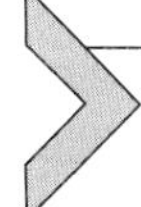

11. STEP 2.2 TRANSFECT CELLS

11.1. Overview

Transfect the cells using the optimal ratio or PEI to DNA as determined above.

11.2. Duration

45 min active time, expression time as determined above

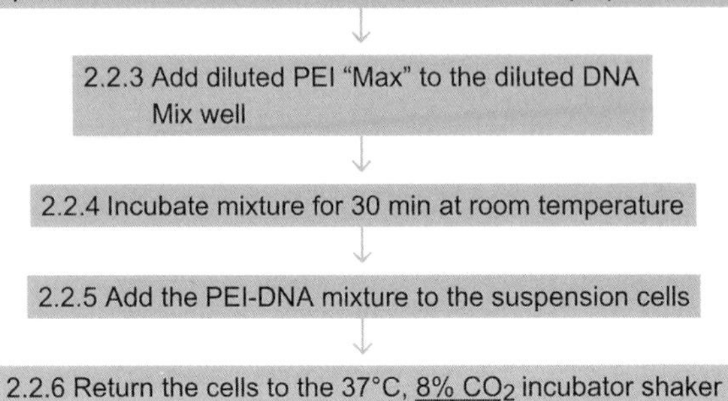

Figure 18.6 Flowchart of Step 2.2.

2.2.1 Dilute 400 μg of DNA in a total volume of 20 ml of Hybridoma SFM (without serum).

2.2.2 Dilute the appropriate amount of PEI 'Max,' as determined in Step 1, into a total volume of 20 ml of Hybridoma SFM (without serum).

2.2.3 Add the diluted PEI 'Max' to the diluted DNA and mix.

2.2.4 Incubate the mixture at room temperature for 30 min.

2.2.5 Add the PEI-DNA mixture to the suspension cells from Step 2.1.8.

2.2.6 Return the cells to the 37 °C incubator shaker, shaking at 130 rpm.

See Fig. 18.6 for the flowchart of Step 2.2.

12. STEP 2.3 HARVEST CELLS AND PROCESS PROTEIN AS NEEDED

12.1. Overview

Harvest the cells (or medium for a secreted protein). Purify the protein or process as needed for downstream applications (see Salting out of proteins using ammonium sulfate precipitation, Using ion exchange chromatography to purify a recombinantly expressed protein, Gel filtration chromatography (Size exclusion chromatography) of proteins, Use and Application of Hydrophobic Interaction Chromatography for Protein Purification or

Hydroxyapatite Chromatography: Purification Strategies for Recombinant Proteins, or look up the chapters on affinity purification if tags have been added to the protein: Purification of His-tagged proteins, Affinity purification of a recombinant protein expressed as a fusion with the maltose-binding protein (MBP) tag, Purification of GST-tagged proteins, Protein Affinity Purification using Intein/Chitin Binding Protein Tags, Immunoaffinity purification of proteins or Strep-tagged protein purification).

12.2. Duration

About 1 h

2.3.1 After the appropriate amount of time, as determined in Step 1, centrifuge the cells at 200 × g for 5 min at room temperature.

2.3.2 If the protein is secreted, collect the medium. Centrifuge the cells at 200 × g for 5 min at room temperature and sterile-filter the medium through a 0.22-μm filter. Add sodium azide to 0.02%. The medium can be stored at 4 °C for months until needed.

2.3.3 If the protein is to be purified, the cell pellet should be flash-frozen in liquid nitrogen and stored at −80 °C until needed.

12.3. Tip

In suspension cultures, single cells are transfected more efficiently than cells that have clumped together during growth. Growth conditions may need to be optimized for single cell growth.

12.4. Tip

Square bottles should be autoclaved in two consecutive dry cycles (45 min each, dry 15 min each) with the lids as loose as possible without falling off. They should be allowed to cool completely in the laminar flow hood before tightening the lids. If the bottles collapse inward, cells will not grow well.

12.5. Tip

To generate more protein, we have found that 24 h posttransfection the cells can be diluted between 1:2 and 1:5 in the appropriate media. The effect of dilution and optimal dilution ratios should be determined empirically.

See Fig. 18.7 for the flowchart of Step 2.3.

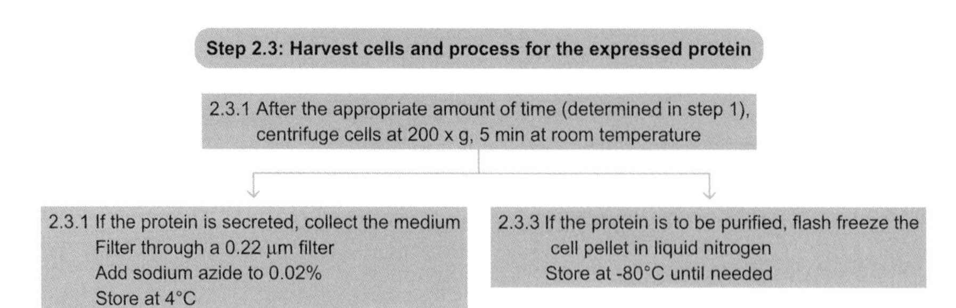

Figure 18.7 Flowchart of Step 2.3.

REFERENCES

Referenced Literature

Boussif, O., et al. (1995). A versatile vector for gene and oligonucleotide transfer into cells in culture and in vivo: Polyethylenimine. *Proceedings of the National Academy of Sciences of the United States of America*, *92*(16), 7297–7301.

Sonawane, N. D., Szoka, F. C., Jr., & Verkman, A. S. (2003). Chloride accumulation and swelling in endosomes enhances DNA transfer by polyamine-DNA polyplexes. *The Journal of Biological Chemistry*, *278*(45), 44826–44831.

Referenced Protocols in Methods Navigator

Small-scale Expression of Proteins in *E. coli*

Rapid creation of stable mammalian cell lines for regulated expression of proteins using the Gateway® Recombination Cloning Technology and Flp-In T-REx® lines

Generating mammalian stable cell lines by electroporation

Molecular Cloning

Transformation of Chemically Competent *E. coli*

Transformation of *E. coli* via electroporation

Isolation of plasmid DNA from bacteria

Explanatory Chapter: Nucleic Acid Concentration Determination

Lysis of mammalian and Sf9 cells

Western Blotting using Chemiluminescent Substrates

Salting out of proteins using ammonium sulfate precipitation

Using ion exchange chromatography to purify a recombinantly expressed protein

Gel filtration chromatography (Size exclusion chromatography) of proteins

Use and Application of Hydrophobic Interaction Chromatography for Protein Purification

Hydroxyapatite Chromatography: Purification Strategies for Recombinant Proteins

Purification of His-tagged proteins

Affinity purification of a recombinant protein expressed as a fusion with the maltose-binding protein (MBP) tag

Purification of GST-tagged proteins

Protein Affinity Purification using Intein/Chitin Binding Protein Tags

Immunoaffinity purification of proteins

Strep-tagged protein purification

CHAPTER NINTEEN

Site-Directed Mutagenesis

Julia Bachman[1]
Department of Neuroscience, Johns Hopkins University School of Medicine, Baltimore, MD, USA
[1]Corresponding author: e-mail address: jbachma9@gmail.com

Contents

1. Theory 242
2. Equipment 242
3. Materials 242
   3.1 Solutions & buffers 243
4. Protocol 243
   4.1 Preparation 243
   4.2 Duration 243
5. Step 1 Setting up and Run the PCR 243
   5.1 Overview 243
   5.2 Duration 244
   5.3 Tip 245
   5.4 Tip 245
   5.5 Tip 245
6. Step 2 Digestion of Template DNA 246
   6.1 Overview 246
   6.2 Duration 246
7. Step 3 Transformation into Chemically Competent *E. coli* 246
   7.1 Overview 246
   7.2 Duration 246
   7.3 Tip 247
   7.4 Tip 247
8. Step 4 Colony Screening 248
   8.1 Overview 248
   8.2 Duration 248
   8.3 Tip 248
References 248

Abstract

Site-directed mutagenesis is a PCR-based method to mutate specified nucleotides of a sequence within a plasmid vector. This technique allows one to study the relative importance of a particular amino acid for protein structure and function. Typical mutations are designed to disrupt or map protein–protein interactions, mimic or block posttranslational modifications, or to silence enzymatic activity. Alternatively, noncoding changes are often used to generate rescue constructs that are resistant to knockdown via RNAi.

Methods in Enzymology, Volume 529
ISSN 0076-6879
http://dx.doi.org/10.1016/B978-0-12-418687-3.00019-7

1. THEORY

Forward and reverse primers are centered on the desired base change(s) and overlap completely in both directions. Extension of these primers around the entire plasmid creates a copy of the template with the inserted mutation. This mutated plasmid then serves as a template for further PCR amplification. Since plasmids grown in *E. coli* are *dam* methylated, the restriction endonuclease Dpn I, which is specific to methylated and hemimethylated DNA, is used to digest the original parent template. The remaining plasmid is then transformed into *E. coli* for selection and DNA isolation.

2. EQUIPMENT

PCR Thermocycler
Water bath (37 and 42 °C)
Shaking incubator
Incubator
Micropipettors
Micropipettor tips
0.2-ml thin-walled PCR tubes
1.5-ml polypropylene tubes
15-ml sterile snap-cap tubes
Petri plates

3. MATERIALS

PCR primers
Plasmid DNA (with desired sequence to be mutated)
dNTPs (100 mM each of dATP, dCTP, dGTP, and dTTP)
10× High-fidelity DNA polymerase buffer
High-fidelity *Pfu*-based DNA polymerase
Ultrapure water
DpnI
Chemically competent cells (e.g., XL 10-Gold ultracompentent cells, Stratagene)
Lysogeny Broth (LB media)
Petri plates

3.1. Solutions & buffers

Step 1 10 mM dNTP mix

Component	Final Concentration	Stock	Amount
dATP	10 mM	100 mM	100 μl
dCTP	10 mM	100 mM	100 μl
dGTP	10 mM	100 mM	100 μl
dTTP	10 mM	100 mM	100 μl

Add purified water to 1 ml

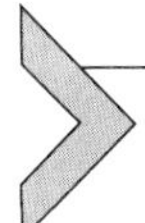

4. PROTOCOL

4.1. Preparation

PCR primers must be designed and synthesized prior to starting this protocol (see Explanatory chapter: PCR -Primer design).

Design forward and reverse primers containing the desired mutation(s). Primers should be 25–45 nucleotides long leaving at least 10–15 nucleotides of perfect matching sequence at both ends. In general, more point mutations being made will require longer primers. For example, a single base change probably requires only 25-mers. If three bases are to be mutated, the primers should be about 35–38 bases in length. Ideally, terminate primers with C or G and maintain a minimum GC content of 40%. Additional primers are needed for sequencing through the insert in order to verify mutations made and check for nonspecific mutations.

4.2. Duration

Preparation	Variable
Protocol	5–6 h + overnight incubation
	2 h + DNA sequencing

See Fig. 19.1 for the flowchart of the complete protocol.

5. STEP 1 SETTING UP AND RUN THE PCR

5.1. Overview

Set up and run the PCR using the primers containing the mutation(s) and a high-fidelity thermostable DNA polymerase.

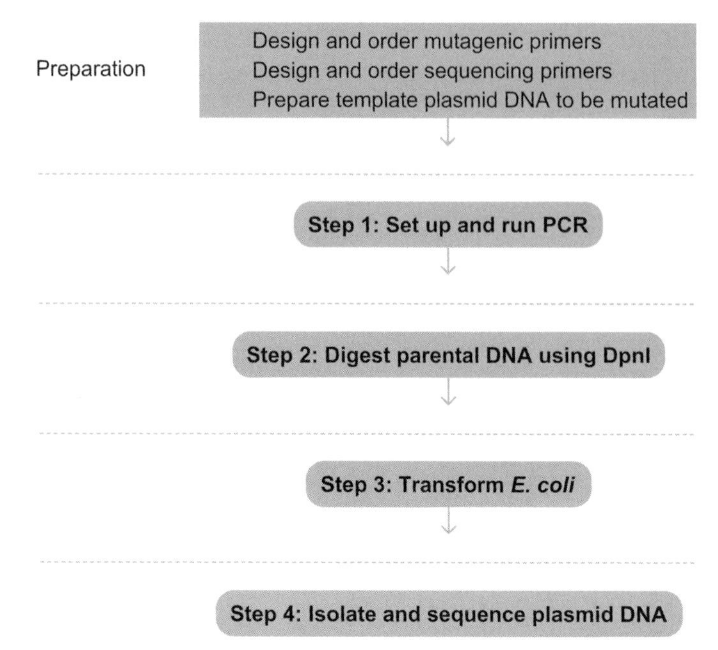

Figure 19.1 Flowchart of the complete protocol, including preparation.

5.2. Duration

about 2 h

1.1 Set up the reaction mixture on ice. Add to a 0.2-ml thin-walled PCR tube:

1 μl template (10 ng μl^{-1})
5 μl 10× Polymerase Buffer
2.5 μl Forward primer (50 ng μl^{-1})
2.5 μl Reverse primer (50 ng μl^{-1})
1 μl dNTP mix (10 mM)
37.5 μl purified water
0.5 μl DNA Polymerase

1.2 Run the PCR with the following cycling parameters:

At 98 °C for 30 s (Initial denaturation of the template)
Repeat for 18 cycles:
  At 98 °C for 10 s (Denature)
  At 60 °C for 30 s (Anneal)
  At 68 °C for 1 min per kb length of entire plasmid (Extension)

5.3. Tip

Add the enzyme last, immediately before starting the thermocycler. If setting up more than one reaction, make a master mix and dispense into tubes already containing the specific primers and/or template.

5.4. Tip

The annealing temperature can vary from 55 to 60 °C according to the primer T_m.

5.5. Tip

The extension temperature for Taq DNA polymerase is 72 °C. However, the recommended extension temperature for many high-fidelity polymerases is 68 °C. Check the product information sheet of the high-fidelity thermostable DNA polymerase being used.

See Fig. 19.2 for the flowchart of Steps 1 and 2.

Step 1: Set up and run PCR

1.1 Add to a 0.2 ml thin-walled PCR tube:

Template DNA (10 ng/μl)	1 μl
10X Polymerase buffer	5 μl
Primer 1 (50 ng/μl)	2.5 μl
Primer 2 (50 ng/μl)	2.5 μl
dNTP mix (10 mM)	1 μl
Ultrapure water	37.5 μl
Thermostable DNA polymerase	0.5 μl

↓

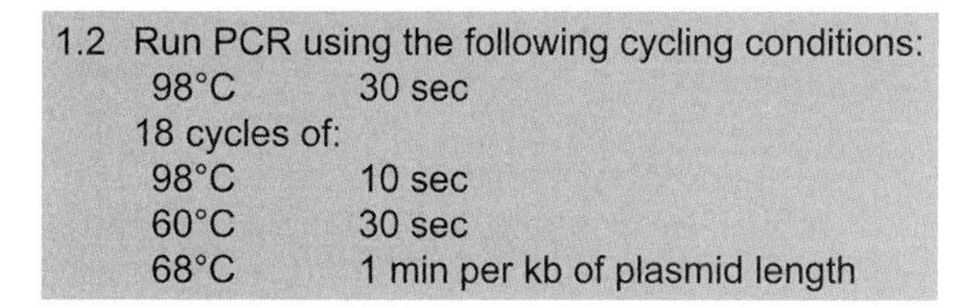

1.2 Run PCR using the following cycling conditions:
98°C 30 sec
18 cycles of:
98°C 10 sec
60°C 30 sec
68°C 1 min per kb of plasmid length

Step 2: Digest parental DNA using DpnI

2.1 Add 1 μl DpnI (10-20 units)

↓

2.2 Incubate at 37°C, 1-2 h

Figure 19.2 Flowchart of Steps 1 and 2.

6. STEP 2 DIGESTION OF TEMPLATE DNA

6.1. Overview

Parental strands of DNA will be digested using DpnI, which cuts methylated and hemimethylated DNA.

6.2. Duration

1–2 h

2.1 Add 1 μl of Dpn I restriction enzyme (10–20 units).

2.2 Incubate at 37 °C for 1–2 h.

7. STEP 3 TRANSFORMATION INTO CHEMICALLY COMPETENT *E. COLI*

7.1. Overview

The mutated plasmid is transferred into chemically competent *E. coli* via heat-shock (see Transformation of Chemically Competent *E. coli*, or alternatively see Transformation of *E. coli* via electroporation if you are using electrocompetent cells). The transformed bacteria are then grown on media containing the appropriate antibiotic for the plasmid.

7.2. Duration

2 h + overnight incubation

3.1 Let the competent cells thaw on ice.

3.2 Add 5 μl of the reaction to 50 μl of chemically competent cells. Mix gently to avoid damaging the cells (do not pipette up and down).

3.3 Incubate on ice for 30 min.

3.4 Heat shock at 42 °C for 1 min.

3.5 Incubate tube on ice for 2 min.

3.6 Add 1 ml of room temperature LB (or other medium, e.g., SOC).

3.7 Incubate at 37 °C for 1 h with shaking (100–150 rpm).

3.8 Centrifuge bacteria at low speed (~7500 rpm) for 2 min.

3.9 Aspirate all but 100 μl of the LB.

3.10 Carefully resuspend the pellet and spread all the 100 μl evenly onto a selective plate.

3.11 Incubate overnight at 37 °C with the plate turned upside down.

7.3. Tip

XL 10-Gold ultracompetent cells are recommended for their high transformation efficiency.

7.4. Tip

Check the supplier's instructions for the optimal length of time for the heat shock.

See Fig. 19.3 for the flowchart of Steps 3 and 4.

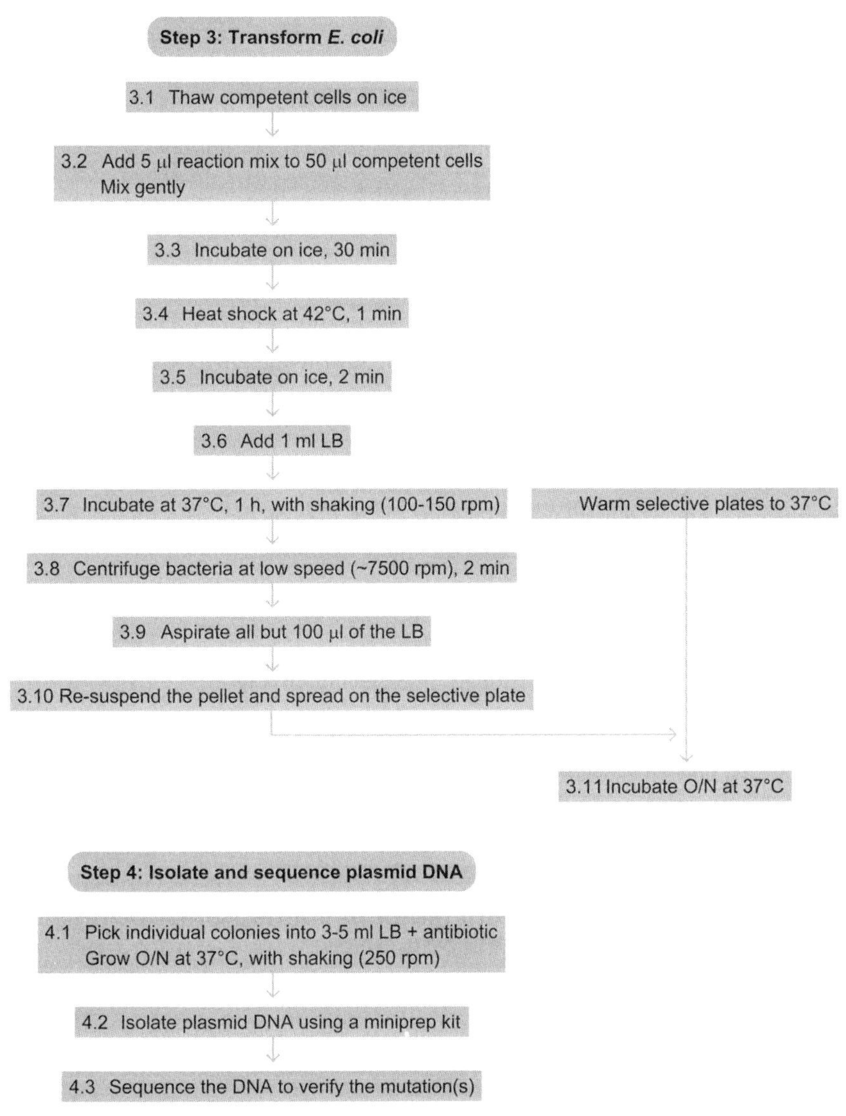

Figure 19.3 Flowchart of Steps 3 and 4.

8. STEP 4 COLONY SCREENING

8.1. Overview

Pick single colonies, grow overnight, and isolate plasmid DNA using a miniprep kit (alternatively see Isolation of plasmid DNA from bacteria). Sequence the DNA to verify that the mutations have been made.

8.2. Duration

15 min + overnight growth

1 h + DNA sequencing

4.1 Pick single colonies into 3–5 ml LB + antibiotic in a 14-ml snap-cap tube. Grow at 37 °C overnight with shaking (250 rpm).

4.2 Isolate plasmid DNA using a plasmid miniprep kit according to the manufacturer's instructions.

4.3 Sequence the DNA to verify that appropriate base changes have been made.

8.3. Tip

Since PCR is used to amplify the whole plasmid, it is important to sequence the entire insert in order to check for PCR-induced mutations. Using a high-fidelity polymerase will minimize this possibility. Alternatively, one could sequence the smaller region containing the desired mutations and subclone this region back into the original vector.

REFERENCES

Referenced Protocols in Methods Navigator

Explanatory chapter: PCR -Primer design
Transformation of Chemically Competent *E. coli*
Transformation of *E. coli* via electroporation
Isolation of plasmid DNA from bacteria

CHAPTER TWENTY

PCR-Based Random Mutagenesis

Jessica S. Dymond[1]

The High Throughput Biology Center and Department of Molecular Biology and Genetics, Johns Hopkins University School of Medicine, Baltimore, MD, USA

[1]Corresponding author: e-mail address: jsiege16@jhmi.edu

Contents

1. Theory 250
2. Equipment 250
3. Materials 250
   3.1 Solutions & buffers 251
4. Protocol 251
   4.1 Preparation 251
   4.2 Duration 251
5. Step 1 PCR Setup 252
   5.1 Overview 252
   5.2 Duration 252
   5.3 Tip 252
6. Step 2 Generation of Additional Template for Mutagenesis 253
   6.1 Overview 253
   6.2 Duration 253
   6.3 Tip 254
7. Step 3 Mutagenic PCR 254
   7.1 Overview 254
   7.2 Duration 254
   7.3 Tip 255
8. Step 4 Subclone and Sequence the PCR Products 255
   8.1 Overview 255
   8.2 Duration 255
   8.3 Tip 256
   8.4 Tip 256
   8.5 Tip 256
   8.6 Tip 256
   8.7 Tip 256
References 257

Abstract

Random PCR mutagenesis enables the rapid and inexpensive construction of a library of mutant genetic elements.

Methods in Enzymology, Volume 529
ISSN 0076-6879
http://dx.doi.org/10.1016/B978-0-12-418687-3.00020-3

1. THEORY

PCR is routinely performed with standard polymerases, yielding a final PCR product with few errors. Although polymerases vary in fidelity, the inherent mutation rate of PCR is insufficient to generate the high frequency of mutations necessary to generate a mutant library. The addition of manganese to the PCR reaction reduces polymerase fidelity, resulting in error-prone PCR and an increased substitution frequency in the PCR products. By performing error-prone PCR on a genetic element of interest and cloning PCR products, a mutant library can be constructed and characterized by sequencing. Manganese-induced error-prone PCR favors transitions (purine-to-purine and pyrimidine-to-pyrimidine mutations); should a broad spectrum of mutations be desired, the relative concentrations of the dNTPs may be altered (Cadwell and Joyce, 1992). Additionally, error-prone PCR may generate no or multiple substitutions in a given PCR product, and additional screening and/or subcloning may be required to isolate single mutations.

2. EQUIPMENT

PCR thermocycler
Micropipettors
Micropipettor tips
0.2-ml thin-walled PCR tubes or plates

3. MATERIALS

PCR Primers
Template DNA
Thermostable DNA polymerase
10× Polymerase Buffer (supplied with enzyme)
dNTP mix (or dATP, dTTP, dCTP, and dGTP)
Manganese chloride ($MnCl_2$)
Sterile water

3.1. Solutions & buffers

Step 1 dNTP mix

Component	Final concentration	Stock	Amount
dATP	2.0 mM	100 mM	20 μl
dTTP	2.0 mM	100 mM	20 μl
dCTP	2.0 mM	100 mM	20 μl
dGTP	2.0 mM	100 mM	20 μl
H_2O			920 μl

Step 3 $MnCl_2$

Component	Final concentration	Stock	Amount
$MnCl_2$	500 μM	1 M	50 μl
H_2O			9.95 ml

4. PROTOCOL

4.1. Preparation

Prepare template DNA to be mutagenized. Appropriate templates include genomic DNA, plasmids, PCR products, etc.

PCR primers should be designed and optimized prior to beginning this protocol (see Explanatory chapter: PCR -Primer design). The primers should flank only the region to be mutagenized as mutations will occur with equal frequency throughout the PCR product. A nonmutagenic PCR should be performed to optimize annealing temperatures and extension times for each oligo pair to be used in this protocol.

4.2. Duration

Preparation	About 1 day
Protocol	About 2–3 h

See Fig. 20.1 for the flowchart of the complete protocol.

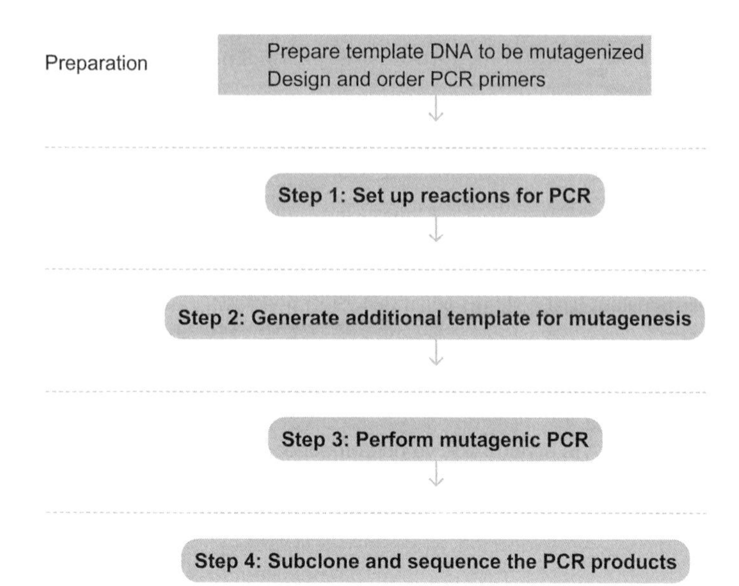
Preparation
Prepare template DNA to be mutagenized
Design and order PCR primers
Step 1: Set up reactions for PCR
Step 2: Generate additional template for mutagenesis
Step 3: Perform mutagenic PCR
Step 4: Subclone and sequence the PCR products

Figure 20.1 Flowchart of the complete protocol, including preparation.

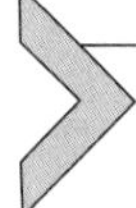

5. STEP 1 PCR SETUP

5.1. Overview

The PCR reaction is assembled.

5.2. Duration

5 min

1.1 Place a PCR tube on ice. Add the following to the PCR tube:

100 ng Template DNA (in a volume of 69 μl or less)
10 μl 10× Polymerase buffer
10 μl 2 mM dNTP Mix
1 μl 100 μM Forward primer
1 μl 100 μM Reverse primer
1 μl 5 u μl^{-1} Polymerase
H_2O to 100 μl

5.3. Tip

If the volume of template DNA is large, it should be resuspended in H_2O, not TE, to avoid significantly altering the concentration of salt and EDTA in the reaction.

See Fig. 20.2 for the flowchart of Steps 1–3.

Step 1: Set up reactions for PCR

1.1 Add to a 0.2 ml PCR tube on ice:
100 ng template DNA
10 μl 10X polymerase buffer
10 μl 2 mM dNTP mix
1 μl 100 μM Forward primer
1 μl 100 μM Reverse primer
1 μl 5 U/μl polymerase
water to 100 μl

Step 2: Generate additional template for mutagenesis

2.1 Run PCR with the following conditions:

94°C	5 min
10 cycles:	
94°C	30 sec
55-65°C	30 sec
72°C	30 sec to 5 min

Step 3: Perform mutagenic PCR

3.1 Add 1 μl 500 μM $MnCl_2$ to the PCR reaction
Mix well

↓

3.2 Return tube to thermocycler and continue PCR with the following conditions:

30 cycles:	
94°C	30 sec
55-66°C	30 sec
72°C	30 sec to 5 min
72°C	5-10 min
4°C	hold

Figure 20.2 Flowchart of Steps 1–3.

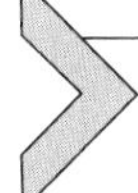

6. STEP 2 GENERATION OF ADDITIONAL TEMPLATE FOR MUTAGENESIS

6.1. Overview

The template DNA is amplified in a normal PCR to generate a large pool of template DNA for mutagenesis.

6.2. Duration

15 min – 1 h

2.1 Put the PCR mixture in the thermocycler and run the following program:

At 94 °C for	5 min
10 cycles:	
At 94 °C	for 30 s
At 55–65 °C for	30 s
At 72 °C	for 30 s – 5 min

6.3. Tip

The annealing temperature (55–65 °C) must be empirically determined prior to the mutagenic PCR. The extension time (72 °C incubation) is specific to each enzyme, but 1 min kb^{-1} is a good estimate; extension times should not fall below 30 s regardless of the amplicon size.

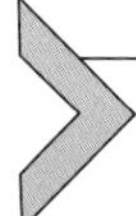

7. STEP 3 MUTAGENIC PCR

7.1. Overview

The PCR is continued in the presence of manganese to increase the mutation rate.

7.2. Duration

1–4 h

3.1 Add 1 μl 500 μM $MnCl_2$ to the PCR reaction; mix well.
3.2 Put the mixture in the thermocycler and run the following program:

30 cycles:	
At 94 °C	for 30 s
At 55–65 °C	for 30 s
At 72 °C	for 30 s – 5 min
At 72 °C	for 5–10 min
4 °C	hold

7.3. Tip

The conditions for the two sets of PCR should be the same.

8. STEP 4 SUBCLONE AND SEQUENCE THE PCR PRODUCTS

8.1. Overview

The PCR products from the mutagenic PCR will be subcloned into a plasmid vector (see Molecular Cloning), transformed into *E. coli* (see Transformation of Chemically Competent *E. coli* or Transformation of *E. coli* via electroporation), and DNA from individual colonies will be sequenced (see Isolation of plasmid DNA from bacteria).

8.2. Duration

3–6 days

4.1 Run 10 μl of the PCR products on an agarose gel to determine the purity of the products (see Agarose Gel Electrophoresis).

4.2 If there is a single band of the correct size, use a PCR purification kit to clean it up. If not, run the remainder of the reaction on an agarose gel and isolate the correct band using a DNA gel extraction kit.

4.3 Digest the purified PCR product and vector DNA with the appropriate restriction enyzmes. Add to a 1.5-ml microcentrifuge tube:

DNA	1–2 μg
10× restriction enzyme buffer	5 μl
10× BSA (if required)	5 μl
Restriction enzyme(s)	1 μl
Water	up to 50 μl

Incubate at 37 °C (or appropriate temperature) for 1–2 h.

4.4 Run the digested DNAs on an agarose gel and isolate them using a DNA gel extraction kit.

4.5 Ligate the digested PCR product and vector. Add to a 1.5-ml microcentrifuge tube:

10× T4 DNA ligase buffer	2 μl
Digested vector DNA	50–100 ng
Digested PCR products	100–300 ng
T4 DNA ligase	1 μl
Water	up to 20 μl

Incubate at room temperature for 1–2 h or overnight at 16 °C.

4.6 Transform competent *E. coli* and plate cells on selective plates. Grow overnight at 37 °C.

4.7 Pick single colonies into 5 ml LB containing antibiotic and grow overnight at 37 °C, with shaking.

4.8 Isolate plasmid DNA using a Plasmid miniprep kit and sequence the inserts to determine whether they contain mutations.

8.3. Tip

Digests of the purified PCR product and vector DNA should be set up in separate tubes.

8.4. Tip

BSA is required by some restriction enzymes. Consult the manufacturer's product material to determine whether it is necessary. BSA does not exhibit an inhibitory effect on restriction enzymes and may be routinely included in digests if so desired.

8.5. Tip

The incubation time and temperature may need to be adjusted depending on the enzyme(s) to be used. Consult the manufacturer's product material for the optimal conditions for a specific enzyme.

8.6. Tip

If you can screen genetically for a phenotype or have a functional assay, use this to prescreen for potentially interesting mutants prior to DNA sequencing.

8.7. Tip

If possible, transform E. coli via electroporation since this will give the greatest number of transformants, increasing the chances of obtaining interesting mutations.

See Fig. 20.3 for the flowchart of Step 4.

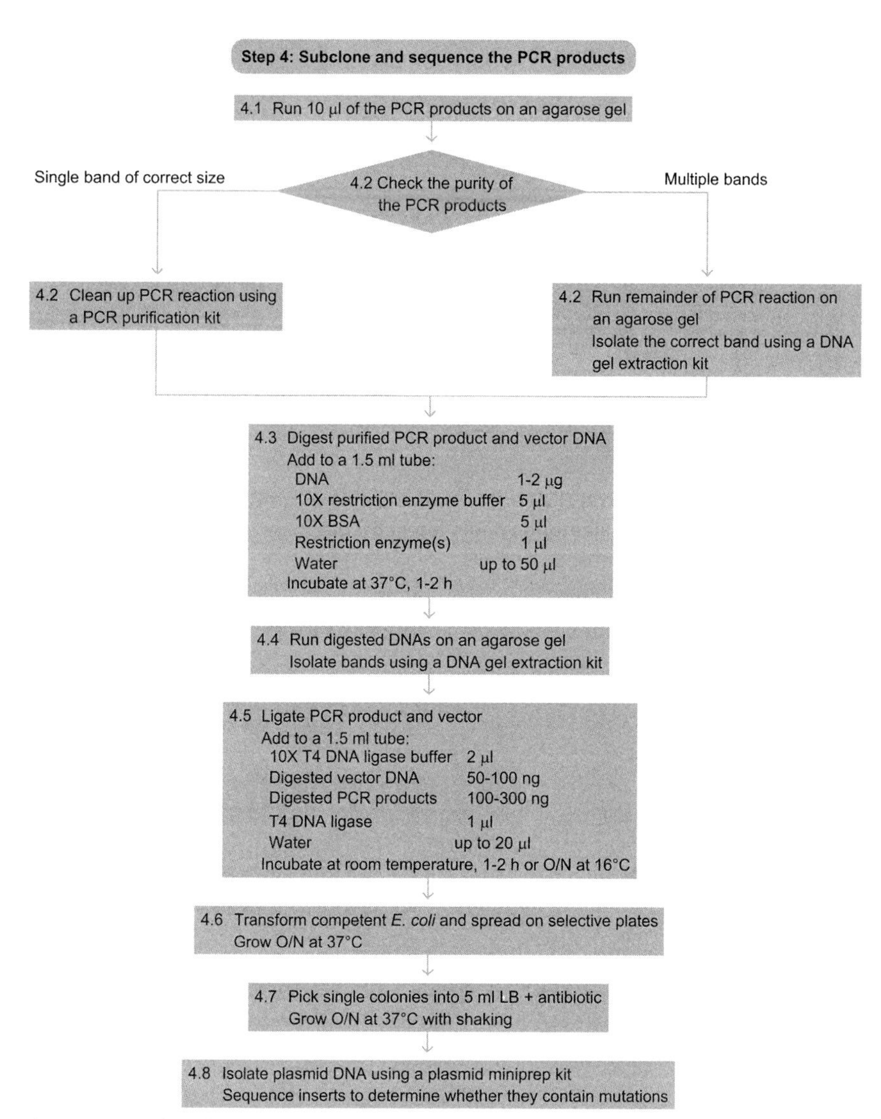

Figure 20.3 Flowchart of Step 4.

REFERENCES

Referenced Literature

Cadwell, R. C., & Joyce, G. F. (1992). Randomization of genes by PCR mutagenesis. *PCR Methods and Applications*, *2*, 28–33.

Referenced Protocols in Methods Navigator

Explanatory chapter: PCR -Primer design
Molecular Cloning

Transformation of Chemically Competent *E. coli*
Transformation of *E. coli* via electroporation
Isolation of plasmid DNA from bacteria
Agarose Gel Electrophoresis

CHAPTER TWENTY ONE

Megaprimer Method for Mutagenesis of DNA

Craig W. Vander Kooi[1]
Department of Molecular and Cellular Biochemistry and Center for Structural Biology, University of Kentucky, Lexington, KY, USA
[1]Corresponding author: e-mail address: craig.vanderkooi@uky.edu

Contents

1. Theory 260
2. Equipment 260
3. Materials 261
   3.1 Solutions & Buffers 261
4. Protocol 261
   4.1 Duration 261
   4.2 Preparation 261
5. Step 1 Design Mutagenic Primer 262
   5.1 Overview 262
   5.2 Duration 262
   5.3 Tip 263
6. Step 2 First Round of PCR 263
   6.1 Overview 263
   6.2 Duration 263
   6.3 Tip 264
7. Step 3 Gel Purification of the Megaprimer 265
   7.1 Overview 265
   7.2 Duration 265
   7.3 Tip 265
   7.4 Caution 265
8. Step 4 Second Round of PCR 265
   8.1 Overview 265
   8.2 Duration 265
   8.3 Tip 267
   8.4 Tip 267
9. Step 5 Gel Purification of the Mutagenized Gene 267
   9.1 Overview 267
   9.2 Duration 267
   9.3 Tip 267

Methods in Enzymology, Volume 529
ISSN 0076-6879
http://dx.doi.org/10.1016/B978-0-12-418687-3.00021-5

Abstract

To create mutations in a gene of interest, including point mutations, short insertions, or deletions.

1. THEORY

Alteration of a gene of interest is a fundamental tool to study the structure and function of DNA and proteins in biomedical research. Point mutations are often produced to study the role of specific amino acid residues in protein activity and interactions. Insertions and deletions can also be incorporated to study protein function.

One commonly used method for producing these alterations is the megaprimer method (Sarkar and Sommer, 1990), which utilizes a two-step PCR-based protocol to incorporate changes followed by incorporation into a target plasmid. This is a flexible and robust technique allowing a wide range of alterations to be incorporated into a gene of interest.

Another commonly used method involves replication of the complete plasmid and gene of interest while incorporating the mutation (e.g., Quikchange mutagenesis, Stratagene).

2. EQUIPMENT

PCR Thermocycler
Microcentrifuge
Micropipettors
Agarose gel electrophoresis equipment
UV light box
0.2-ml thin-walled PCR tubes
1.5-ml microcentrifuge tubes
Scalpel
Micropipettor tips

3. MATERIALS

High-fidelity DNA polymerase [e.g., Phusion (New England Biolabs)]
dNTP mix
Agarose
Tris base
Glacial acetic acid
EDTA
Ethidium bromide
DNA gel extraction kit

3.1. Solutions & Buffers

Steps 3 & 5 50× TAE agarose gel running buffer

Component	Stock	Amount
Tris base		242 g
Glacial acetic acid		57.1 ml
EDTA, pH 8	0.5 M	100 ml

Add water to 1 l

Ethidium bromide stock solution (10 mg ml^{-1})
Dissolve 10 mg ethidium bromide in 1 ml water
Use an amber colored 1.5-ml tube or wrap tube with aluminum foil

4. PROTOCOL

4.1. Duration

Preparation	About 1 h
Protocol	About 8 h

4.2. Preparation

Obtain or purify DNA containing the gene of interest.

Design and order standard (outside) forward and reverse primers for amplification of the gene with appropriate restriction endonuclease

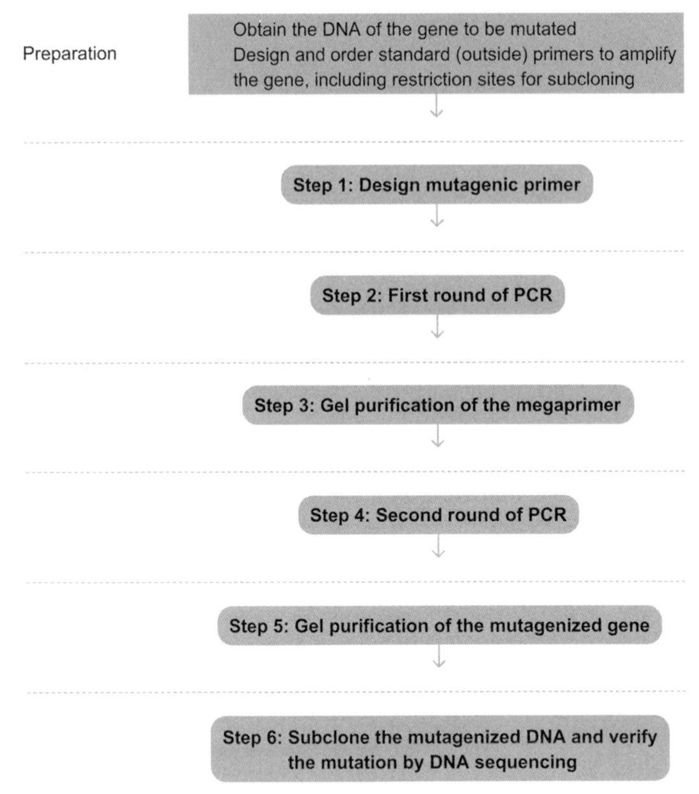
Preparation
Obtain the DNA of the gene to be mutated
Design and order standard (outside) primers to amplify the gene, including restriction sites for subcloning
Step 1: Design mutagenic primer
Step 2: First round of PCR
Step 3: Gel purification of the megaprimer
Step 4: Second round of PCR
Step 5: Gel purification of the mutagenized gene
Step 6: Subclone the mutagenized DNA and verify the mutation by DNA sequencing

Figure 21.1 Flowchart of the complete protocol, including preparation.

recognition sequences and overhangs for subsequent subcloning (see Explanatory chapter: PCR -Primer design).

See Fig. 21.1 for the flowchart of the complete protocol.

5. STEP 1 DESIGN MUTAGENIC PRIMER

5.1. Overview

Design a primer that incorporates the mutation of interest along with flanking regions overlapping the gene of interest.

5.2. Duration

1 h

1.1 Determine the nucleic acid sequence for the desired mutation.

1.2 Select regions of DNA upstream and downstream of the mutation site that each have a T_m ~45 °C.

1.3 If the mutation lies within the first half of the gene, order the reverse complement of the above sequence.

1.4 If the mutation lies within the second half of the gene, order the designed sequence.

5.3. Tip

The mutagenic primer can be purified as needed (see Purification of DNA Oligos by Denaturing Polyacrylamide Gel Electrophoresis (PAGE)).

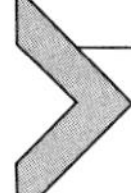

6. STEP 2 FIRST ROUND OF PCR

6.1. Overview

Generate the megaprimer by PCR, in this example using Phusion DNA polymerase (New England Biolabs).

6.2. Duration

2 h

2.1 If the mutation is in the first half of the gene, use the standard (outside) forward primer and reverse mutagenic primer. If the mutation is in the second half of the gene, use the forward mutagenic primer and the standard (outside) reverse primer. Dilute all primers to a working stock concentration of 25 μM.

2.2 Add to a 0.2-ml thin-walled PCR tube:

5× Phusion HF buffer	4 μl
10 mM dNTP mix	0.4 μl
Template DNA (200 ng ml^{-1})	0.5 μl
Primer A	0.4 μl
Primer B	0.4 μl
Phusion DNA polymerase	0.2 μl
Water	to 20 μl

2.3 Run PCR with the following cycling conditions:

At 98 °C for	30 s
20–35 cycles:	
At 98 °C for	5–10 s
At 55 °C	for 10–30 s
At 72 °C	for 15–30 s/1 kb
At 72 °C	for 5–10 min
At 4 °C	hold

6.3. Tip

Cast agarose gel while PCR is running.

See Fig. 21.2 for the flowchart of Steps 2 and 4.

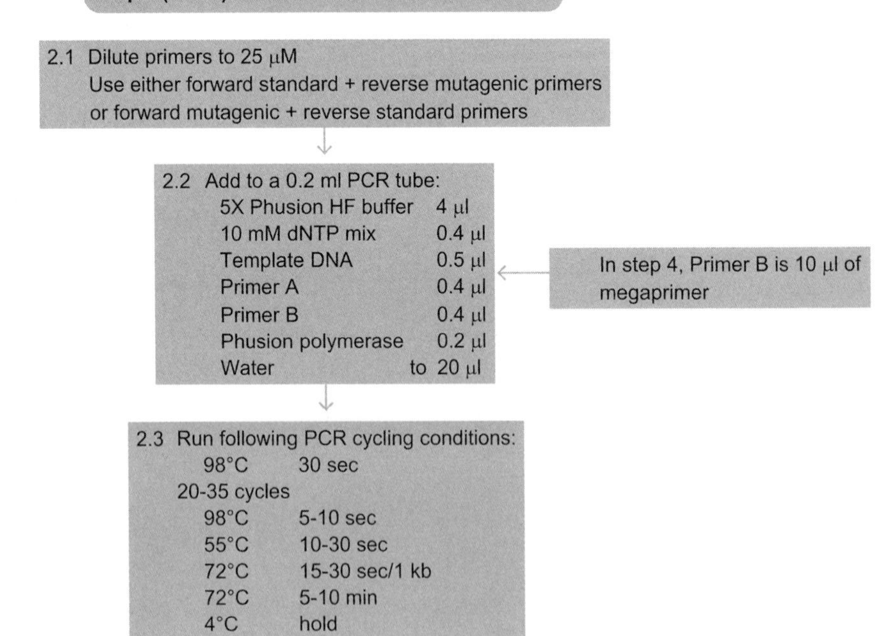

Figure 21.2 Flowchart of Step 2 (and 4).

7. STEP 3 GEL PURIFICATION OF THE MEGAPRIMER

7.1. Overview

Run an agarose gel (see Agarose Gel Electrophoresis) to purify the megaprimer to use in subsequent PCR.

7.2. Duration

1.5h

3.1 Prepare a 2% (w/v) agarose gel. Add 2 g agarose to 100 ml 1× TAE.

3.2 Microwave in glass flask or beaker to dissolve agarose. Swirl flask occasionally to prevent boiling over. Cool on bench for 10 min.

3.3 Add 2 μl of ethidium bromide and swirl to mix. Pour the gel into gel casting box and allow gel to fully harden.

3.4 Load the PCR sample from Step 2 and run the agarose gel at 100 V for ~30 min.

3.5 Examine gel under a UV light box, excise the megaprimer using a scalpel, and put it in a 1.5-ml microcentrifuge tube.

3.6 Dissolve the gel fragment and isolate the megaprimer using a gel extraction kit.

7.3. Tip

Work quickly when excising gel band to minimize UV exposure. Alternatively, use a handheld UV light source.

7.4. Caution

Ethidium bromide is a mutagen. Wear gloves and handle with care. Dispose of gel and buffer appropriately according to local regulations.

See Fig. 21.3 for the flowchart of Steps 3 and 5.

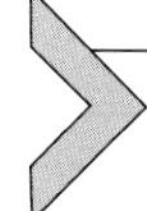

8. STEP 4 SECOND ROUND OF PCR

8.1. Overview

Utilize the megaprimer to produce the mutated gene by PCR.

8.2. Duration

2 h

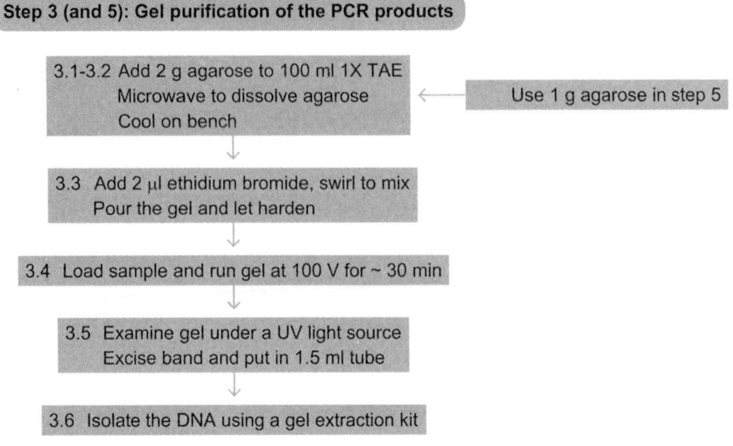

Figure 21.3 Flowchart of Step 3 (and 5).

4.1 If the mutation is in the first half of the gene, use the (forward) megaprimer (10 μl) and the standard (outside) reverse primer (0.4 μl).

If the mutation is in the second half of the gene, use the standard (outside) forward primer (0.4 μl) and (reverse) megaprimer (10 μl).

4.2 Add to a 0.2-ml thin-walled PCR tube:

5× Phusion HF buffer	4 μl
10 mM dNTP mix	0.4 μl
Template DNA (200 ng ml^{-1})	0.5 μl
Primer A	0.4 μl
Primer B (megaprimer)	10 μl
Phusion DNA polymerase	0.2 μl
Water	to 20 μl

4.3 Run PCR with the following cycling conditions:

98 °C	30 s
20–35 cycles:	
At 98 °C for	5–10 s
At 55 °C for	10–30 s
At 72 °C for	15–30 s/1 kb
At 72 °C for	5–10 min
At 4 °C	hold

8.3. Tip

The protocol is designed to utilize excess megaprimer for the second round of PCR, which significantly improves the efficiency and allowable megaprimer size (Smith and Klugman, 1997). As needed, the amount of megaprimer and template can be varied to maximize production of the desired product.

8.4. Tip

Cast agarose gel while PCR is running.

See Fig. 21.2 for the flowchart of Steps 2 and 4.

9. STEP 5 GEL PURIFICATION OF THE MUTAGENIZED GENE

9.1. Overview

Run an agarose gel to purify the mutagenized gene (see Agarose Gel Electrophoresis).

9.2. Duration

1.5 hours

5.1 Prepare a 1% (w/v) agarose gel. Add 1 g agarose to 100 ml 1× TAE.

5.2 Microwave in glass flask or beaker to dissolve agarose. Swirl flask occasionally to prevent boiling over. Cool on bench for 10 min.

5.3 Add 2 μl of ethidium bromide and swirl to mix. Pour the gel into gel casting box and allow gel to fully harden.

5.4 Load the PCR sample from Step 4 and run the agarose gel at 100 V for ~30 min.

5.5 Examine gel under a UV light box, excise the DNA (gene with incorporated mutation) using a scalpel, and put it in a 1.5-ml microcentrifuge tube.

5.6 Dissolve the gel fragment and isolate the DNA using a gel extraction kit.

9.3. Tip

Work quickly when excising gel band to minimize UV exposure.

See Fig. 21.3 for the flowchart of Steps 3 and 5.

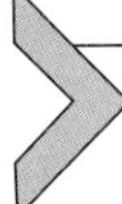

10. STEP 6 SUBCLONE THE MUTAGENIZED DNA AND VERIFY THE MUTATION BY DNA SEQUENCING

10.1. Overview

Use standard protocols to digest the vector DNA and the PCR product with the appropriate restriction enzymes and gel-purify the DNA. Ligate the insert into the digested vector (see Molecular Cloning), transform *E. coli* (see Transformation of Chemically Competent *E. coli* or Transformation of *E. coli* via electroporation), screen colonies for the presence of the insert (see Colony PCR), purify (see Isolation of plasmid DNA from bacteria) and sequence the plasmids to verify the mutation.

10.2. Duration

4–7 days

6.1 Digest the PCR product and 1 μg of the desired vector DNA with the chosen restriction enzymes.

6.2 Prepare a 1% (w/v) agarose gel (as above). Run digested DNAs on gel at 100 V for ~30 min and gel-purify the DNA (as earlier).

6.3 Add to a 1.5-ml microcentrifuge tube:

10× T4 DNA ligase buffer	2 μl
Prepared vector	50–100 ng
Annealed insert	100–300 ng
T4 DNA ligase	1 μl
Water	up to 20 μl

Incubate at room temperature for 1–2 h or overnight at 16 °C (a PCR theyrmocycler can be used).

6.4 Transform competent *E. coli* of choice. Plate cells on selective media and incubate overnight.

6.5 Pick colonies and screen for the presence of the insert by colony PCR or by restriction enzyme digestion.

6.6 Purify the insert-containing plasmids and subject them to DNA sequencing to verify the presence of the mutation.

See Fig. 21.4 for the flowchart of Step 6.

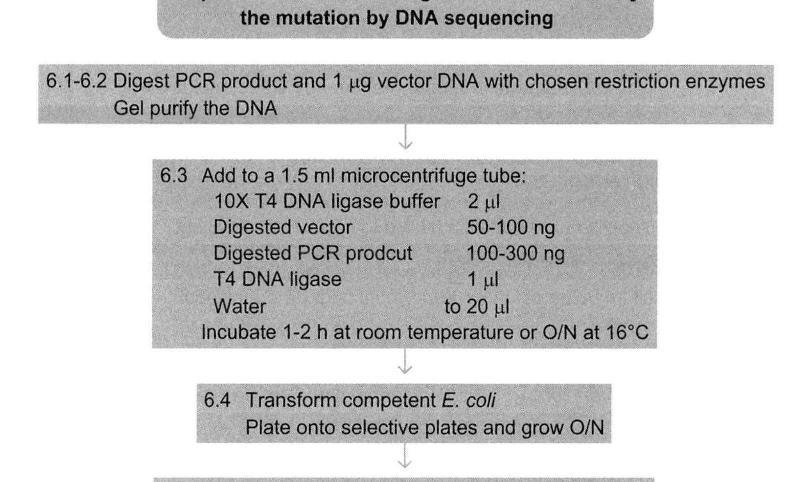

Figure 21.4 Flowchart of Step 6.

REFERENCES

Referenced Literature

Sarkar, G., & Sommer, S. S. (1990). The "megaprimer" method of site-directed mutagenesis. *BioTechniques*, *8*(4), 404–407.

Smith, A. M., & Klugman, K. P. (1997). "Megaprimer" method of PCR-based mutagenesis: the concentration of megaprimer is a critical factor. *BioTechniques*, *22*(3), 438–442.

Referenced Protocols in Methods Navigator

Explanatory chapter: PCR -Primer design
Transformation of Chemically Competent *E. coli*
Transformation of *E. coli* via electroporation
Colony PCR
Agarose Gel Electrophoresis
Molecular Cloning
Isolation of plasmid DNA from bacteria
Purification of DNA Oligos by Denaturing Polyacrylamide Gel Electrophoresis (PAGE)

CHAPTER TWENTY TWO

Explanatory Chapter: Troubleshooting PCR

Kirstie Canene-Adams[1]
Department of Pathology, Johns Hopkins University School of Medicine, Baltimore, MD, USA
[1]Corresponding author: e-mail address: kirstieadams26@yahoo.com

Contents

Abstract

The purpose of this chapter is to assist with any problems you are having with the polymerase chain reaction (PCR) protocol from General PCR.

1. THEORY

The simple theory behind PCR is to use custom-designed primers, template DNA, a thermostable DNA polymerase, and nucleotides (dNTPs) to amplify a particular region of DNA so that it can be easily detected or manipulated. There are many applications for PCR including gene cloning, manipulation, mutagenesis, genotyping, detection and diagnosis of pathogens, DNA sequencing, and analyzing expression levels of a gene.

PCR comprises three main steps:

- *Denaturation*: this separates the double-stranded DNA into single-stranded DNA.
- *Annealing*: the primers bind to the complementary target sequence in the template DNA.
- *Extension*: the thermostable DNA polymerase synthesizes the complementary DNA strand, starting from the 3′-end of the primer.

Methods in Enzymology, Volume 529
ISSN 0076-6879
http://dx.doi.org/10.1016/B978-0-12-418687-3.00022-7

All PCR should be evaluated to confirm that a PCR product of the expected size was made. This is most commonly done by agarose gel electrophoresis (see Agarose Gel Electrophoresis). This chapter assists in the troubleshooting of PCR in the event that there was no product produced or a product of the wrong size was produced.

2. EQUIPMENT

PCR thermal cycler
Micropipettors
Aerosol barrier pipet tips
Sterile microcentrifuge tubes
0.2-ml thin-walled PCR reaction tubes or plates
Surface decontamination wipe (e.g., DNA AWAY®, Molecular Bioproducts)
Agarose gel electrophoresis equipment

3. MATERIALS

Template DNA
Primers
dNTPs
Taq DNA polymerase (or other thermostable DNA polymerase)
10× PCR buffer
Magnesium chloride ($MgCl_2$, 25 mM)
Sterile ultra pure water
Ethidium bromide
Normal melting agarose
TAE buffer
DNA ladder

4. PROTOCOL

Rather than being a standard protocol, this chapter lists common problems with PCR, the reasons why they might have occurred, and steps that can be taken to prevent these problems in the future. Most often, after performing PCR, an agarose gel is run with the PCR product alongside a DNA ladder to verify the size of the product. Two problems can be immediately apparent: first, there is no PCR product, and second, there is a PCR

Table 22.1 Troubleshooting the lack of a PCR product seen on an agarose gel

PCR problem	Possible reasons	Fixes
No product formed	• Problem with reagents	• Repeat the experiment, as a reagent may have been unintentionally left out • Check that the reagents had been fully thawed and mixed thoroughly • Try a new vial of dNTPs as they can be damaged by repeated freeze–thaw cycles • Use a different polymerase • Try using an additive such as DMSO or glycerol
	• Quality of the template DNA is poor	• Use a NanoDrop spectrophotometer to check the quantity and quality of template DNA
	• Too much or too little starting template	• Remake template DNA. Older stocks can degrade, particularly for genomic DNA • Set up a series of reactions with varying amounts of template (between 10 and 200 ng DNA)
	• Primers are not working	• Check that the primers have been diluted to the correct concentration • Make sure that the sequence of the primer is what you expected it to be synthesized as. Typos are common! • Increase primer concentration • Redesign the primers
	• PCR inhibitors are present in the template DNA	• Demonstrate that a control gene or another DNA sequence can be effectively amplified using the same template DNA. If it can, there is another problem with the PCR reaction
	• Problem with the settings on thermal cycler	• Check the cycler – Are the temperatures and times as you expect? • Change the annealing temperature. Find the optimum temperature by using a gradient cycler and test a range from the lowest primer T_m to 10 °C below the T_m

Continued

Table 22.1 Troubleshooting the lack of a PCR product seen on an agarose gel—cont'd

PCR problem	Possible reasons	Fixes
		• Try the reaction in another cycler – the calibration of the one you are using may be off
	• The template DNA is GC-rich	• Use a polymerase buffer intended for GC-rich templates. Test a full range of the additive (e.g., GC melt)

Table 22.2 Troubleshooting PCR products which are either too long or too short

PCR problem	Possible reasons	Fixes
Wrong sized product formed	• Cross-contamination	• Use the following order when setting up the reactions: actual samples, positive controls, and then negative controls
	• One or both primers anneal to an off-target site on the template	• Use aerosol barrier pipettor tips • Pre- and postamplification samples should be manipulated in physically separated areas. Set up a separate lab bench area and a set of pipettors for running gels
Long nonspecific products	• One or both primers anneal to an off-target site on the template	• Decrease the annealing and/or extension time • Decrease the extension temperature to 62–68 °C • Increase the annealing temperature. • Increase the $MgCl_2$ concentration to 3–4.5 mM, while keeping the dNTP concentration constant • Use less primers, DNA template, and/or polymerase • If none of the above works, BLAST the primer for repetitive sequences and try designing new primer(s)

Table 22.2 Troubleshooting PCR products which are either too long or too short—cont'd

PCR problem	Possible reasons	Fixes
Short nonspecific products or 'primer dimers'	• Small PCR bands about the size of one primer or both together, which are formed by the annealing of a primer to itself or to the other primer	• Increase the annealing temperature and/or time. Try to find the optimal annealing temperature using a gradient PCR machine
	• One or both primers anneal to an off-target site on the template, resulting in a very short PCR product	• Increase the extension time and/or temperature to 74–78 °C • Titrate the $MgCl_2$ concentration to 3–4.5 mM, while keeping the dNTP concentration constant • Use less primer or Taq polymerase • Try a 'hot start' polymerase instead of a standard polymerase • Increase the amount of the DNA template • If none of the above works, BLAST the primer for repetitive sequences and try designing new primer(s)

product(s) but it is the wrong size. Table 22.1 gives some reasons and possible fixes for when there is no PCR product produced. Table 22.2 gives some reasons and possible fixes for when there is a PCR product, but it is the wrong size.

Since there are five main reagents in a PCR, you can alter any of them in order to optimize the reaction for your particular gene of interest and product. It is preferred to only change one aspect of the reaction at a time. The following gives examples of how each reagent, $MgCl_2$, Taq DNA polymerase, the primers, dNTPs, and template DNA, can cause problems in PCR and acceptable ranges of concentrations for each reagent.

$MgCl_2$: Magnesium plays several roles in PCR: it is an essential cofactor for thermostable DNA polymerases, it stabilizes double-stranded DNA, and it elevates the T_m. A deficient or low Mg^{2+} concentration requires stricter base pairing when the primers and DNA anneal, and can cause a small or no

yield of PCR product. If Mg^{2+} ions are in excess, there can be an increased yield of nonspecific products, and the reliability of DNA polymerases can be reduced, promoting the misincorporation of nucleotides and the samples appearing as a smear on the gel. Since dNTPs sequester Mg^{2+} ions, any alteration in dNTP concentrations will require additional changes in the concentration of $MgCl_2$. Normally, the $MgCl_2$ concentration is best between 1 and 4 mM.

Taq DNA polymerase: Low levels of Taq polymerase can cause incomplete primer elongation or untimely termination of the PCR product synthesis during the extension step. Too much polymerase will result in an excessive background of unwanted DNA fragments, which will look like a smear on the gel. A markedly overabundant amount of polymerase can cause the reaction to completely fail with no product formed at all. A polymerase concentration of 1 unit per 25 μl of reaction mix is sufficient to generate a clean PCR product.

Primers: Low primer concentration generally results in a cleaner product, yet raising the primer concentration does not lead to a corresponding increase in the amount of product formed. Using elevated concentrations of primers can result in the creation of primer-dimers, nonspecific primer binding, and the generation of undesirable PCR products. Conversely, to amplify short target sequences, such as 100 base pairs, a higher number of molecules of PCR product are required to provide a specified amount of amplified DNA (in nanograms). In this case, a higher primer concentration could be beneficial. The suggested primer concentration is between 0.1 and 1 μM of each primer. Primer base composition and priming sites can also dramatically affect the PCR performance, see chapter Explanatory chapter: PCR -Primer design.

dNTPs: A suboptimal concentration of nucleotides can lead to incomplete primer elongation or untimely termination of synthesis at some point in the extension step. Excessive dNTP concentrations can inhibit PCR, preventing product formation. The usual dNTP concentration is between 40 and 200 μM for each of the four dNTPs. To amplify longer DNA fragments, a higher dNTP concentration may be required.

DNA template: The concentration of the template DNA should be balanced with the number of cycles in the reaction. When the total amount of DNA is extremely small, there is increased likelihood of loss, contaminating DNA from impurities, and/or degradation. Using too much DNA can lead to off-target annealing of the primers as well as poor DNA synthesis due to the obstructed diffusion of polymerase. However, reducing the number of cycles may help to resolve these problems. Contamination can come from unlikely sources such as the dust floating in the air or particles of skin or hair from your body, which can carry both DNA and DNA-degrading nucleases. To prevent this, autoclave the tubes, wear gloves, and clean the working space using an oxidizing substance (e.g., 6% H_2O_2), 100% ethanol, or a surface decontamination wipe (e.g., DNA AWAY®, Molecular Bioproducts).

Weak amplification of your target can be improved by:

- Increasing the amount of primers, DNA template, and/or polymerase.
- Checking primer sequences for mismatches and/or increasing the primer length by five nucleotides (see also Explanatory chapter: PCR -Primer design).
- Increasing the annealing time or decreasing the annealing temperature.
- Increasing the number of cycles.
- Trying an additive of 5% (v/v, final concentration) DMSO or glycerol.
- Extracting the weak PCR product from an agarose gel using a DNA gel purification kit and using this as the template in a new reaction.

Nonspecific band amplification can be fixed by:

- Repeating the reaction with a negative control, such as water. The non-specific bands are most likely from contaminating foreign DNA. If this is the problem, use new stocks for all reagents. Also, remember to always use autoclaved PCR tubes and wear gloves.
- Using less DNA template.
- Increasing the annealing time if the nonspecific products are shorter than your target. If they are longer than your target, reduce the annealing time.
- Increasing the annealing temperature.
- Redesigning the primer(s) to increase the length at its 3′-end as the extra bands may be from similar sequences to your target. Increasing the 3′ primer length by adding further matching base pairs will make the primer more specific for your target because correct sequence binding will be enhanced while the extension of the nonspecific sequences will be blocked. When trying to eliminate aberrant bands, it is usually not a good idea to increase the length of a primer at its 5′-end, because this will increase its ability to anneal to the off-target sequences. In some cases shorter primers will work better because they do not base pair as well to the off-target sequences.

REFERENCES

Related Literature

Thermo Scientific "PCR Troubleshooting Guide" Assessed May–June 2010.
http://www.thermo.com/eThermo/CMA/PDFs/Product/productPDF_9174.pdf.
Bitesize Bio Beta, Brain Food for Scientists "The Essential PCR Troubleshooting Checklist" Nick Oswald, PhD; Assessed May–June 2010.
http://bitesizebio.com/2008/01/23/the-essential-pcr-troubleshooting-checklist/.
"PCR Troubleshooting: The Essential Guide" Michael L. Altshuler; Caister Academic Press; March 2006, Review online and Assessed May–June 2010.
http://www.highveld.com/pages/pcr-troubleshooting.html.

Referenced Protocols in Methods Navigator

General PCR
Agarose Gel Electrophoresis
Explanatory chapter: PCR -Primer design

CHAPTER TWENTY THREE

Explanatory Chapter: Quantitative PCR

Jessica S. Dymond[1]
The High Throughput Biology Center and Department of Molecular Biology and Genetics, Johns Hopkins University School of Medicine, Baltimore, MD, USA
[1]Corresponding author: e-mail address: jsiege16@jhmi.edu

Contents

1. Theory 280
2. Terminology 281
3. Equipment 281
4. Materials 282
5. Protocol 282
   5.1 Preparation 282
   5.2 Duration 282
6. Step 1 Primer Design 282
7. Step 2 Control Gene 283
8. Step 3 Amplification Efficiency 283
9. Step 4 Probe Choice 284
   9.1 SYBR green I 285
   9.2 TaqMan probes 286
10. Step 5 Data Analysis 287
   10.1 Normalization 287
   10.2 Absolute quantification 287
   10.3 Relative quantification 287
   10.4 Summary 288
References 288

Abstract

Quantitative PCR (qPCR), also called real-time PCR or quantitative real-time PCR, is a PCR-based technique that couples amplification of a target DNA sequence with quantification of the concentration of that DNA species in the reaction. This method enables calculation of the starting template concentration and is therefore a frequently used analytical tool in evaluating DNA copy number, viral load, SNP detection, and allelic discrimination. When preceded by reverse-transcription PCR, qPCR is a powerful tool to measure mRNA expression and is the gold standard for microarray gene expression data confirmation. Given the broad applications of qPCR and the many technical variations that have been developed, a brief survey of qPCR, including technical background,

Methods in Enzymology, Volume 529
ISSN 0076-6879
http://dx.doi.org/10.1016/B978-0-12-418687-3.00023-9

available chemistries, and data analysis techniques will provide a framework for both experimental design and evaluation.

1. THEORY

The defining feature of qPCR is 'real-time' measurement of the concentration of a DNA species of interest during a PCR, rather than end-point measurement at the completion of the amplification series. Although end-point analysis is extremely valuable in determining the presence or absence of template DNA in a reaction, the results can be very misleading. Much like other enzyme-catalyzed reactions, product formation in a PCR reaction follows a saturation curve (Fig. 23.1). Early amplification cycles generate product in an exponential fashion; the PCR is working at top efficiency. As the PCR progresses, the efficiency decreases because of lowered concentrations of consumables such as dNTPs and primers, and enzyme degradation, and product formation enters a linear phase. Finally, the reaction progresses into a plateau phase where product formation is greatly decreased because of the depletion of consumables. The PCR cycle at which each phase transition occurs is unique to each reaction and is largely dependent on the concentration of template DNA. PCRs containing a large template concentration will generally plateau in earlier cycles than PCRs containing a lesser amount of template. By end-

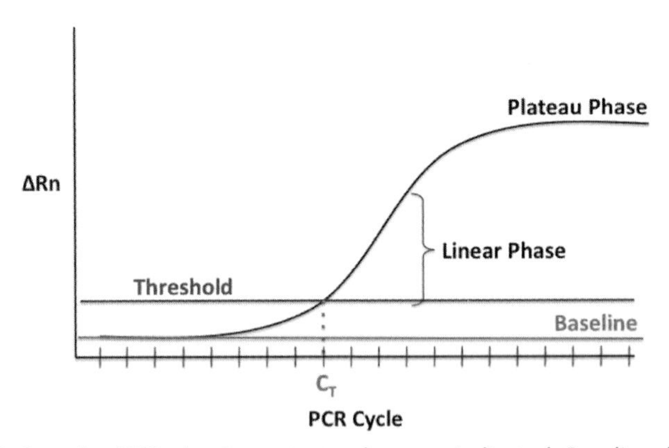

Figure 23.1 Sample qPCR plot. Important values are indicated. Baseline: background fluorescence prior to product accumulation. Threshold: 10 times the standard deviation of baseline fluorescence for the first 15 cycles. C_T: Amplification cycle at which fluorescence exceeds the threshold. ΔRn = normalized fluorescence (subtracted background). Linear phase: The most efficient amplification cycles when the rate of product accumulation is greatest. Plateau phase: Product accumulation slows then ceases as the reaction is exhausted.

point analysis, however, these two conditions are indistinguishable if both reactions have reached the plateau phase by the last amplification cycle performed. For this reason, measurement of PCR product accumulation during the PCR is a much more sensitive assay than end-point analysis.

Analysis of product accumulation in qPCR is accomplished through use of fluorescent dyes that specifically label dsDNA or a specific DNA species of interest; the amount of fluorescence is proportional to the quantity of DNA present. Although many models of qPCR thermocyclers are in production, all share common features: a standard thermocycler platform (96- or 384-well) is coupled with an excitation source (usually a laser or tungsten lamp), a camera for fluorescence detection, and computer and software for data processing. Depending on the available excitation source and detection filters, a variety of fluorescent dyes may be used in qPCR.

2. TERMINOLOGY

Pioneering studies by Heid et al. (1996); Gibson et al. (1996) defined many of the important parameters in qPCR (Fig. 23.1).

Baseline: During cycles 1–15 of the qPCR, fluorescence is below the limits of detection, but it is increasing.

Rn_B: The fluorescence level at the baseline.

Rn_F: The fluorescence level at a given point in the qPCR.

ΔRn: The normalized fluorescence level at a given point in the qPCR; Rn_F/Rn_B.

Threshold: A ΔRn value defined as 10 times the standard deviation of baseline fluorescence during the first 15 amplification cycles. The threshold may be adjusted for a group of different reactions so that it is in the linear amplification range for each, but should be held constant across samples to be compared.

C_T: The amplification cycle at which the ΔRn exceeds the threshold. The C_T is a quantitative description of the starting concentration of template.

3. EQUIPMENT

qPCR thermocycler
qPCR plates (depends on thermocycler used)
Micropipettors
Multichannel pipettor
Micropipettor tips

4. MATERIALS

Primers
Fluorescent probe
TaqMan probe

5. PROTOCOL

5.1. Preparation

Several factors must be considered when designing a qPCR experiment. Once a target of interest has been identified, PCR primers must be designed to amplify that sequence, a detection method must be chosen, control reactions must be planned, and the method of data analysis established. Each of these points is addressed individually in the subsequent sections.

5.2. Duration

Preparation	1–2 days
Protocol	About 1 day

6. STEP 1 PRIMER DESIGN

Primer quality (see Explanatory chapter: PCR -Primer design) plays an essential role in qPCR: well-designed primers with minimal self-complementarity and high specificity give rise to reproducible results; primers with a tendency toward dimerization or mishybridization will greatly reduce PCR efficiency and confound results. PCR amplicons should be 50–150 bp. In mRNA expression studies, it is desirable for the PCR product to cross an exon–exon border to reduce background amplification from contaminating genomic DNA and/or splice variants. Primer T_m should be 58–60 °C, length should be 18–30 bases, and G-C content at 40–60%. To increase specificity, an adenine or thymine at the 3′ position should be avoided. The use of software to design primers is preferable as it optimizes these parameters and matches primers very efficiently. Primer3 is an excellent open-source program that is available over the internet (Rozen and Skaletsky, 2000). Commercially available software for qPCR primer design exists as well. Prior to using newly designed primers in a qPCR reaction, their specificity can be tested by performing a standard PCR; if spurious amplification is observed, the primers should be rejected and the design process restarted.

7. STEP 2 CONTROL GENE

qPCR is an extremely sensitive technique. Small differences in reaction composition may be magnified over the course of the PCR. In order to correct for this eventuality, normalization against an internal control is critical. Performing additional qPCR reactions using the same template and identical setup but instead measuring the expression (or copy number) of a housekeeping gene will provide a 'loading control' against which experimental data may be normalized. These control reactions should be performed simultaneously alongside experimental reactions. The most commonly used control genes include β-actin, glyceraldehyde 3-phosphate dehydrogenase (GAPDH), and ribosomal RNA. These genes have become well established as controls, since they are ubiquitously expressed and do not appear to vary dramatically under many experimental conditions. Care should be taken when selecting a control gene, however, as there have been reports of unexpected changes in expression of control genes, particularly GAPDH (reviewed in Suzuki et al., 2000). A thorough literature search with planned experimental conditions in mind is necessary for the best choice of control gene. Once a control has been decided upon, amplification primers should be designed in a fashion identical to that for the experimental genes.

8. STEP 3 AMPLIFICATION EFFICIENCY

In order to compare expression or copy number data between two samples and/or genes, they must first be normalized against a control gene. One assumption of normalization of experimental data against a control gene is that they have relatively equal amplification efficiencies (AE). PCR efficiency may vary because of primer hybridization, GC-content of the amplicon, and disparities in amplicon size. If the amplification efficiencies are equal, varying the starting template concentration will have an equal effect on amplification of both genes and the control gene remains an appropriate control. If, however, AEs differ, an increase in template concentration may manifest as variable increases in product accumulation between the experimental and control genes, and the results are no longer directly comparable. Therefore, AEs must be matched as closely as possible.

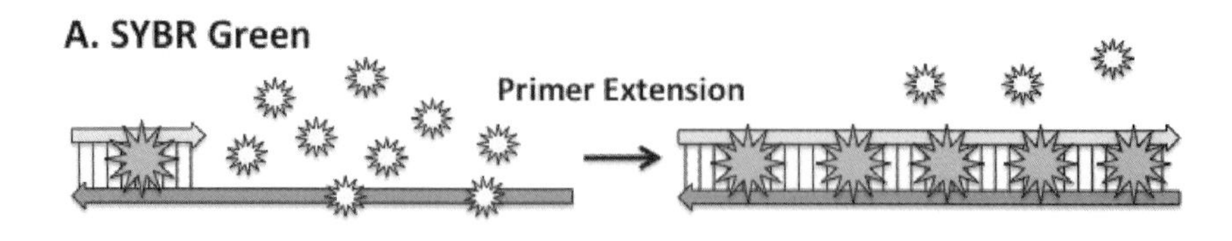

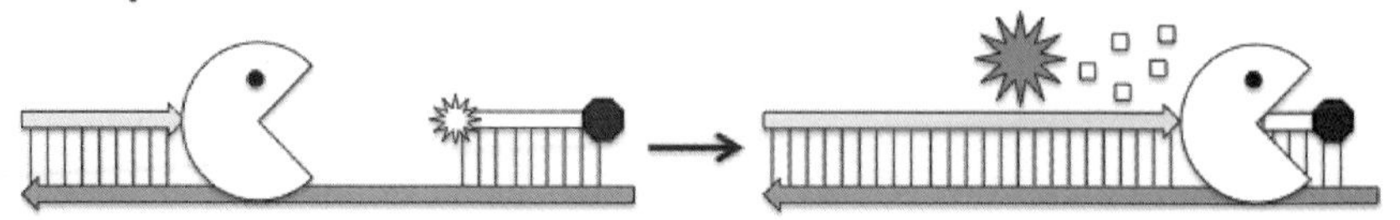

Figure 23.2 Standard curve for starting template concentration. Five template concentrations falling in the linear range are shown. The trend line and corresponding equation are also shown. Each point represents the average of three replicates. In this case, the efficiency is calculated as 0.99.

If necessary, however, corrections that consider mismatched AEs are available (Pfaffl, 2001; Yuan et al., 2001).

The standard method to determine the AE is to generate a standard curve. qPCR is performed with a series of dilutions of the template; the C_T for each reaction is then plotted against the log template concentration, and the slope of the resulting line is calculated (Fig. 23.2). The AE is defined as

$$AE = 10^{(-1/\text{slope})} - 1$$

(Rasmussen, 2001). An AE of 1.0, corresponding to an efficiency of 100% and a standard curve slope of −3.33, is optimal. A template concentration in the middle of those falling within the −3.33 slope should be chosen for future experiments. Efficiencies of 0.9–1.1 (corresponding to 90–110%) are acceptable. In order to compare two AEs, a plot of template dilution versus ΔC_T where

$$\Delta C_T = C_{T,\ \text{target gene}} - C_{T,\ \text{reference gene}}$$

should be constructed; a slope of approximately zero indicates equivalent AEs.

9. STEP 4 PROBE CHOICE

Multiple technical variations exist to detect the DNA species of interest in a qPCR reaction, but all rely upon fluorescent labeling of the DNA of interest and subsequent measurement of fluorescence during each cycle of the PCR. The most commonly used methods, DNA intercalating agents (specifically SYBR Green) and TaqMan probes, are described here. A more comprehensive review has been compiled by Mackay and Landt (2007).

9.1. SYBR green I

DNA intercalating agents have long been used to label double-stranded (ds) DNA. Although ethidium bromide has been the most common labeling agent in the past, SYBR Green I (SG), which does not carry the same exposure risks for humans and exhibits brighter fluorescence, has largely supplanted ethidium bromide in qPCR applications. SG is a nonspecific minor groove-binding molecule, and exhibits a large increase in fluorescence when in the dsDNA-bound form (Zipper et al., 2004; Fig. 23.3(a)). One major advantage of SG is its nonspecific nature, which allows it to be used for multiple samples with little or no preparation. This

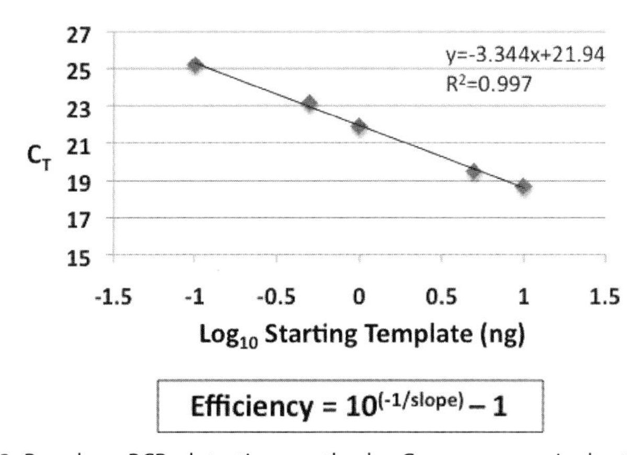

Figure 23.3 Popular qPCR detection methods. Gray arrows: single-stranded DNA (ssDNA); arrowheads: 3′ ends; vertical bars: hydrogen bonds. (a) SYBR Green I (SG) is a DNA intercalating agent that binds dsDNA with high affinity. Free or ssDNA-bound SG (light green star) has a low level of intrinsic fluorescence; however, when bound to dsDNA, SG fluorescence increases greatly (bright green star). Generation of dsDNA over the course of a PCR results in a higher proportion of dsDNA-bound SG and an increase in fluorescence. (b) TaqMan probes are short primers (white) that are dual-labeled with a 5′ fluorophore (white and blue stars) and a 3′ quencher (black octagon). The probe hybridizes to complementary ssDNA, holding the fluorophore and quencher in close proximity. Due to FRET between the two dyes, there is little fluorescence (white star). As primer extension occurs, Taq polymerase (chomper) encounters the TaqMan probe, and the enzyme's 5′ exonuclease activity cleaves the probe, liberating the 5′ fluorophore from the 3′ quencher. Fluorescence increases as a result (blue star).

same feature, however, may be a disadvantage if primers are not carefully designed; SG will bind any dsDNA, including primer dimers and nonspecific PCR products. As a result, the fluorescence due to formation of the product of interest may be masked, making analysis extremely difficult. Use of primer-design software typically ameliorates these complications.

9.2. TaqMan probes

Taq polymerase, frequently used for standard PCR applications, exhibits 5′-exonuclease activity: if the enzyme encounters a double-stranded region during extension, the polymerase will nick the nontemplate strand, freeing its constituent nucleotides. This property is exploited by the TaqMan probe (Lee et al., 1993; Fig. 23.3(b)). A pair of PCR primers is designed, and an additional primer is designed within the amplicon. This internal primer is the foundation of the TaqMan probe and is labeled at the 5′ end with one fluorescent reporter dye, typically FAM, TET, VIC, or HEX, and the 3′ end is labeled with a quencher dye, often TAMRA, DABCYL, or a black hole quencher (BHQ). As this technology relies upon FRET (fluorescence resonance energy transfer) between the reporter and quencher dyes, the choice of reporter and quencher dye is not arbitrary; the two dyes must have overlapping excitation/emission spectra. When the reporter and quencher are in close proximity (labeling the intact primer), FRET occurs and there is no fluorescence observable. As Taq polymerase extends through this region and encounters the probe, the enzyme's exonuclease activity nicks the 5′ end of the probe, liberating the reporter dye from the primer anchoring the quencher. As the reporter and quencher are no longer in close proximity, FRET does not occur and fluorescence is observed. A major advantage of TaqMan probes is the added specificity conferred by the probe's hybridization between the two PCR primers. One drawback is that a separate probe must be designed for each gene of interest; however, the multiple available combinations of reporters and quenchers in concert with sequence specificity conferred by the probe itself enable multiplexing, that is, the combination of several probes in one reaction (Gibson et al., 1996).

When designing TaqMan probes, several guidelines must be kept in mind: the probe should be 15–30 nucleotides long, be ~30–80% GC, have a T_m 10 °C higher than the amplification primers, and be free of nucleotide runs. It is also desirable to have fewer than 2 GCs at the 3′ end to introduce a small degree of instability to prevent mispriming. Additionally, a G in the 5′ position should be avoided as it may quench fluorescence (Arya et al., 2005).

If maximal specificity is desired, such as for allelic discrimination, a non-fluorescent quencher (such as a BHQ) may be coupled with a minor groove binder (MGB). MGBs increase T_m, allowing shorter probes, and show higher specificity for single base mismatches (Kutyavin et al., 2000).

10. STEP 5 DATA ANALYSIS

Two methods are widely used to analyze qPCR data: absolute and relative quantification. Regardless of the method of analysis to be employed, all qPCR reactions should be performed in triplicate simultaneously to further protect against error.

10.1. Normalization

Small variations in the composition of qPCR reactions may be magnified by PCR amplification and increase variability both within and between samples. For this reason, normalization against an internal control is necessary. qPCR on control genes should be performed on the same samples and at the same time as experimental runs.

10.2. Absolute quantification

To determine the exact concentration of starting template, a standard curve may be generated from reference DNA (or cDNA) and used to quantify experimental samples. Ideally, the reference DNA source will be the gene of interest on a plasmid so that its concentration can be accurately calculated by spectrophotometry and copy number can be extrapolated. Copy number versus C_T is plotted to generate a standard curve; alternatively, the log starting template concentration may be used. Experimental data may then be compared with this curve to calculate copy number. In both constructing a standard curve and defining experimental samples, the raw C_T average should be normalized against an internal control gene.

10.3. Relative quantification

A relative quantification, or 'fold-change' description, is typically sufficient for most applications. The most commonly used method of relative quantification is the $\Delta\Delta C_T$ analysis: The C_T of the target gene in a sample is normalized against the reference gene, yielding a ΔC_T value for that gene and sample. The experimental strain is then compared with the reference strain,

$$\text{Ratio} = \frac{(AE_{target})^{\Delta C_{T,target(control\text{-}sample)}}}{(AE_{reference})^{\Delta C_{T,reference(control\text{-}sample)}}}$$

Figure 23.4 $\Delta\Delta C_T$ analysis for the relative quantification of qPCR data. The efficiency can be considered in calculating the ratio of gene expression and/or copy number.

generating a $\Delta\Delta C_T$. Finally, the fold-change is calculated by raising 2 to $-\Delta\Delta C_T$ (Livak and Schmittgen, 2001).

$$\Delta C_T = C_{T,\text{ target gene}} - C_{T,\text{ reference gene}}$$
$$\Delta\Delta C_T = [\Delta C_T]_{\text{experimental}} - [\Delta C_T]_{\text{reference}}$$
$$2^{-\Delta\Delta CT} = \text{fold change}$$

This method, however, assumes equal AEs, which is not always the case. Because of the exponential nature of PCR, differences in AE can be dramatically magnified after repeated rounds of amplification, resulting in an inability to directly compare experimental and control data. If necessary, the efficiency can be considered in calculating the ratio of gene expression and/or copy number (see Fig. 23.4):

10.4. Summary

Quantitative PCR is a powerful tool for determining copy number and relative concentration of nucleic acids. Proper choice and use of a detection method and data analysis will ensure accuracy and reproducibility. Although qPCR technologies are rapidly evolving, the basic concepts covered here provide a foundation upon which to evaluate and make appropriate decisions regarding qPCR experimental design and execution.

REFERENCES

Referenced Literature

Arya, M., Shergill, I. S., Williamson, M., Gommersall, L., Arya, N., & Patel, H. R. H. (2005). Basic principles of real-time quantitative PCR. *Expert Reviews of Molecular Diagnostics*, *5*(2), 209–219.

Gibson, U. E., Heid, C. A., & Williams, P. M. (1996). A novel method for real time quantitative RT-PCR. *Genome Research*, *6*, 995–1001.

Heid, C. A., Stevens, J., Livak, K. J., & Williams, P. M. (1996). Real time quantitative PCR. *Genome Research*, *6*, 986–994.

Kutyavin, I. V., Afonina, I. A., Mills, A., et al. (2000). 3′-Minor groove binder-DNA probes increase sequence specificity at PCR extension temperatures. *Nucleic Acids Research*, *28*(2), 655–661.

Lee, L. G., Connell, C. R., & Bloch, W. (1993). Allelic discrimination by nick-translation PCR with fluorogenic probes. *Nucleic Acids Research*, *21*(16), 3761–3766.

Livak, K. J., & Schmittgen, T. D. (2001). Analysis of relative gene expression data using real-time quantitative PCR and the 2(-ΔΔ*CT*) method. *Methods*, *25*, 402–408.

Mackay, J., & Landt, O. (2007). Real-time PCR fluorescent chemistries. In E. Hilario, & J. Mackay (Eds.), Methods in Molecular Biology, vol. 353: Protocols for Nucleic Acid Analysis by Nonradioactive Proves (pp. 237–261) (2nd ed.). Totowa, NJ: Humana Press.

Pfaffl, M. W. (2001). A new mathematical model for relative quantification in real-time RT-PCR. *Nucleic Acids Research*, *29*(9), 2002–2007.

Rasmussen, R. (2001). Quantification on the LightCycler. In S. Meuer, C. Wittwer, & K. Nakagawara (Eds.), *Rapid Cycle Real-Time PCR: Methods and Applications* (pp. 21–34). Heidelberg: Springer.

Rozen, S., & Skaletsky, H. J. (2000). Primer3 on the WWW for general users and for biology programmers. In S. Krawetz & S. Misener (Eds.), *Bioinformatics Methods and Protocols: Methods in Molecular Biology* (pp. 365–386). Totowa, NJ: Humana Press.

Suzuki, T., Higgins, P. J., & Crawford, D. R. (2000). Control selection for RNA quantitation. *BioTechniques*, *29*(2), 332–337.

Yuan, J. S., Wang, D., & Stewart, C. N., Jr. (2001). Statistical methods for efficiency adjusted real-time PCR quantification. *Biotechnology Journal*, *3*, 112–123.

Zipper, H., Brunner, H., Bernhagen, J., & Vitzthum, F. (2004). Investigations on DNA intercalation and surface binding by SYBR Green I, its structure determination and methodological implications. *Nucleic Acids Research*, *32*(12), e103.

Referenced Protocols in Methods Navigator

Explanatory chapter: PCR -Primer design

CHAPTER TWENTY FOUR

General PCR

Kirstie Canene-Adams[1]
Department of Pathology, Johns Hopkins University School of Medicine, Baltimore, MD, USA
[1]Corresponding author: e-mail address: Kirstieadams26@yahoo.com

Contents

1. Theory 292
2. Equipment 293
3. Materials 293
   3.1 Solutions & buffers 293
4. Protocol 294
   4.1 Preparation 294
   4.2 Duration 294
   4.3 Caution 294
5. Step 1 Polymerase Chain Reaction 295
   5.1 Overview 295
   5.2 Duration 295
   5.3 Tip 296
   5.4 Tip 296
   5.5 Tip 296
   5.6 Tip 296
   5.7 Tip 296
6. Step 2 Analyze PCR Products 297
   6.1 Overview 297
   6.2 Duration 297
   6.3 Tip 298
References 298

Abstract

The primary purpose of polymerase chain reaction (PCR) is to rapidly make many copies of a specific region of DNA or RNA so that it can be adequately detected, often by agarose gel electrophoresis. PCR is routinely used to amplify, modify, and clone genes for expression studies. There are many other applications for PCR, including paternity testing, biological relationships, mouse genotyping, diagnosing genetic diseases, forensics, and finding bacteria and viruses.

Methods in Enzymology, Volume 529
ISSN 0076-6879
http://dx.doi.org/10.1016/B978-0-12-418687-3.00024-0

1. THEORY

There are four main components of a PCR reaction: primers, the DNA template, thermostable DNA polymerase, and nucleotides (dNTPs). Primers are custom designed, short oligonucleotides that are complementary to the ends of the piece of DNA you are interested in amplifying. The template DNA provides the starting material for the reaction and contains the region of DNA you want to amplify. In the case of RT-PCR, the template is cDNA made from the RNA. The thermostable DNA polymerase copies the DNA using the oligonucleotides as primers in a cycle that is repeated many times, resulting in the exponential amplification of the DNA between the primers. This can continue until the primers and/or the free nucleotides are used up or the double-stranded DNA reaches a reaction-inhibiting concentration.

There are three main steps in a PCR reaction: denaturation, annealing, and extension. These are repeated using an automated thermal cycler until sufficient amplification has taken place (often > 25 cycles). Denaturation occurs at 94 °C and it separates the double-stranded DNA template into single-stranded DNA. Annealing occurs at a lower temperature, usually between 45 and 60 °C. Annealing allows the primers to base pair with their complementary sites on the template DNA. This formation of double-stranded DNA provides a place for the polymerase to attach and begin the process of copying the template. This last phase, known as extension, usually takes place at 72 °C. Because both strands of DNA are copied during the PCR process, there is an exponential growth in the number of gene copies.

The PCR reaction must be analyzed to confirm that the desired product was made. This is most commonly done using an agarose gel (see Agarose Gel Electrophoresis). For some applications, it is important to know the sequence as well. PCR products can be sequenced directly or cloned into a plasmid vector (see Molecular Cloning) and individual clones sequenced (see sequencing methods on Sanger Dideoxy Sequencing of DNA and Explanatory Chapter: Next Generation Sequencing). If there are problems with the PCR, you can troubleshoot the protocol (see Explanatory Chapter: Troubleshooting PCR).

2. EQUIPMENT

PCR thermal cycler
Agarose gel electrophoresis equipment
Micropipettors
Aerosol barrier micropipettor tips
Sterile 1.5-ml microcentrifuge tubes
0.2-ml thin-walled PCR tubes (or plates)

3. MATERIALS

Template DNA
Primers
dNTPs (100 mM each dATP, dCTP, dGTP, dTTP)
Thermostable DNA polymerase (e.g., Taq polymerase or a high fidelity enzyme)
10× PCR buffer
Tris base
Hydrochloric acid (HCl)
Ammonium sulfate [$(NH_4)_2SO_4$]
Tween-20
Magnesium chloride ($MgCl_2$)
Agarose
50× TAE buffer
Ethidium bromide
6× DNA gel loading dye
Sterile ultra pure water

3.1. Solutions & buffers

Step 1 10× PCR buffer

Component	Final concentration	Stock	Amount
Tris–HCl, pH 8.8	750 mM	1 M	7.5 ml
$(NH_4)_2SO_4$	200 mM	1 M	2.0 ml
Tween-20	0.1% (v/v)	10%	100 µl

Add purified water to 10 ml

dNTP mix

Component	Final concentration	Stock	Amount
dATP	2.5 mM	100 mM	25 μl
dCTP	2.5 mM	100 mM	25 μl
dGTP	2.5 mM	100 mM	25 μl
dTTP	2.5 mM	100 mM	25 μl

Add purified water to 1 ml

25 mM $MgCl_2$
Dilute 25 μl 1 M $MgCl_2$ in 975 μl sterile ultra pure water

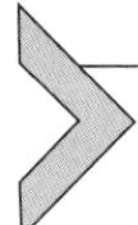

4. PROTOCOL

4.1. Preparation

Isolate DNA from the source of choice, which can be cells, human or lab animal tissue, or another origin. Kits can be purchased to isolate DNA from a variety of sources or standard protocols can be used. If you wish to isolate genomic DNA, see Preparation of Genomic DNA from Bacteria, Preparation of genomic DNA from *Saccharomyces cerevisiae*, or Isolation of Genomic DNA from Mammalian cells.

Design and order PCR primers (see Explanatory chapter: PCR - Primer design).

Dilute PCR primers in purified water to a working concentration of 10 μM.

4.2. Duration

Preparation	Variable, 1–2 days
Protocol	~4 h

4.3. Caution

If using a phenol or phenol-chloroform extraction technique, make sure to wear the appropriate gloves, eye protection, and use a hood ventilation system.

See Fig. 24.1 for the flowchart of the complete protocol.

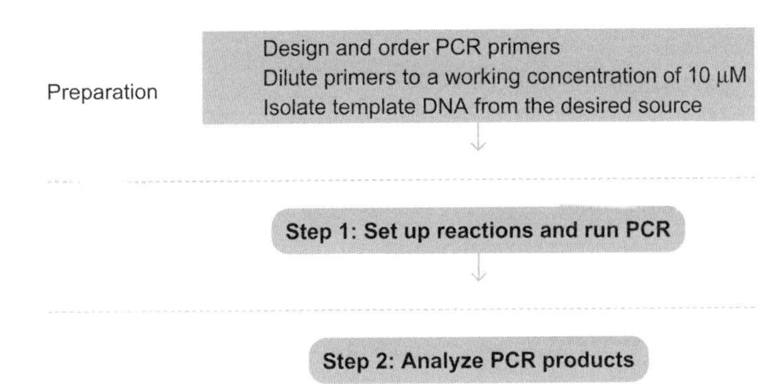

Figure 24.1 Flowchart of the complete protocol, including preparation.

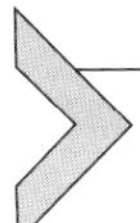

5. STEP 1 POLYMERASE CHAIN REACTION

5.1. Overview

Set up the reactions and run the PCR in a thermal cycler.

5.2. Duration

3–4 h

1.1 Prepare a PCR Master Mix. On ice, combine the following reagents in a sterile 1.5-ml microcentrifuge tube. Note that these amounts are for a single PCR reaction; the volumes should be multiplied by the total number of reactions you are performing.

10× PCR buffer	2.5 μl
Primer 1 (10 μM)	1.25 μl
Primer 2 (10 μM)	1.25 μl
dNTP mix (2.5 mM each)	2.5 μl
$MgCl_2$ (25 mM)	1.6 μl
Taq DNA polymerase	0.1 μl
Sterile ultra pure water	10.8 μl

1.2 Dispense 20 μl of the reaction mixture into 0.2-ml thin-walled PCR tubes. Add 5 μl of the template DNA (50–200 ng) for a total reaction volume of 25 μl.

1.3 Quickly vortex samples and centrifuge for a short time to ensure there are no air bubbles in the reaction mix. Alternatively, mix by gently pipetting up and down.

1.4 Place tubes in a thermal cycler with the following cycling conditions:

Temperature	Time	Purpose	# Cycles
94 °C	1 min	1	Initial denaturing
94 °C	1 min	25–35	(Denaturing)
55 °C	1 min		(Annealing)
72 °C	1 min (per kilobase PCR product)		(Extension)
72 °C	4 min	1	Final extension
4 °C	Hold	1	Stop/hold

5.3. Tip

Use a DNA contamination control product (e.g., DNA Away™ wipes, Molecular BioProducts) to clean your bench top and pipettors to prevent contamination of your PCR. These are particularly useful when performing nested PCR and degenerate PCR.

5.4. Tip

Thaw samples and reagents on ice!

5.5. Tip

Use aerosol barrier pipette tips to reduce contamination. Always use a new tip when pipetting samples or reagents.

5.6. Tip

Run a negative and positive control reaction with each PCR.

5.7. Tip

Some 10× PCR buffers supplied with the enzyme may already contain $MgCl_2$. Check the recipe of your buffer.

See Fig. 24.2 for the flowchart of Step 1.

Step 1: Set up reactions and run PCR

1.1 Prepare a PCR Master Mix. For each reaction, add to a 1.5 ml tube:

10X PCR buffer	2.5 μl
Primer 1 (10 μM)	1.25 μl
Primer 2 (10 μM)	1.25 μl
dNTP mix (2.5 mM each)	2.5 μl
$MgCl_2$ (25 mM)	1.6 μl
Taq polymerase	0.1 μl
Sterile ultra pure water	10.8 μl

↓

1.2 Dispense 20 μl Master Mix to 0.2 ml thin-walled PCR tubes
Add 5 μl template DNA (50-200 ng)

↓

1.3 Run PCR using the following cycling conditions:

94°C 1 min
25-35 cycles of:
94°C 1 min
55°C 1 min
72°C 1 min per kilobase of PCR product
72°C 4 min
4°C Hold

Figure 24.2 Flowchart of Step 1.

6. STEP 2 ANALYZE PCR PRODUCTS

6.1. Overview

Run a portion of the PCR reaction on an agarose gel to assess the quantity and quality of the PCR product (see Agarose Gel Electrophoresis).

6.2. Duration

About 2 h

2.1 Prepare a 1% agarose gel in 1× TAE buffer plus ethidium bromide.

2.2 Mix 5 μl PCR reaction and 1 μl.

2.3 Load samples and a DNA ladder on the gel. Run gel at 100 V for 30–45 min (depending on the expected size of the PCR product).

2.4 Visualize bands using a UV gel documentation system.

2.5 If appropriate, clone the PCR product into a plasmid vector and/or directly sequence the PCR product.

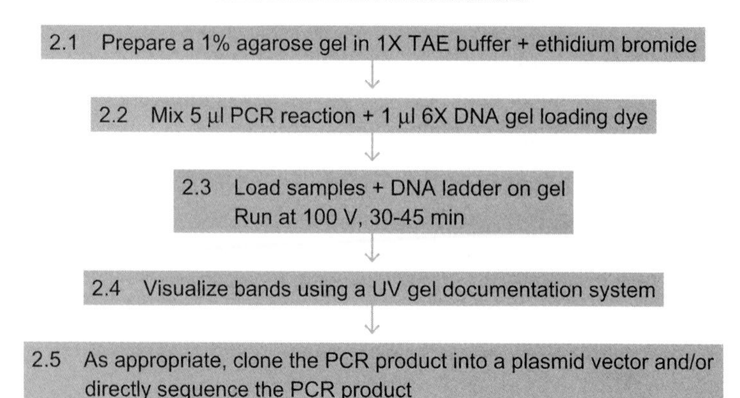

Figure 24.3 Flowchart of Step 2.

6.3. Tip

If there is little or no PCR product, the conditions may have to be optimized (see Explanatory Chapter: Troubleshooting PCR).

See Fig. 24.3 for the flowchart of Step 2.

REFERENCES

Related Literature

Genetic Science Learning Center. (2010). *PCR Virtual Lab*. http://learn.genetics.utah.edu/content/labs/pcr/, Learn.GeneticsTM, The University of Utah.

Referenced Protocols in Methods Navigator

Agarose Gel Electrophoresis
Molecular Cloning
Sanger Dideoxy Sequencing of DNA
Explanatory Chapter: Next Generation Sequencing
Explanatory Chapter: Troubleshooting PCR
Preparation of Genomic DNA from Bacteria
Preparation of Genomic DNA from *Saccharomyces cerevisiae*
Isolation of Genomic DNA from Mammalian cells
Explanatory chapter: PCR -Primer design

CHAPTER TWENTY FIVE

Colony PCR

Megan Bergkessel, Christine Guthrie[1]
Department of Biochemistry and Biophysics, University of California, San Francisco, CA, USA
[1]Corresponding author: e-mail address: christineguthrie@gmail.com

Contents

1. Theory 300
2. Equipment 300
3. Materials 301
   3.1 Solutions & buffers 301
4. Protocol 303
   4.1 Duration 303
   4.2 Preparation 303
5. Step 1 Extraction of DNA from a Colony 304
   5.1 Overview 304
   5.2 Duration 305
   5.3 Tip 305
   5.4 Tip 306
   5.5 Tip 306
6. Step 2 PCR 307
   6.1 Overview 307
   6.2 Duration 307
   6.3 Tip 307
   6.4 Tip 308
   6.5 Tip 308
   6.6 Tip 308
7. Step 3 Visualization of Product on an Agarose Gel 308
   7.1 Overview 308
   7.2 Duration 308
   7.3 Tip 309
References 309

Abstract

Colony PCR is a method for rapidly screening colonies of yeast or bacteria that have grown up on selective media following a transformation step, to verify that the desired genetic construct is present, or to amplify a portion of the construct.

Methods in Enzymology, Volume 529
ISSN 0076-6879
http://dx.doi.org/10.1016/B978-0-12-418687-3.00025-2

1. THEORY

Strategies for genetic manipulation of yeast and bacteria usually require linking a desired modification with a selectable marker on a construct that is transformed into the organism. The construct can be either a linear piece of DNA that must then undergo homologous recombination in order to be stably integrated into the genome, or a circular plasmid that can be maintained under selection. Following transformation, the organism is plated onto selective media. A successful transformation will result in the growth of one to a few hundred colonies on the plate. However, there are many possible scenarios that can lead to retention of the selectable marker and subsequent growth of a colony in the absence of the desired genetic construct. Therefore, a screening method is required to distinguish colonies carrying the correct construct from colonies carrying the selective marker in some other context. Colony PCR involves designing PCR primers that will yield a specific product of known size only if the desired construct is present, and ideally, a product of a different size if the site of the desired manipulation remains unaltered. This technique takes advantage of the high sensitivity of PCR – the small amount of template DNA that is required to give an easily visualized band on an agarose gel following PCR amplification can be recovered from a very crude preparation of cells. Colony PCR is thus a powerful tool for rapidly and easily screening through potentially large numbers of colonies to distinguish true positives from false positives. In most cases it is a superior alternative to the older strategy of growing small cultures from several colonies, preparing microgram quantities of DNA from each culture, and performing restriction digests to verify that the desired construct is present. Additionally, in the case of a cloning strategy that utilizes a step in one organism to generate a construct that is then recovered, amplified by PCR, and transformed into a different organism, colony PCR can be used directly to generate the product for the second transformation.

2. EQUIPMENT

Benchtop microcentrifuge
Vortex mixer
PCR Thermocycler
Microwave oven
Gel electrophoresis equipment

Transilluminator, or other UV light source
Petri dishes, containing colonies to be tested
Micropipettors
Micropipettor tips
0.2-ml thin-walled PCR tubes

3. MATERIALS

Selective culture media, on which colonies are growing
Sodium hydroxide (NaOH)
Purified water (can be distilled and autoclaved, or filtered by Milli-Q or similar filtration system)
PCR primers
Taq polymerase (or similar thermostable DNA polymerase)
dNTP mix (may be supplied with Taq enzyme, or contains:
  dATP
  dGTP
  dTTP
  dCTP)
PCR buffer (may be supplied with Taq enzyme, or contains:
  Tris–HCl, pH 8.8
  $(NH_4)_2SO_4$
  Tween 20
  $MgCl_2$)
Agarose
Tris Base
Boric Acid (H_3BO_3)
EDTA
Ethidium Bromide
Ficoll 400
Bromophenol blue
Xylene Cyanol
DNA ladder

3.1. Solutions & buffers

Step 1 Sodium Hydroxide (20 mM)

Component	Stock	Amount/10 ml
NaOH	10 N	20 µl

Add purified water to 10 ml

Step 2 dNTP mix

Component	Final concentration	Stock	Amount/1 ml
dATP	2.5 mM	100 mM	25 μl
dCTP	2.5 mM	100 mM	25 μl
dGTP	2.5 mM	100 mM	25 μl
dTTP	2.5 mM	100 mM	25 μl

Add purified water to 1 ml

10× PCR buffer

Component	Final concentration	Stock	Amount/10 ml
Tris–HCl (pH 8.8)	750 mM	1 M	7.5 ml
$(NH_4)_2SO_4$	200 mM	1 M	2.0 ml
Tween 20	0.1% (v/v)	10%	100 μl
$MgCl_2$	15 mM	1 M	150 μl

Add purified water to 10 ml

PCR primers

Amounts given for PCR primers assume a working stock concentration of 10 μM.

Step 3 5× TBE buffer

Component	Final concentration	Stock	Amount/liter
Tris base	445 mM	N/A	54 g
Boric Acid	445 mM	N/A	27.5 g
EDTA	10 mM	0.5 M	20 ml

Add purified water to 1 l

1% Agarose Gel Mix

Component	Final concentration	Stock	Amount/100 ml
Agarose	1% w/v	N/A	1 g
TBE buffer	1×	5×	20 ml
Ethidium Bromide	0.5 μg ml^{-1}	10 mg ml^{-1}	5 μl

Add water to 100 ml and heat to boiling in microwave

6× DNA loading dye

Component	Final concentration	Stock	Amount/10 ml
Ficoll	15% w/v	N/A	1.5 ml
Bromophenol Blue	0.25% w/v	N/A	25 mg
Xylene Cyanol FF	0.25% w/v	N/A	25 mg

Add water to 10 ml

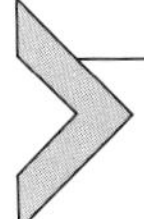

4. PROTOCOL

4.1. Duration

Preparation	variable
Protocol	3–5 h

4.2. Preparation

PCR primers must be designed and synthesized prior to starting this protocol. Commercial primer synthesis has become very economical and widely available, and the design of appropriate primers is critical to the success of colony PCR. It is usually worthwhile to design primers specifically for this purpose rather than use suboptimal primers that may have been designed for other purposes. Ideally, primers designed to verify the integration of a construct by homologous recombination should include one primer against the construct and one against the genome adjacent to the desired site of integration. Colony PCR is most effective for products less than 1 kb in length, especially when the organism is *S. cerevisiae*. Larger products have been successfully obtained, especially from *E. coli* colonies, but if the goal is simply to verify an integration event, smaller products are usually better. It is best to design primers such that a product is obtained whether the desired construct is present or not, but the product sizes are unique in each case. Sometimes it is necessary to design more than one primer set to accomplish this. At a minimum, primers that give a product only if the desired construct is present can be used, but will not distinguish between a failure to obtain the desired construct and a failure of the PCR reaction in the case that no product is obtained. See Fig. 25.1 for an example.

See Fig. 25.2 for the flowchart of the complete protocol.

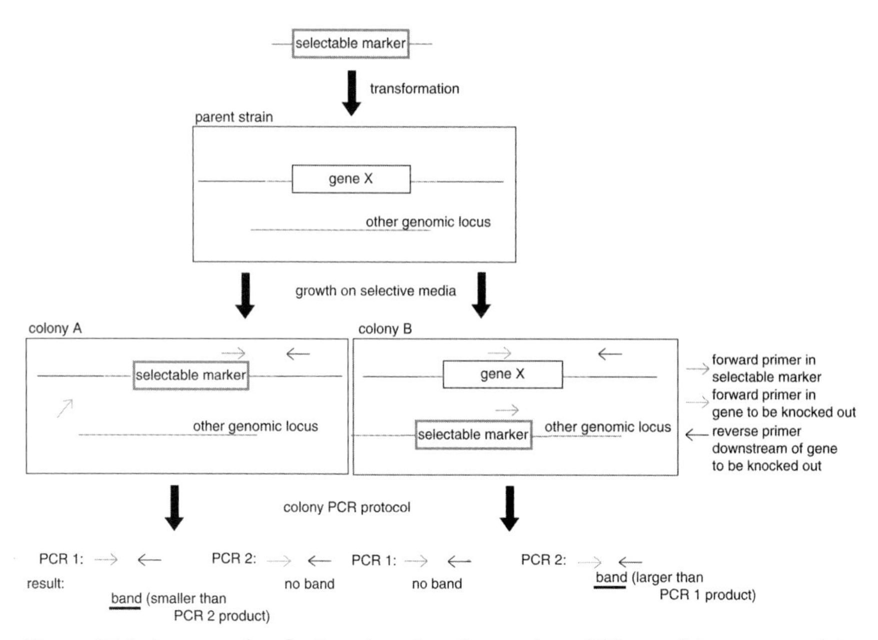

Figure 25.1 An example of primer locations for a colony PCR to validate a gene deletion. A construct consisting of a selectable marker flanked by sequences upstream and downstream of a gene to be deleted is transformed into the organism. Colonies that grow on selective media presumably express the selectable marker, but some colonies may have integrated the marker into a random genomic locus (as in colony B) rather than into the locus of the gene to be deleted (as in colony A). A forward primer against sequence within the selectable marker combined with a reverse primer against sequence just downstream of the gene to be deleted (PCR 1) will yield a product only if the marker has integrated in the appropriate locus. A forward primer against sequence within the gene to be deleted, combined with a reverse primer against sequence just downstream of that gene (PCR 2) will yield a product only if the correct integration has not occurred. Thus, a colony with the correct genotype should yield a band for PCR 1 but not PCR 2. Each colony should yield a band for either PCR 1 or PCR 2, however – failure of both reactions would indicate a failure of the PCR protocol, rather than a negative result for the presence of the desired construct. All the three PCR primers should have a similar melting temperature, preferably above 60 °C, and should not be complementary to each other. Product sizes should be ~200 bp–1 kb.

5. Step 1 Extraction of DNA from a Colony

5.1. Overview

Although PCR requires only a very small amount of template DNA, it is more successful, especially for *S. cerevisiae*, if cells are treated to break down the cell wall and allow the DNA to escape.

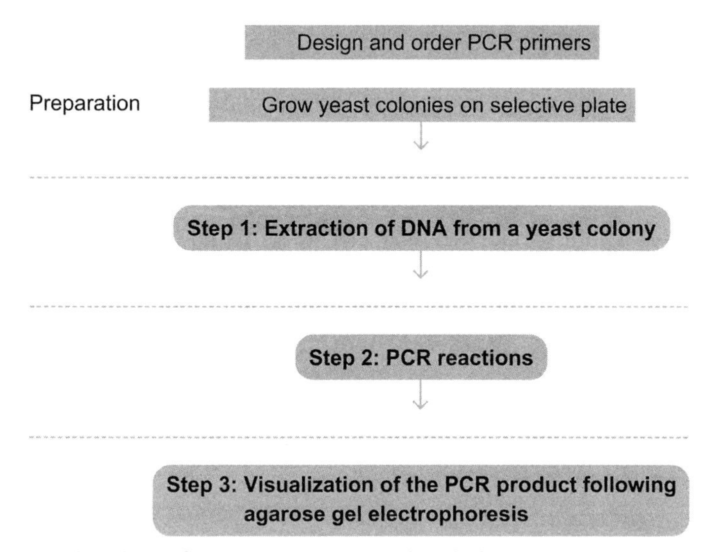

Figure 25.2 Flowchart of the complete protocol, including preparation.

5.2. Duration

15 min

1.1 Start with large (~2 mm) colonies. Colonies can be restreaked into a larger patch on a second plate of selective media if they are very slow-growing, or if the first plate is crowded with colonies.

1.2 For each colony to be screened, pipette 20 µl 20 mM sodium hydroxide into a 0.2-ml thin-walled PCR tube. Using a pipettor tip, scrape up about half of a colony and pipette up and down in the 20 µl sodium hydroxide to disperse the colony in the liquid. Label the colony on the plate so that positive colonies can later be identified.

1.3 Incubate tubes for 8 min at 100 °C in a PCR machine, with a heated lid.

1.4 Briefly spin the tubes in a microfuge to collect liquid at the bottoms of the tubes. Mix each tube vigorously using a vortex mixer at top speed for about 15 s.

1.5 Spin tubes in a microfuge for about 30 s at top speed to pellet the cell debris.

1.6 Use the supernatant as the template DNA in a PCR reaction, as described in Step 2.

5.3. Tip

Because PCR is so sensitive in detecting very small amounts of DNA, it is possible for DNA from dead cells that have failed to grow on the selective media to be amplified in colony PCR. Depending on how the primers for the colony PCR have been designed, this can lead to a false band in a negative control reaction. Usually these false bands are faint. This can be minimized by restreaking colonies onto a fresh plate of selective media and using these cells for colony PCR, but adding this step costs 1–2 days.

5.4. Tip

This step is probably not necessary if the organism is E. coli. *Good results have been observed when a small amount of an* E. coli *colony is scraped up with a pipetor tip and dispersed directly into a PCR reaction mix. This also has been effective with some other yeast species and may even sometimes be effective for* S. cerevisiae. *This step will not decrease the efficacy of the PCR, however, and probably helps the DNA escape the cell wall.*

5.5. Tip

A single E.coli *colony can be used for simultaneous PCR amplification and overnight growth. First, disperse some of the scraped colony into the PCR master mix. Disperse the remainder into a tube containing 1–2 ml of LB medium plus the appropriate antibiotic and grow overnight at 37 °C with shaking.*

See Fig. 25.3 for the flowchart of Step 1.

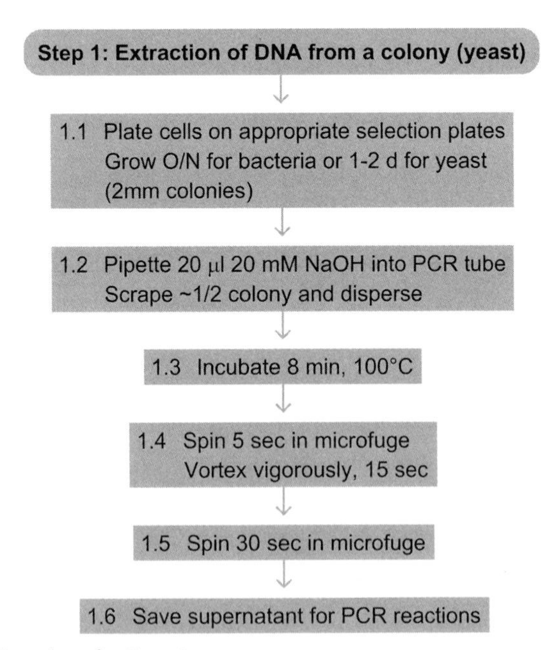

Figure 25.3 Flowchart for Step 1.

6. STEP 2 PCR

6.1. Overview

A standard PCR reaction is used to amplify the product of interest from the DNA that has been extracted from the colony in Step 1. A typical set of PCR conditions is described, but any commercially available thermostable polymerase and its supplied buffer should work (see General PCR).

6.2. Duration

2–2.5 h

2.1 Make a PCR master mix, and keep on ice:

per sample:

16.25 μl purified water
2.5 μl 10× PCR buffer
2.5 μl dNTP mix
1.25 μl forward primer
1.25 μl reverse primer
0.25 μl 100× Taq

2.2 Pipette 24 μl of the PCR master mix into 0.2-ml thin-walled PCR tubes. Keep tubes on ice.

2.3 Pipette 0.5–1 μl of the supernatant from each sample into a PCR tube containing PCR master mix.

2.4 Use a thermocycler to carry out PCR. Typical reaction conditions are as follows, for a product up to 1 kb in length:

5 min at 95 °C
repeat 30 times:
15 s at 95 °C
15 s at 57 °C
1 min at 72 °C
finish with:
5 min at 72 °C

6.3. Tip

If skipping Step 1, increase the amount of water per reaction by 1 μl, and pipette 25 μl PCR master mix into a 0.2-ml thin-walled PCR tube for each colony to be tested. Then disperse the colonies directly into this liquid. Be sure to label the colonies on the plate so positives can later be identified.

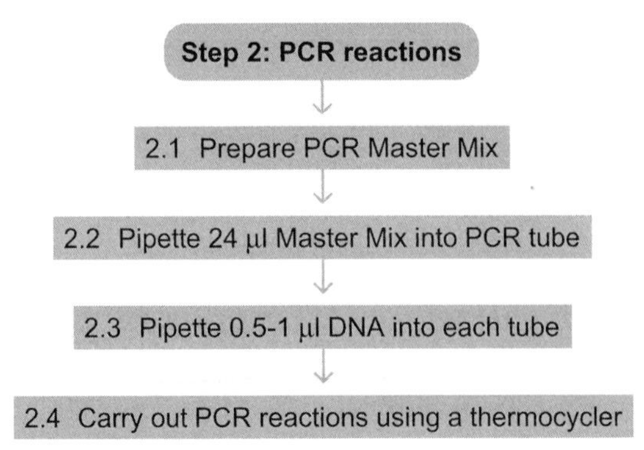

Figure 25.4 Flowchart for Step 2.

6.4. Tip

If using a HotStart Taq polymerase, it is not necessary to place the tubes on ice while setting up the reactions.

6.5. Tip

When preparing a PCR master mix, make enough for the number of reactions plus 0.5–1 extra.

6.6. Tip

If the goal of the colony PCR is to produce a sufficient quantity of a construct for a subsequent transformation step into a different organism, a larger volume PCR can be carried out. It may also be advisable to use a higher fidelity thermostable polymerase, or mixture of polymerases, since any mutations acquired during this PCR step could be deleterious in future cloning steps. Check the supplier's manual if you use a high-fidelity thermostable polymerase. The extension is often carried out at a lower temperature.

See Fig. 25.4 for the flowchart of Step 2.

7. STEP 3 VISUALIZATION OF PRODUCT ON AN AGAROSE GEL

7.1. Overview

The PCR products that have been generated are now separated by size on an agarose gel containing a DNA-intercalating dye such as ethidium bromide (see Agarose Gel Electrophoresis).

7.2. Duration

1.5 h

3.1 Cast the agarose gel mix into a gel tray after boiling in a microwave oven to completely dissolve the agarose. Take care while heating the

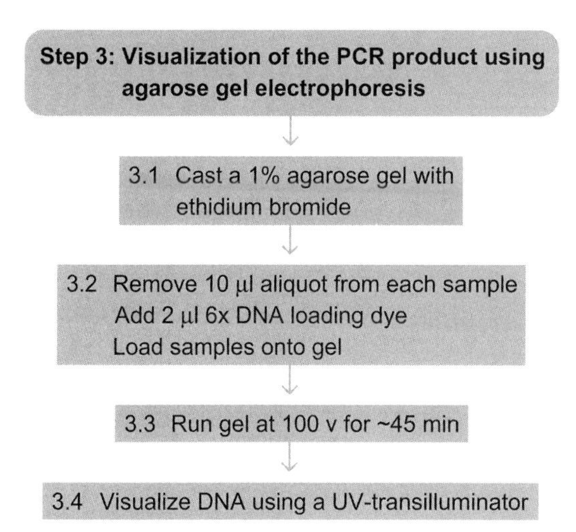

Figure 25.5 Flowchart for Step 3.

gel mix not to allow it to boil over. This can be done while the PCR is running. Fill the gel tank with 1× TBE, the buffer in which the gel will be run. Note that ethidium bromide is toxic. Wear gloves when handling the gel and take care to dispose of it properly.

3.2 Add 2 µl DNA loading dye to 10 µl of each sample. Load onto the gel. Load DNA ladder (as commercially supplied, or about 500 ng DNA in a volume of about 10 µl, containing loading dye) in one lane.

3.3 Run the gel at ~100 V for 45 min, or until xylene cyanolhas run about three quarters the length of the gel.

3.4 Use a UV light box to visualize the DNA bands on the gel.

7.3. Tip

If the goal of the colony PCR is to produce a sufficient quantity of a construct for a future transformation step, run only a small fraction of the PCR product on a gel to verify that the product is the correct size. If a single band is obtained in high enough amounts, the PCR mixture can be used directly for a transformation of some organisms, including S. cerevisiae.

See Fig. 25.5 for the flowchart of Step 3.

REFERENCES

Referenced Protocols in Methods Navigator

Agarose Gel Electrophoresis
General PCR

CHAPTER TWENTY SIX

Chemical Transformation of Yeast

Megan Bergkessel, Christine Guthrie[1]
Department of Biochemistry and Biophysics, University of California, San Francisco, CA, USA
[1]Corresponding author: e-mail address: christineguthrie@gmail.com

Contents

1. Theory 312
2. Equipment 312
3. Materials 313
   3.1 Solutions & buffers 313
4. Protocol 314
   4.1 Preparation 314
   4.2 Duration 315
5. Step 1 Preparation of Competent Cells 315
   5.1 Overview 315
   5.2 Duration 315
   5.3 Tip 316
   5.4 Tip 316
   5.5 Tip 316
6. Step 2 Transformation of Yeast Cells 317
   6.1 Overview 317
   6.2 Duration 317
   6.3 Tip 318
   6.4 Tip 318
   6.5 Tip 319
   6.6 Tip 319
   6.7 Tip 320
References 320

Abstract

Transformation of chemically competent yeast cells is a method for introducing exogenous DNA into living cells. Typically, the DNA is either a plasmid carrying an autonomous replication sequence that allows for propagation or a linear piece of DNA to be integrated into the genome. The DNA usually also carries a marker that allows for selection of successfully transformed cells by plating on the appropriate selective media.

Methods in Enzymology, Volume 529
ISSN 0076-6879
http://dx.doi.org/10.1016/B978-0-12-418687-3.00026-4

1. THEORY

Since the phenomenon was first reported in 1983, transformation of yeast cells by treatment with alkali cations has become the most commonly used method for yeast transformations. Cells are incubated with a mixture of lithium acetate, polyethylene glycol (PEG), single-stranded carrier DNA, and DNA to be transformed. Next, cells are briefly heat-shocked and plated on selective media. The precise mechanism by which treatment with alkali cations allows for exogenous DNA to enter a yeast cell remains unclear, but it is thought that interactions among the negatively charged PEG, the monovalent cations, and the yeast cell surface result in an increase in permeability of the cell wall and plasma membrane, and that membrane potential then helps drive the exogenous DNA across.

2. EQUIPMENT

Roller drum (30 °C)
Rotary shaker (30 °C)
Magnetic stirrer
Magnetic stir bars
Spectrophotometer
Bench top or clinical centrifuge
Microcentrifuge
Vortex mixer
Bunsen burner
Micropipettors
Micropipettor tips
Pipet-aid
Inoculating loop (optional)
250-ml Erlenmeyer flask, autoclaved
Sterile 17 × 100-mm glass culture tubes
Sterile glass or plastic serological pipettes, 5–25 ml
Disposable spectrophotometer cuvettes
50-ml polypropylene conical tubes
1.5-ml polypropylene microcentrifuge tubes, autoclaved
5-mm glass beads, autoclaved (optional)
Sterile 100 × 15 mm disposable Petri plates

3. MATERIALS

YPAD medium (YPD medium supplemented with 400 μM adenine hemisulfate)
Bacto agar
PEG 3350
Lithium acetate (LiOAc·$2H_2O$)
Salmon sperm DNA
Tris base
Hydrochloric acid (HCl)
EDTA
Distilled water

3.1. Solutions & buffers

Prepare 1× YPAD liquid and agar plates, and 2× YPAD liquid medium. Prepare selective medium agar plates, either containing an antibiotic (if an antibiotic resistance marker is used) or lacking an essential nutrient (if a nutritional marker is used). See the yeast media recipes on *Saccharomyces cerevisiae* Growth Media

Step 2 50% PEG 3350

Add 50 g PEG 3350 to about 30 ml distilled water in a 150-ml beaker and stir until completely dissolved, heating gently if necessary. Bring the volume up to 100 ml with water and sterilize by autoclaving. This solution can be stored at room temperature, but is best prepared fresh every few months

1 M Lithium Acetate

Dissolve 10.2 g lithium acetate dihydrate in 100 ml distilled water and sterilize by autoclaving

2 mg ml^{-1} single-stranded carrier DNA

Prepare 100 ml sterile TE by adding 1 ml of 1 M Tris–HCl, pH 8.0 and 200 μl of 0.5 M EDTA to 99 ml sterile water. Dissolve 200 mg of salmon sperm DNA in the 100 ml TE on a magnetic stir plate at 4 °C. This will take several hours. Dispense in 1–5 ml aliquots. Denature carrier DNA by boiling for 5 min and then immediately chilling on ice before use. Small aliquots can be denatured and then stored at −20 °C and used again without reboiling; however, repeated boiling up to 4 times will not adversely affect it. All aliquots should be stored at −20 °C

Transformation Mix

Component	Final concentration	Stock	Amount per transformation
PEG 3350	33%	50%	240 μl
Lithium Acetate	100 mM	1 M	36 μl
Single-stranded carrier DNA	280 $\mu g\ ml^{-1}$	2.0 $mg\ ml^{-1}$	50 μl
DNA to be transformed	1 μg/ transformation	About 40 $ng\ \mu l^{-1}$	24 μl

4. PROTOCOL

4.1. Preparation

Prior to the start of this protocol, the DNA to be transformed into yeast must be prepared (see Molecular Cloning). Depending on the goals of the transformation, the DNA can be of different forms and sources. Often, a plasmid carrying a selectable marker and an autonomous replication sequence is transformed; the plasmid can be prepared by miniprep isolation of plasmid DNA from a bacterial stock carrying the plasmid, if a stock of the plasmid DNA is not already available. Transformation with plasmid DNA is usually highly efficient. Transformation with a linear piece of DNA that is intended to integrate into the genome by homologous recombination is also very common. There are a couple sources of the linear piece of DNA. It is usually PCR amplified from a cassette carrying a selectable marker using primers that carry sequences to direct homologous recombination. Alternatively, a plasmid carrying a selectable marker and other sequences of interest can be digested with restriction endonucleases. PCR product can be used directly for transformation if the reaction yields a single product in sufficient quantity. If multiple PCR products are present, or in the case of a fragment that has been cut out of a plasmid, the desired linear piece of DNA should be gel-purified away from the other pieces. Because homologous recombination is itself a somewhat low-frequency event, and is required for successful maintenance of the selectable marker, transformation by a linear piece of DNA that must undergo homologous recombination is usually much less efficient than transformation by plasmid DNA. Thus, it is recommended that the higher efficiency version of the protocol is followed for linear pieces of DNA, and that a full 1 μg of DNA is used per transformation (i.e., follow

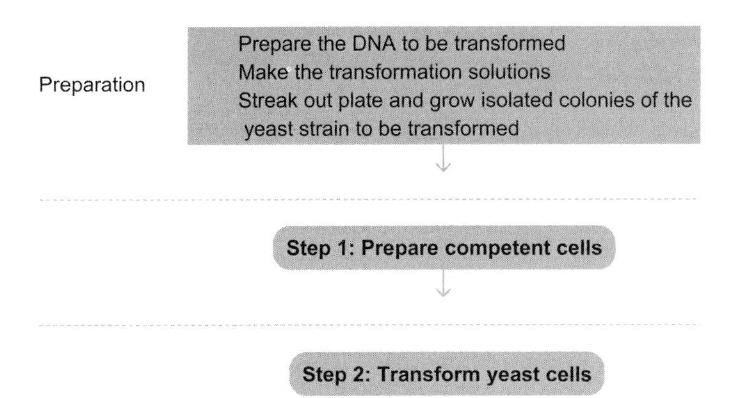

Figure 26.1 Flowchart of the complete protocol, including preparation.

all of Step 1). For plasmid transformations, the lower efficiency protocol and less DNA can usually be used successfully (see Tip after Step 1).

Make solutions (dissolving salmon sperm DNA can take quite a while).

Grow the yeast strain that is to be transformed as patches on an YPAD agar plate.

4.2. Duration

Preparation	Variable
Protocol	About 1 day + 3–4 days for transformed yeast colonies to grow

See Fig. 26.1 for the flowchart of the complete protocol.

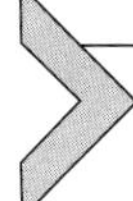

5. STEP 1 PREPARATION OF COMPETENT CELLS

5.1. Overview

The efficiency of transformation correlates with the growth status of the yeast cells used for the transformation – cells in log phase growth transform most efficiently. The protocol described below gives the highest transformation efficiency. If high efficiency is not required (for a transformation using plasmid DNA, for example), see the tip below for a faster version of this step.

5.2. Duration

1 day

1.1 Set up a 5 ml starter culture by scraping up a colony of the yeast strain to be transformed and dispersing it into 5 ml of YPAD medium in a

17 × 100-mm glass culture tube. Work near a Bunsen burner flame with aseptic technique.

1.2 Allow starter to grow to saturation overnight at 30 °C (or room temperature if the strain is temperature sensitive) on a roller drum.

1.3 The next morning, measure the OD at 600 nm of the starter culture, using a spectrophotometer. Dilute the culture 1:20 with YPAD medium in a plastic cuvette before making a measurement, in order to bring the OD within the linear range of the spectrophotometer.

1.4 Use the starter culture to inoculate a 50 ml culture in 2× YPAD in a sterile 250-ml Erlenmeyer flask at an OD_{600} of 0.2.

1.5 Allow the culture to grow at 30 °C on a rotary shaker for 4–5 h, or until the OD_{600} reaches ~1.0–1.5.

1.6 Transfer the culture to a 50-ml conical centrifuge tube. Harvest the yeast cells by spinning at 3000 rpm at room temperature for 5 min in a benchtop centrifuge.

1.7 Discard the supernatant and resuspend the pellet in 20 ml of sterile, room temperature water.

1.8 Pellet the cells by spinning again at 3000 rpm for 5 min.

1.9 Resuspend the pellet in 1 ml sterile water and transfer it to a 1.5-ml microcentrifuge tube. Use 100 μl of this cell suspension per transformation.

5.3. Tip

Remember to take into account the 1:20 dilution of the overnight starter culture. For example, if the OD_{600} reading is 0.4, then the actual OD_{600} of the starter culture is 8. Usually this will require adding ~1–2 ml of the starter culture to 48–49 ml of 2× YPAD media, although the exact amount needed will vary greatly with the growth rate of the strain and the amount of time it had been left to grow.

5.4. Tip

If a very high efficiency of transformation is not required, cells can be used directly from a fresh patch on an agar plate. The patch should be no more than a few days old. Scrape approximately a 50-μl volume of cells off of the plate and disperse them in 100 μl of sterile water in a microcentrifuge tube. Follow the rest of the protocol as written.

5.5. Tip

Cells harvested by this protocol can be stored at 4 °C for up to a week with only minor losses in transformation efficiency. If planning to store cells, pellet them out of the sterile water suspension and resuspend them in 100 mM lithium acetate instead.

See Fig. 26.2 for the flowchart of Step 1.

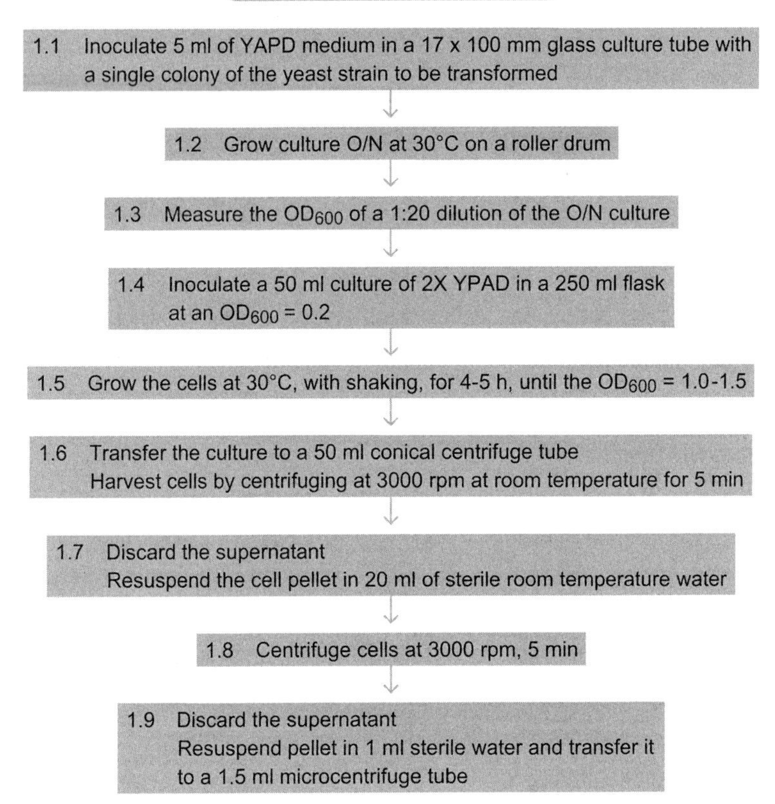

Figure 26.2 Flowchart of Step 1.

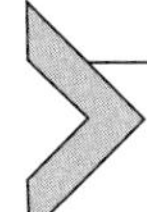

6. STEP 2 TRANSFORMATION OF YEAST CELLS

6.1. Overview

An exogenous piece of DNA is introduced to the cells in the presence of lithium acetate and polyethylene glycol. The addition of single-stranded salmon sperm DNA as a carrier dramatically increases the efficiency of transformation.

6.2. Duration

1.5–2 h

2.1 Transfer 100 μl of the cell suspension into a 1.5-ml microcentrifuge tube. Centrifuge the cells at 10 000 rpm for 1 min in a microcentrifuge. Discard the supernatant.

2.2 Add 360 μl Transformation Mix and resuspend the cells by mixing vigorously using a vortex mixer.

2.3 Incubate 15–30 min at 30 °C. Ensure mixing by taping the tubes to a roller drum.

2.4 Heat-shock cells by incubating for 15–40 min in a 42 °C water bath.

2.5 Collect cells by spinning at 10 000 rpm for 1 min in a microcentrifuge.

2.6 Discard supernatant and gently resuspend the cell pellet in 500 μl of sterile water.

2.7 Pipette 20–200 μl cell suspension onto the surface of an agar plate, depending on expected transformation efficiency. For plasmid transformations, generally less is required, while for transformations with linear DNA more cells should be plated.

2.8 Add 5–20 autoclaved 5 mm glass beads to each plate by shaking them out of their bottle. Replace the lids on the plates and shake back and forth briefly to disperse the cells over the surface of the plate. A whole stack of plates can be shaken simultaneously. After the transformation reaction has been spread over the surface of the plate, dump the glass beads into a receptacle containing 70% ethanol. They can be washed, autoclaved, and reused.

2.9 Incubate plates at 30 °C, or at room temperature for temperature-sensitive strains. For antibiotic resistance markers, replica plate to antibiotic-containing plates the day after plating to YPAD. It should take 3–4 days for sizeable yeast colonies to grow up following transformation.

6.3. Tip

The optimal heat shock time varies from strain to strain. Generally, longer heat shock times improve transformation efficiency, but may lead to increased cell death, especially for temperature-sensitive strains. Extremely temperature-sensitive strains can be transformed by incubating overnight at room temperature rather than at 30 and 42 °C as described above.

6.4. Tip

Some commonly used antibiotics and their concentrations to use for selection include Geneticin (G418, 0.2 mg ml^{-1}), nourseothricin (NAT, 0.2 mg ml^{-1}), and hygromicin B (HPH, 0.3 mg ml^{-1}).

6.5. Tip

If the yeast strain being transformed is unfamiliar, it is advisable to perform a mock transformation with water being added to the transformation mix instead of DNA, to use as a negative control.

6.6. Tip

If unsure what efficiency is expected, plate 20 μl on one plate and 200 μl on a second plate. For plating 20 μl, add 100–200 μl of sterile water directly to the plate. If the transformation efficiency is anticipated to be very low, plate the entire transformation reaction, splitting it between at least two plates. Increased plating density lowers the transformation efficiency. If using a nutritional marker for selection, plate directly onto

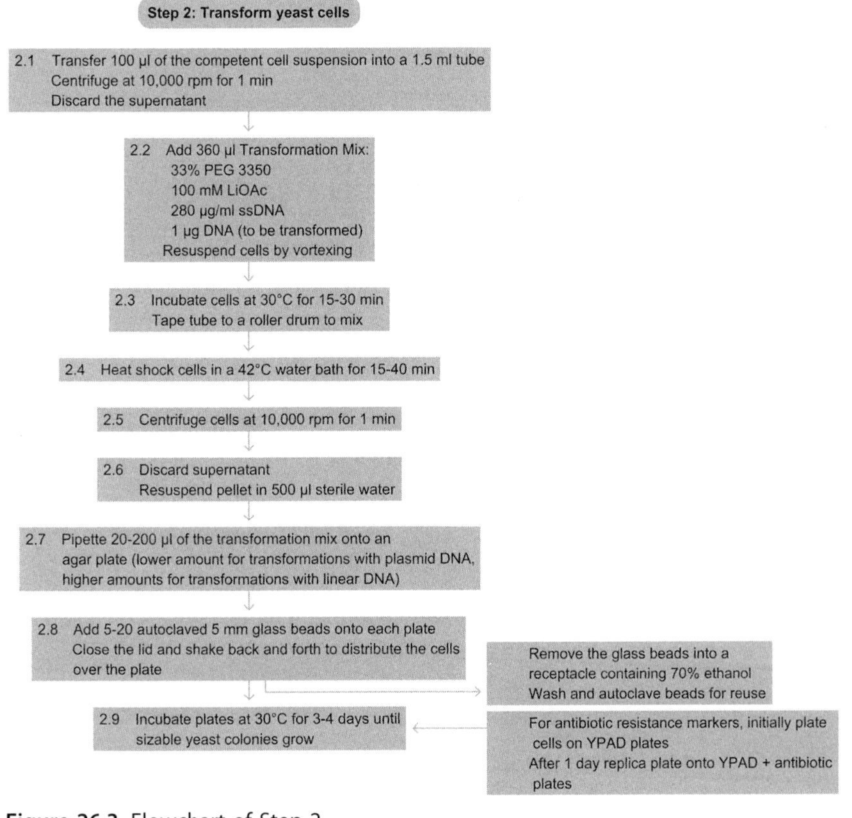

Figure 26.3 Flowchart of Step 2.

selective media. If using an antibiotic resistance marker, plate onto an YPAD agar plate.

6.7. Tip

Alternatively, a bent glass rod that has been sterilized by dipping in ethanol and flaming can be used to spread the transformation reaction over the surface of the plate (see how to flame-bend a glass Pasteur pipette and spread the yeast on Pouring Agar Plates and Streaking or Spreading to Isolate Individual Colonies).

See Fig. 26.3 for the flowchart of Step 2.

REFERENCES

Related Literature

Gietz, R. D., & Schiestl, R. H. (2007). High-efficiency yeast transformation using the LiAc/SS carrier DNA/PEG method. *Nature Protocols*, *2*(1), 31–34.

Gietz, R. D., & Schiestl, R. H. (2007). Quick and easy yeast transformation using the LiAc/SS carrier DNA/PEG method. *Nature Protocols*, *2*(1), 35–37.

Ito, H., et al. (1983). Transformation of intact yeast cells treated with alkali cations. *Journal of Bacteriology*, *153*(1), 163–168.

Referenced Protocols in Methods Navigator

Saccharomyces cerevisiae Growth Media

Molecular Cloning

Pouring Agar Plates and Streaking or Spreading to Isolate Individual Colonies

CHAPTER TWENTY SEVEN

Transformation of *E. coli* Via Electroporation

Juliane C. Lessard[1]
Department of Biochemistry and Molecular Biology, Johns Hopkins School of Public Health, Baltimore, MD, USA
[1]Corresponding author: e-mail address: jkellne2@jhmi.edu

Contents

1. Theory 321
2. Equipment 322
3. Materials 322
4. Protocol 322
   4.1 Preparation 322
   4.2 Duration 323
   4.3 Tip 323
   4.4 Tip 323
5. Step 1 Create Electro-Competent *E. coli* 324
   5.1 Overview 324
   5.2 Duration 324
   5.3 Tip 324
   5.4 Tip 324
6. Step 2 Electroporation of *E. coli* 324
   6.1 Overview 324
   6.2 Duration 325
   6.3 Tip 327
   6.4 Tip 327
References 327

Abstract

To create electro-competent *E. coli* and transform them with a plasmid of choice via electroporation.

1. THEORY

Electroporation of *E. coli* is a popular alternative to traditional heat-shock transformation of chemically competent cells. A high-voltage current

Methods in Enzymology, Volume 529
ISSN 0076-6879
http://dx.doi.org/10.1016/B978-0-12-418687-3.00027-6

is applied to the cells, which temporarily permeabilizes the plasma membrane and allows DNA or other small molecules to enter. The main advantages of electroporation over heat-shock transformation are the higher efficiency in the uptake of plasmid DNA and a faster, less involved production of competent cells.

2. EQUIPMENT

Electroporator for bacteria (e.g., Gen Pulser Xcell Electroporation System, Bio-Rad)
Refrigerated centrifuge
Refrigerated microcentrifuge
Shaking incubator (37 °C)
Incubator (37 °C)
15-ml sterile polypropylene snap-cap tubes
Electroporation cuvette, 0.1 cm gap
15-ml glass centrifuge tube
Rubber adaptors (to fit glass centrifuge tube into floor centrifuge rotor)
Pipettes
Micropipettors
Micropipettor tips
1.5-ml microcentrifuge tubes
Kimwipes

3. MATERIALS

Plasmid DNA (to be transformed)
LB agar plates (selective plates containing appropriate antibiotic)
Lysogeny broth (LB)
Bacto agar
Sterile ddH_2O
Glycerol (optional)

4. PROTOCOL

4.1. Preparation

Inoculate 5 ml of LB with *E. coli*. Grow overnight at 37 °C with shaking.

Chill the 15-ml glass centrifuge tube, a 1.5-ml microcentrifuge tube, the electroporation cuvette, and the ddH_2O on ice. Run the floor centrifuge for a few minutes to cool the chamber to 4 °C.

4.2. Duration

Preparation	15 min + overnight
Protocol	About 2 h

4.3. Tip

For most simple plasmid transformations, it is not necessary to harvest bacteria at early to mid log phase. However, if the transformation efficiency is low, dilute 1 ml of the overnight culture in 20 ml of LB and grow at 37 °C with shaking until the OD_{600} measures between 0.4 and 0.6 (1–1.5 h). Use 10 ml to make 50 μl of competent cells as described subsequently.

4.4. Tip

For all steps, keep reagents, tubes, and bacteria on ice at all times to ensure the production of high-quality electro-competent cells. Keeping the bacteria at a low temperature during the electric pulse also helps prevent electrically induced heating and thus increases cell survival.

See Fig. 27.1 for the flowchart of the complete protocol.

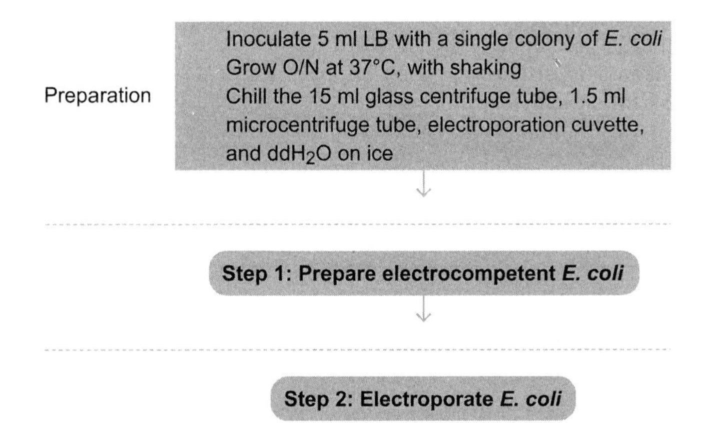

Figure 27.1 Flowchart of the complete protocol, including preparation.

5. STEP 1 CREATE ELECTRO-COMPETENT *E. COLI*

5.1. Overview

E. coli are washed several times with ice-cold water to prepare them for the uptake of plasmid DNA during electroporation.

5.2. Duration

20–30 min

1.1 Pour the bacterial culture into the prechilled 15-ml glass centrifuge tube on ice.
1.2 Centrifuge at 7000 rpm at 4 °C, for 5 min.
1.3 Discard the supernatant.
1.4 Resuspend the pellet in 1 ml of sterile, ice-cold ddH_2O and transfer into a chilled 1.5-ml microcentrifuge tube.
1.5 Spin in a microcentrifuge at 7000 rpm at 4 °C, for 5 min.
1.6 Discard the supernatant.
1.7 Resuspend the pellet in 1 ml of ice-cold ddH_2O.
1.8 Repeat steps 1.5–1.7 two more times for a total of three washes.
1.9 Spin a final time at 7000 rpm at 4 °C, for 5 min, discard the supernatant, resuspend the pellet in 50 μl of ice-cold ddH_2O, and put the cells on ice.

5.3. Tip

If you want to store the electro-competent cells at −80 °C for later use, substitute 10% glycerol for ddH_2O in Step 1.4 and in all subsequent steps.

5.4. Tip

This protocol can be scaled up to make larger quantities of competent cells.

See Fig. 27.2 for the flowchart of Step 1.

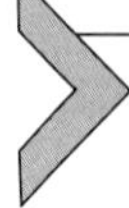

6. STEP 2 ELECTROPORATION OF *E. COLI*

6.1. Overview

Plasmid DNA is added to electro-competent *E. coli*. An electric pulse mediates the uptake of the DNA by the bacteria.

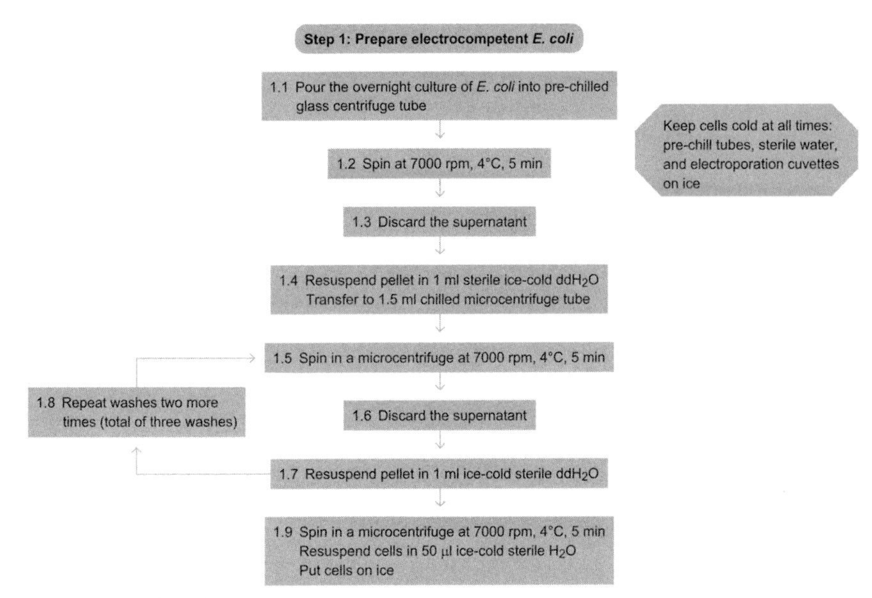

Figure 27.2 Flowchart of Step 1.

6.2. Duration

1 h and 10 min

2.1 Add 50–100 ng of supercoiled plasmid to the electro-competent *E. coli* and mix gently (do not pipette up and down).

2.2 Transfer the bacteria to a chilled electroporation cuvette. Be careful to pipette straight in between the metal plates and avoid introducing any bubbles.

2.3 Cap the cuvette and tap it lightly on the bench to settle the bacteria/DNA mix.

2.4 Put the cuvette back on ice and carry it to the electroporator.

2.5 Turn on the electroporator and set it to 1.8 kV, 25 μF, 200 Ω. This is a standard setting for most *E. coli* strains. Other bacterial strains may require an adjustment of the electroporation conditions.

2.6 Wipe the cuvette briefly with a Kimwipe to remove any residual water or ice, and then place it in the electroporation chamber.

2.7 Push the pulse button. The time constant displayed should be around 4 ms.

2.8 *Immediately* after the pulse has been delivered, add 1 ml of LB (or other growth medium, e.g., SOC) to the cuvette and pipette quickly but gently up and down. Be aware that the transformation efficiency decreases proportionally to the lag time between the electric pulse and the addition of media.

2.9 Transfer the mixture to a fresh 1.5-ml microcentrifuge tube.

2.10 Incubate at 37 °C for 1 h with shaking.

2.11 Evenly spread 100 μl of your transformation onto a selective plate. Electroporation is highly efficient and often yields a very large number of colonies. To ensure that individual, medium-sized colonies can be picked the next day, mix 10 μl of the transformation with 90 μl of LB and spread this 1:10 dilution evenly onto another

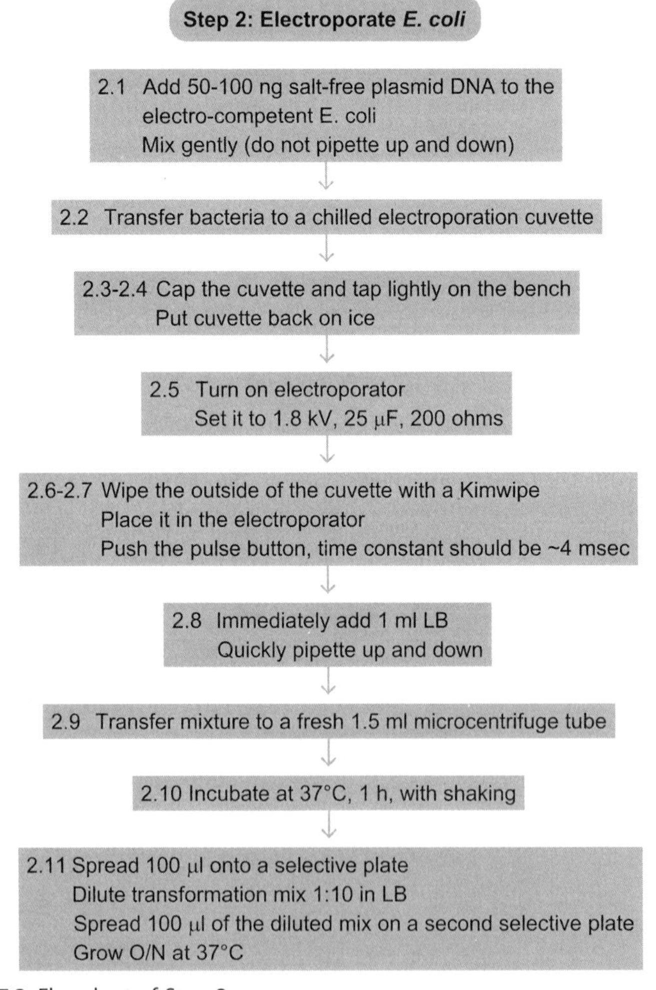

Figure 27.3 Flowchart of Step 2.

selective plate. Incubate both plates upside down overnight at 37 °C.

6.3. Tip

Using low amounts of well-purified DNA is key to a successful electroporation. Even small amounts of residual salt from DNA preparations may interfere with the electric current and cause arcing, resulting in excessive cell death.

6.4. Tip

Warm up the selective plates at 37 °C while the bacteria are recovering. Cold plates lead to lower transformation efficiency.

See Fig. 27.3 for the flowchart of Step 2.

REFERENCES

Related Literature

"Gene Pulser Xcell Electroporation System" Instruction Manual, BioRad

CHAPTER TWENTY EIGHT

Transformation of Chemically Competent *E. coli*

Rachel Green[1], Elizabeth J. Rogers
Johns Hopkins School of Medicine/HHMI, Molecular Biology and Genetics, Baltimore, MD, USA
[1]Corresponding author: e-mail address: ragreen@jhmi.edu

Contents

1. Theory 330
2. Equipment 330
3. Materials 330
   3.1 Solutions & buffers 331
4. Protocol 332
   4.1 Preparation 332
   4.2 Duration 332
5. Step 1 Prepare Competent Cells 333
   5.1 Overview 333
   5.2 Duration 333
   5.3 Tip 333
   5.4 Tip 334
   5.5 Tip 334
   5.6 Tip 334
   5.7 Tip 334
6. Step 2 Transform Competent Cells 334
   6.1 Overview 334
   6.2 Duration 334
   6.3 Tip 335
   6.4 Tip 336
   6.5 Tip 336
   6.6 Tip 336
References 336

Abstract

To introduce DNA into *E. coli* cells.

Methods in Enzymology, Volume 529
ISSN 0076-6879
http://dx.doi.org/10.1016/B978-0-12-418687-3.00028-8

1. THEORY

Transformation of *E. coli* is an important step that allows the introduction of heterologous DNA using plasmid vectors or introducing mutations via homologous recombination events.

2. EQUIPMENT

Shaking incubator (37 °C)
UV/Vis spectrophotometer
Refrigerated low-speed centrifuge (4 °C)
Water bath (42 °C)
Incubator (37 °C)
Erlenmeyer flask, 500 ml (sterile)
10-cm Petri plates
0.2-μm filters
14-ml sterile polypropylene snap-cap tubes
Disposable cuvettes
50-ml sterile polypropylene tubes (e.g., Corning 430829)
2-ml sterile screw-capped conical tubes with no skirt (e.g., Phenix SCS-02 S)
5–10-ml glass vials or autoclavable screw-capped tubes
Glass spreader

3. MATERIALS

Antibiotic of choice
Bacto agar
Bacto tryptone
Bacto yeast extract
Calcium chloride ($CaCl_2$)
Magnesium chloride ($MgCl_2$)
Magnesium sulfate ($MgSO_4$)
Manganese chloride ($MnCl_2$)
3-(*N*-Morpholino)-propanesulfonic acid (MOPS)
Potassium acetate (KOAc)
Potassium chloride (KCl)
Rubidium chloride (RbCl)
Sodium chloride (NaCl)

Glucose
Glycerol
Acetic acid
Sodium hydroxide (NaOH)
Potassium hydroxide (KOH)
Dry ice
Ethanol
Liquid nitrogen

3.1. Solutions & buffers

Step 1 Psi Media

Component	Amount
Tryptone	20 g
Yeast extract	5 g
$MgCl_2$	5 g

Dissolve in 900 ml deionzed water. Adjust the pH to 7.6 with KOH. Add water to 1 l and autoclave to sterilize
Note: You can also use LB medium (low salt) supplemented with 4 mM $MgSO_4$ and 10 mM KCl or SOB (also commercially available)

Tfb I (Transformation Buffer I)

Component	Final concentration	Amount
Potassium acetate	30 mM	1.18 g
$RbCl_2$	100 mM	4.84 g
$CaCl_2 \cdot 2H_2O$	10 mM	0.59 g
$MnCl_2 \cdot 4H_2O$	50 mM	3.96 g
Glycerol	15% (v/v)	60 ml

Dissolve in 300 ml water. Adjust the pH to 5.8 with dilute acetic acid. Add water to 400 ml and pass through a 0.2-μm filter to sterilize

Tfb II (Transformation Buffer II)

Component	Final concentration	Amount
MOPS	10 mM	0.21 g
$CaCl_2 \cdot 2H_2O$	75 mM	1.1 g
$RbCl_2$	10 mM	0.12 g
Glycerol	15% (v/v)	15 ml

Dissolve in 75 ml water. Adjust the pH to 6.5 with dilute NaOH. Add water to 100 ml and pass through a 0.2 μm filter to sterilize

Step 2 LB Agar (Miller's high salt)

Component	Amount
Tryptone	10 g
Yeast extract	5 g
NaCl	10 g
Agar	15 g

Dissolve in 900 ml water. Adjust the pH to 7.2 with ~0.2 ml 5 N NaOH. Add water to 1 l and autoclave to sterilize. Cool to ~55 °C, add appropriate antibiotic and pour plates
Note: LB Agar is available commercially as a premixed powder

SOC media

Component	Final concentration	Stock	Amount
Yeast extract	0.5%	–	0.5 g
Tryptone	2%	–	2.0 g
NaCl	10 mM	3 M	0.33 ml
KCl	2.5 mM	1 M	0.25 ml
$MgCl_2$	10 mM	1 M	1 ml
$MgSO_4$	10 mM	1 M	1 ml
Glucose	20 mM	1.1 M	1.82 ml

Dissolve in 90 ml of deionized water and bring the volume to 100 ml. Dispense into 5–10 ml aliquots and autoclave to sterilize

4. PROTOCOL

4.1. Preparation

Pour selective agar plates and let harden.

Pick a single colony from a freshly streaked plate (without antibiotics!) and inoculate a small culture (2–5 ml) of Psi media in a sterile 14-ml snap-cap tube. Grow overnight at 37 °C with shaking at 250 rpm.

Store TfbI and TfbII at 4 °C to make sure they are chilled.

4.2. Duration

Preparation	About 15 min
Protocol	About 4–6 h

See Fig. 28.1 for the flowchart of the complete protocol.

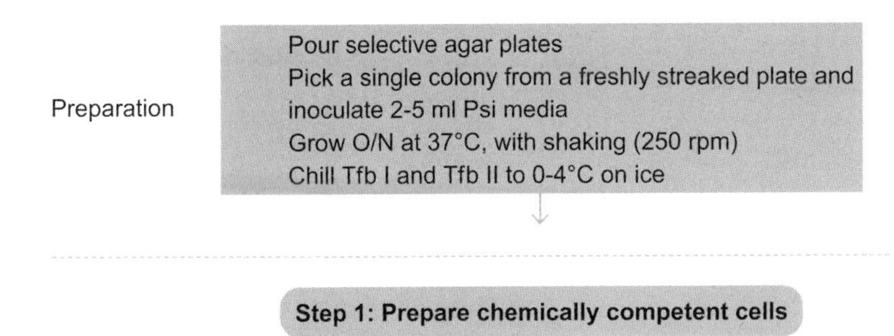

Figure 28.1 Flowchart of the complete protocol, including preparation.

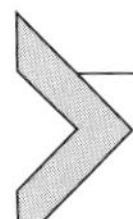

5. STEP 1 PREPARE COMPETENT CELLS

5.1. Overview

Grow cells to mid-log phase and make competent by chemical treatment.

5.2. Duration

3–5 h

1.1 Inoculate 100 ml of Psi media with 0.5 ml of the overnight culture and incubate at 37 °C with vigorous shaking.

1.2 When A_{600} reaches 0.4–0.5, place on ice and chill for 5–10 min.

1.3 Transfer the cells to 50-ml sterile chilled polypropylene centrifuge tubes. Pellet cells at 5000 × *g*, 4 °C for 5 min.

1.4 Discard supernatant carefully and gently resuspend the cell pellet in 0.4 volume of ice-cold Tfb I (20 ml for each 50 ml tube). Do not vortex; keep on ice while resuspending.

1.5 Incubate the cells on ice for 15 min.

1.6 Pellet the cells at 2000 × *g*, 4 °C for 10 min.

1.7 Discard the supernatant carefully and gently resuspend in 0.02 volume of ice-cold TfbI II (1 ml for 50 ml of culture). Keep the tube on ice.

1.8 Aliquot 50 μl into 2-ml prechilled sterile screw-capped tubes with conical bottoms.

1.9 Flash-freeze in a dry ice-ethanol bath or liquid nitrogen and store at −80 °C.

5.3. Tip

Use sterile technique at all times; the bacteria are not antibiotic-resistant at this point.

5.4. Tip

If reusable tubes are used to pellet the cells, they must be very clean and free of soap residue.

5.5. Tip

Cells and transformation buffers should be kept cold at all times. It is also preferable to use chilled pipettes and do everything in the cold room if possible.

5.6. Tip

The incubation in Tfb I can vary from 5 min to 1–2 h without any harm. The cells must be kept on ice.

5.7. Tip

Do not use standard 1.5-ml conical microcentrifuge tubes – they do not work well in the heat shock step of the transformation.

See Fig. 28.2 for the flowchart of Step 1.

6. STEP 2 TRANSFORM COMPETENT CELLS

6.1. Overview

Introduce DNA into competent cells and select for antibiotic-resistant bacteria.

6.2. Duration

2 h

2.1 Equilibrate a water bath to 42 °C. A dry heating block will work if the tube fits snugly, but it is not as good as a water bath.
2.2 Thaw one vial of competent cells on ice for each transformation. Handle gently since cells are sensitive to temperature changes and mechanical lysis.
2.3 Add 1–5 μl of DNA (10 pg to 100 ng) to a vial of thawed competent cells. DO NOT VORTEX OR PIPETTE UP AND DOWN.
2.4 Incubate on ice for 30 min.
2.5 Heat-shock cells for 30 s at 42 °C. Do not go any longer or shake the tube.
2.6 Remove the tube from the water bath and place on ice for 2 min.
2.7 Add 250 μl of SOC media to each vial.
2.8 Make sure that the cap is tight and incubate the tube on its side in a 37 °C shaking incubator (200–250 rpm) for 1 h.

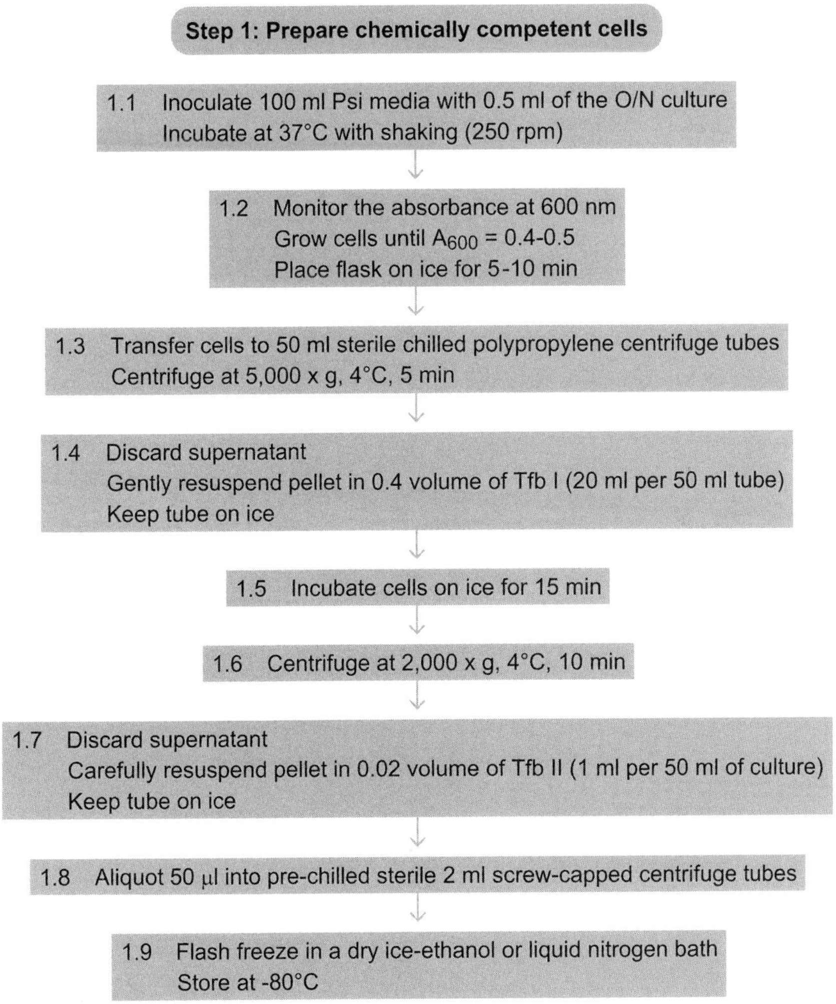

Step 1: Prepare chemically competent cells

1.1 Inoculate 100 ml Psi media with 0.5 ml of the O/N culture
Incubate at 37°C with shaking (250 rpm)

↓

1.2 Monitor the absorbance at 600 nm
Grow cells until A_{600} = 0.4-0.5
Place flask on ice for 5-10 min

↓

1.3 Transfer cells to 50 ml sterile chilled polypropylene centrifuge tubes
Centrifuge at 5,000 x g, 4°C, 5 min

↓

1.4 Discard supernatant
Gently resuspend pellet in 0.4 volume of Tfb I (20 ml per 50 ml tube)
Keep tube on ice

↓

1.5 Incubate cells on ice for 15 min

↓

1.6 Centrifuge at 2,000 x g, 4°C, 10 min

↓

1.7 Discard supernatant
Carefully resuspend pellet in 0.02 volume of Tfb II (1 ml per 50 ml of culture)
Keep tube on ice

↓

1.8 Aliquot 50 μl into pre-chilled sterile 2 ml screw-capped centrifuge tubes

↓

1.9 Flash freeze in a dry ice-ethanol or liquid nitrogen bath
Store at -80°C

Figure 28.2 Flowchart of Step 1.

2.9 Spread from 20 to 200 μl on an appropriate selective plate. The plates should be at room temperature or prewarmed to 37 °C. Incubate overnight at 37 °C.

6.3. Tip

Validate the chemically competent cells by plating untransformed cells on LB plates (without antibiotic – should have a lawn of cells) and LB plates containing antibiotic (should not have any colonies growing on them).

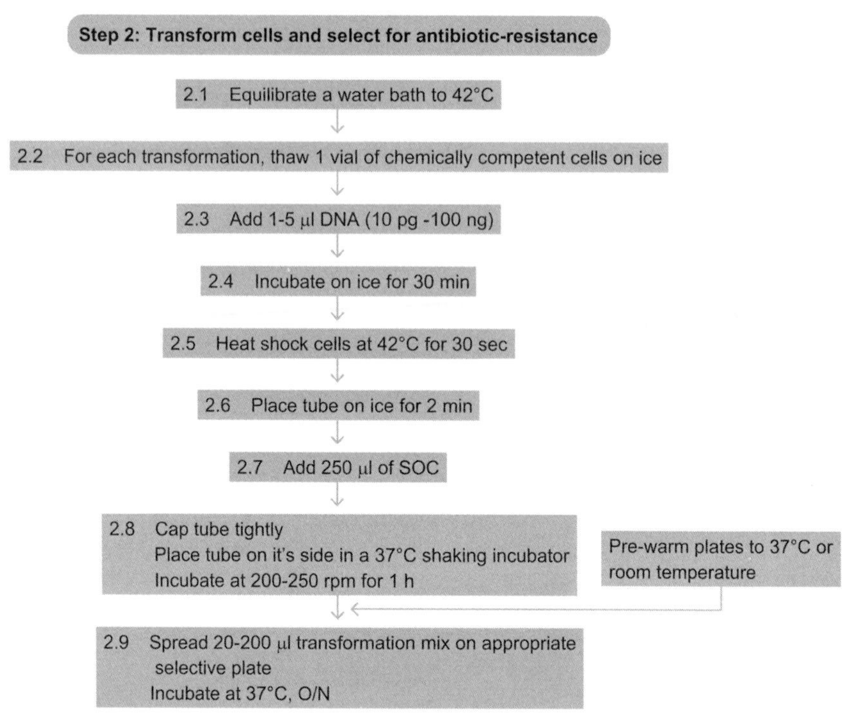

Step 2: Transform cells and select for antibiotic-resistance
2.1 Equilibrate a water bath to 42°C
2.2 For each transformation, thaw 1 vial of chemically competent cells on ice
2.3 Add 1-5 μl DNA (10 pg -100 ng)
2.4 Incubate on ice for 30 min
2.5 Heat shock cells at 42°C for 30 sec
2.6 Place tube on ice for 2 min
2.7 Add 250 μl of SOC
2.8 Cap tube tightly
Place tube on it's side in a 37°C shaking incubator
Incubate at 200-250 rpm for 1 h
Pre-warm plates to 37°C or room temperature
2.9 Spread 20-200 μl transformation mix on appropriate selective plate
Incubate at 37°C, O/N

Figure 28.3 Flowchart of Step 2.

6.4. Tip

Transform the cells with 10, 30, and 100 pg of supercoiled plasmid DNA and determine the transformation efficiency (# colonies per microgram DNA).

6.5. Tip

Bacteria are transformed more efficiently using supercoiled DNA than ligated DNA.

6.6. Tip

Transformed cells can be stored at 4 °C for 24–48 h with minimal loss of viability. Transformation efficiency varies depending on DNA.

See Fig. 28.3 for the flowchart of Step 2.

REFERENCES

Related Literature

Hanahan, D. (1983). Studies on transformation of *Escherichia coli* with plasmids. *Journal of Molecular Biology*, *166*, 557–580.

Hanahan, D., Jessee, J., & Bloom, F. R. (1991). Plasmid transformation of *Escherichia coli* and other bacteria. *Methods in Enzymology*, *204*, 63–114.

AUTHOR INDEX

Note: Page numbers followed by "*f*" indicate figures, and "*t*" indicate tables.

A

Affourtit, J. P., 204
Afonina, I. A., 286–287
Albert, T. J., 203–204
Allison, L. A., 85–98
Altman, W. E., 205, 206
Andersson, S., 103–104
Andreou, L. V., 137*f*, 150*f*
Arya, M., 286–287
Arya, N., 286–287
Atkinson, T., 68
Ausubel, F. M., 161–170

B

Babcock, H., 205
Berens, C., 103–104
Berg, P., 210–211
Berk, A., 85–98
Bernhagen, J., 285–286
Birnboim, H. C., 136
Bisseling, T., 18
Bloch, W., 286
Boeke, J. D., 154
Boussif, O., 228
Boyce, R., 18
Brasch, M. A., 100
Bretscher, A., 85–98
Brownley, A., 18
Brunner, H., 285–286
Bubunenko, M., 17
Buhler, J., 18
Burger, L., 99–124
Butler, J. M., 18
Buzby, P. R., 205

C

Cadwell, R. C., 250
Callow, M. J., 204–205
Canceill, D., 5
Cate, R. L., 48
Chamankhah, M., 14, 18
Chamberlain, J. S., 12
Chilana, P., 18
Connell, C. R., 286
Costa, G. L., 204
Costantino, N., 16–17
Court, D. L., 16–17
Crawford, D. R., 283

D

Dahlback, H., 103–104
Datta, S., 17
Davis, D. L., 103–104
Dieffenbach, C. W., 5
Doly, J., 136
Dorschner, M., 2, 18
Dressman, D., 205
Drmanac, R., 204–205
Dveksler, G. S., 5

E

Edwards, G., 137*f*
Egholm, M., 205, 206
Ehrenfels, C. W., 48
Ehrlich, S. D., 5
Engler-Blum, G., 48

F

Fink, G. R., 154
Frank, J., 48
Fuller, C. W., 172

G

Gaidatzis, D., 99–124
Garfinkel, D. J., 154
Geurts, R., 18
Gibbs, R. A., 12
Gibson, U. E., 281, 286
Glusman, G., 204–205
Gnirke, A., 203–204
Gomez-Sanchez, C. E., 8–9
Gommersall, L., 286–287
Griffin, A. M., 172
Griffin, H. G., 172
Griffith, F., 29

H

Hafner, M., 99–124
Harris, T. D., 205
Hecker, K. H., 67
Heid, C. A., 281, 286
Henikoff, J. G., 18
Henikoff, S., 18
Higgins, P. J., 283
Hillen, W., 103–104
Holmes, D. S., 136
Humbert, R., 2, 18

J

Jornvall, H., 103–104
Joyce, G. F., 250

K

Kaiser, C. A., 85–98
Karsten Tischer, B., 16–17
Kaufer, B., 16–17
Keggins, K. M., 144
Kinzler, K. W., 205
Klugman, K. P., 267
Korbel, J. O., 204
Krieger, M., 85–98
Kushner, S. R., 137*f*
Kutyavin, I. V., 286–287

L

Landt, O., 284–285
Landthaler, M., 99–124
Lee, L. G., 286
Leunissen, J. A. M., 18
Li, K., 18
Linhart, C., 14, 18
Liu, Y.-G., 13–14
Livak, K. J., 281, 287–288
Lodish, H., 85–98
Loker, E. S., 62*f*
Lovett, P. S., 144
Lowe, T. M., 5

M

Mackay, J., 284–285
Maguire, J., 203–204
Maher, B., 203–204
Maierhofer, C., 203–204
Malik, S., 100
Mann, T., 2, 18
Maples, V. F., 137*f*
Margulies, M., 205, 206
Markham, N. R., 18
McArdle, B. F., 172
McDonell, M. W., 69
McKernan, K. J., 204
Medrano, J. F., 12
Meier, M., 48
Melnikov, A., 203–204
Metzker, M. L., 186
Mills, A., 286–287
Molla, M. N., 203–204
Morishita, S., 18
Mülhardt, C., 8
Muller, G. A., 48
Muzny, D. M., 203–204

N

Najafabadi, H. S., 14, 18
Nguyen, P. N., 12
Nijveen, H., 18
Noble, W. S., 2, 18

O

Osterrieder, N., 16–17

P

Patel, H. R. H., 286–287
Peckham, H. E., 204
Pfaffl, M. W., 283–284

Q

Quigley, M., 136

R

Rainer, J. E., 12
Rao, X., 18
Rasmussen, R., 284
Rill, R. L., 67
Roach, J. C., 204–205
Roeder, R. G., 100
Rose, T. M., 18
Rothballer, A., 99–124
Rozen, S., 18, 282–283
Russell, D. W., 103–104
Rychlik, W., 4, 5

S

Saberi, A., 14, 18
SantaLucia, J. Jr., 2, 3–4, 6, 12, 13
Sarkar, G., 260
Sawadogo, M., 72–73
Sawitzke, J. A., 17
Schmittgen, T. D., 287–288
Scott, M. P., 85–98
Shamir, R., 14, 18
Sheperd, J., 137*f*
Shergill, I. S., 286–287
Simon, M. N., 69
Skaletsky, H. J., 18, 282–283
Smit, A. F., 204–205
Smith, A. M., 267
Smith, M., 68
Soma, H., 18
Sommer, S. S., 260
Sonawane, N. D., 228
Southern, E. M., 48
Southern, P. J., 210–211
Souvenir, R., 18
Sparks, A. B., 204–205
Speicher, M. R., 203–204
Stamatoyannopoulos, J., 2, 18
Stevens, J., 281
Stewart, C. N. Jr., 283–284
Stockwell, T. B., 18
Stormo, G., 18
Studier, E. W., 69
Styles, C. A., 154
Suzuki, T., 283
Szoka, F. C., Jr., 228

T

Tammi, M. T., 203–204
Temple, G. F., 100
Thomas, C., 12
Thomason, L. C., 17
Torabi, N., 14, 18
Traverso, G., 205

U

Untergasser, A., 18
Urban, A. E., 204
Urlaub, G., 209–226

V

Vallone, P. M., 18
Van Dyke, M. W., 72–73
Verkman, A. S., 228
Viguera, E., 5
Vitzthum, F., 285–286
Vogelstein, B., 205
von Einem, J., 16–17

W

Walhout, A. J., 100
Wang, D., 283–284
Wang, T. L., 203–204
Whittier, R. F., 13–14
Wigler, M., 210–211
Williams, P. M., 281, 286
Williamson, M., 286–287
Wong, M. L., 12
Wysk, M., 48

X

Xie, C., 203–204

Y

Yamada, T., 18
Yan, H., 205
Yuan, J. S., 283–284

Z

Zhang, S. M., 62*f*
Zhang, W., 18
Zhou, M.-Y., 8–9
Zipper, H., 285–286
Zuker, M., 18

SUBJECT INDEX

Note: Page numbers followed by "*f*" indicate figures, and "*t*" indicate tables.

A

Absolute quantification, 287
Adaptor ligation
  duration, 194–196
  flowchart, 196, 196*f*
  purified DNA storage, 196
  T4 DNA ligase, 194
Adherent cells, transfection, 232–233, 233*f*
Agarose gel electrophoresis
  casting, 40–41, 43*f*
  DNA fragments separation, 54
  duration, 54
  equipment, 36
  flowchart, 54, 55*f*
  linear separation, nucleic acids, 36
  loading and running, 42–44, 42*f*
  materials, 36–39
  microbial genomic DNA, 150*f*
  preparation, flowchart, 36, 40*f*
  sample visualization, 44–45, 45*f*
Alkaline lysis method, 136
Amplification
  fragment library
    cycling conditions, 199
    duration, 198–199
    flowchart, 199, 200*f*
    polymerase, 197–198
  primer, 126
Amplification efficiency (AE)
  definition, 284
  template concentration, 284, 284*f*
Analyze PCR products, 297–298, 298*f*
Anneal primer to DNA, 176–177, 177*f*
Applied Biosystems, 202, 203–204, 205, 206–207
AutoDimer, 12

B

Bacteria
  harvesting
    centrifugation, 139
    flowchart, 139, 140*f*
  transformation
    calcium chloride, 30
    electroporation, 30
Butanol precipitation, crude oligonucleotide
  deprotection, 72–73
  flowchart, 74, 74*f*

C

Calcium chloride transformation, 30
Casting, 40–41, 43*f*
Cell harvesting, 155–156, 166*f*
Cell lysis
  bacteria, genomic DNA, 146–147, 147*f*
  DNA isolation, mammalian cells, 166
  and plasmid DNA isolation, bacteria
    flowchart, 141, 141*f*
    RNase, 139
Chemically competent *E. coli* transformation
  antibiotic resistance, 334, 336*f*
  competent cells preparation, 333–334, 335*f*
  duration, 334–335
  efficiency, 336
  equipment, 330
  flowchart, 336*f*
  materials, 330–332
  protocol
    duration, 332
    flowchart, 333*f*
    preparation, 332
  theory, 330
  validation, 335
Chemical transformation, yeast
  antibiotics and concentrations, 318
  competent cells preparation, 315–316, 317*f*
  duration, 317–318
  efficiency, 319–320
  equipment, 312
  exogenous piece of DNA, 317
  flowchart, 319*f*
  materials, 313–314

Chemical transformation, yeast (*Continued*)
optimal heat shock time, 318
protocol
duration, 315
preparation, 314–315
theory, 312
Cloning vector, 128–129, 129*f*
CODEHOPs. *See* COnsensus-DEgenerate Hybrid Oligonucleotide Primers (CODEHOPs)
Colony PCR
DNA extraction, 304–306, 306*f*
duration, 307
equipment, 300–301
HotStart Taq polymerase, 308
materials, 301–303
PCR master mix, 307, 308
protocol
duration, 303
flowchart, 305*f*
preparation, 303, 304*f*
standard PCR reaction, 307
theory, 300
visualization, agarose gel, 308–309, 309*f*
Colony screening, 248
Colorimetric assay, 61
Competent cells preparation, yeast transformation
duration, 315–316
efficiency, 315, 316
flowchart, 317*f*
storage, 316
COnsensus-DEgenerate Hybrid Oligonucleotide Primers (CODEHOPs), 14–15
Control gene, 283

D

Design mutagenic primer, 262–263
Digestion, template DNA, 246
Digoxigenin-11-dUTP, nonradioactive detection, 59, 60*f*
Dilute plasmid DNA
duration, 216
flowchart, 216, 216*f*
selection markers, 211*t*, 216
DNA-based experimental tools, 86
DNA elution, 27
DNA extraction, colony
duration, 305
E. coli, 306
false band, 305–306
flowchart, 306*f*
DNA oligos purification, PAGE
butanol precipitation, 72–74
chemical synthesis, 66
desalted oligonucleotide solution, 67
electroelution method, 69
elution, 68–69, 78–79
enzymatic modification, 67
equipment, 69–70
ethanol precipitation, 82
flowchart, 72, 73*f*
isolation, 76–78
length, 67
materials, 70–72
polyacrylamide gels, 68
preparative polyacrylamide gel electrophoresis, 75–76
resolution range, 68
reverse-phase HPLC, 67–68
DNA purification methods, 144
DNA template, 276–277
DNA visualization, 44–45, 45*f*

E

E. coli transformation
electroporation, 324–327, 326*f*
equipment, 322
materials, 322
protocol
duration, 323
flowchart, 323*f*
high-quality electro-competent cells, 323
preparation, 322–323
theory, 321–322
Electroporation
E. coli
cold plates, 327
duration, 325–327
flowchart, 326*f*
plasmid DNA, 324
well-purified DNA, 327
mammalian stable cell lines, 219, 219*f*

Elution, oligonucleotide from polyacrylamide gel
  diffusion, 78–79, 79f
  electroelution, 80–81, 81f
End-repair, fragmented DNA
  duration, 193–194
  flowchart, 195f
  kit components and procedures, 194
  purified DNA, storage, 194
  T4 DNA polymerase, 193
Enzymatic lysis, 154
Ethanol precipitation
  elution, oligonucleotide, 82
  flowchart, 82, 82f
  mammalian cells, 167–168
Eukaryotic cells
  chemical-based methods, 32
  electroporation, 32
  microinjection, 33
  particle-based approaches, 33
  transient transfection, 31–32
  viral transduction, 33
  western blotting, 31–32
Exogenous DNA, cells
  bacteria transformation, 30
  eukaryotic cells, 31–33
  yeast transformation, 31
Extension, DNA chains, 178–179, 179f

F

Fragment libraries preparation
  adaptor ligation, 194–196, 196f
  amplification, library, 197–199, 200f
  end-repair, fragmented DNA, 193–194, 195f
  equipment, 187–188
  materials, 188–189
  protocol
    caution, 191, 191f
    duration, 190
    preparation, 190
  shear DNA, 191–192, 192f
  size selection, library, 196–197, 198f
  theory, 187

G

Gateway expression vectors. *See* Molecular cloning
Gateway® Recombination Cloning Technology. *See* Mammalian stable cell lines
Gel purification
  megaprimer, 265, 266f
  mutagenized gene, 266f, 267
Gene of interest (GOI)
  cloning, 112
  destination vector, 118
  mammalian cell lines, 100
  pFRT vectors, 102
  subcloning, 114
  tetracycline-regulated expression, 103
Genomic DNA
  isolation, mammalian cells
    cell lysis, 166, 167f
    collection, 165–166, 166f
    detergent-based buffer, 162
    equipment, 162
    ethanol precipitation, 167–168, 169f
    flowchart, preparation, 165, 165f
    materials, 162–164
    organic extraction, 167, 168f
  preparation, bacteria
    applications, 144
    cell lysis, 146–147
    equipment, 144
    materials, 144–145
    organic extraction and ethanol precipitation, 147–148, 149f
    purification methods, 144
    quantity and quality assessment, 149–151
  Saccharomyces cerevisiae
    applications, 154
    cell harvesting, 155–156
    equipment, 154
    flowchart, preparation, 163, 165f
    initial DNA extraction, 156–157, 167f
    materials, 154–155
    purification, crude DNA preparation, 157–159, 168f
GOI. *See* Gene of interest (GOI)

H

Harvest cells and analyze protein expression, 235–236, 236f
HighlY DEgeNerate primers (HYDEN), 15

HotStart Taq polymerase, 308
HYDEN, *See* HighlY DEgeNerate primers (HYDEN)

L

Labeled probe hybridization
  background binding, 59
  flowchart, 60, 61*f*
Labeling, DNA strand, 177–178, 178*f*
Large-scale transient transfection, 236
Ligation
  E.coli, 133, 134*f*
  events identification
    flowchart, 98, 98*f*
    screening strategies, 96
  molecular cloning
    flowchart, 94, 95*f*
    negative control reaction, 94
Liquid nitrogen storage, 122
Lithium acetate transformation, 31
Lyse bacteria, 25–26

M

MAD-DPD. *See* Minimum Accumulative Degeneracy Degenerate Primer Design (MAD-DPD)
Mammalian stable cell lines
  BP reaction, 101
  equipment, 104–105
  flowchart, preparation, 112, 112*f*
  Flp-In host cell line, 104
  generation, electroporation
    cells preparation, 216–218, 218*f*
    dilute plasmid DNA, 215–216, 216*f*
    duration, 219
    equipment, 211–212
    flowchart, 219*f*
    materials, 212–214
    methotrexate amplification, 223–224, 225*f*
    plating, 220, 221*f*
    protocol, 214–215, 215*f*
    single colonies, 221–223, 223*f*
    theory, 210–211
    transfection, 219
  GOI, 100
  Invitrogen Gateway® Recombination Cloning Technology, 101, 101*f*
  LR reaction, 101
  materials, 105–110
  molecular cloning, 112–118
  pENTR4 and pENTR4_GOI, 102, 103*f*
  stable cell lines, 118–122
  tetracycline-regulated expression, GOI, 103
  TetR-expressing cell lines, 104
  vector maps, 101, 102*f*
Mechanical lysis, 154
Megaprimer method
  design mutagenic primer, 262–263
  equipment, 260
  first round PCR, 263–264, 264*f*
  gel purification, 265, 266*f*
  materials, 261
  mutagenized gene
    gel purification, 266*f*, 267
    subclone, 268, 269*f*
  protocol
    duration, 261
    flowchart, 262*f*
    preparation, 261–262
  second round PCR, 264*f*, 265–267
  theory, 260
Methotrexate amplification
  concentrations, 224
  duration, 223–224
  flowchart, 225*f*
  pSV2-*dhfr* vector transfected cells, 223
  selection, 224
Minimum Accumulative Degeneracy Degenerate Primer Design (MAD-DPD), 15
MIPS. *See* Multiple, Iterative Primer Selector (MIPS)
Molecular cloning
  defined, 86
  DNA-based experimental tools, 86
  duration, 112–116
  equipment, 88
  flowchart, 118, 119*f*
  identification, ligation events, 96–98
  ligation, 93–94
  ligation and transformation, 115
  LR reaction and second transformation, 116
  mammalian expression, 112

materials, 88–89
minipreps and restriction digestion, 116
and PCR
primers, 113
product insert, 90
and product purification, 114
restriction digestion, 115
primer design, 90, 90*f*
restriction digests, vector and insert, 91–93
restriction enzymes, 86
sequencing primers, 113
sequencing reaction and analysis, 116
steps, 86, 87*f*
transformation, *E. coli*, 94–96
vectors, 86
Multiple, Iterative Primer Selector (MIPS), 15
Mutagenic PCR, 254–255

N

NetPrimer, 7
Next-generation sequencing
advantage and disadvantage, 186–187
commercialized platforms, 202
description, 186
fragment library
description, 187
preparation (*see* Fragment libraries preparation)
library choice and construction
'bait' oligonucleotides, 203–204
fragment libraries, 203–204
mate-paired library, 204
'SMRT Bell' libraries, 204–205
massively parallel sequencing
Applied Biosystems, 206–207
Illumina and Helicos, 206
Roche 454 system, 206
near-and long-term horizon, 207–208
preparation, libraries, 205
'short-read' sequencing, 186–187
terminology, 202
Nonradioactive detection, 59, 60*f*
Nonspecific band amplification, 277

O

Oligonucleotides isolation, 78, 78*f*
Organic extraction
and ethanol precipitation, genomic DNA preparation, 147–148, 149*f*
mammalian cells, 167

P

PAGE. *See* Polyacrylamide gel electrophoresis (PAGE)
Pairwise Alignment for Multiple Primer Selection (PAMPS), 14
PAMPS. *See* Pairwise Alignment for Multiple Primer Selection (PAMPS)
PCR. *See* Polymerase chain reaction (PCR)
PCR-based random mutagenesis
additional template generation, 253–254
equipment, 250
materials, 250–251
mutagenic PCR, 254–255
PCR setup, 252, 253*f*
protocol
duration, 251
flowchart, 252*f*
preparation, 251
subclone and sequence, 255–256, 257*f*
theory, 250
PCR primer design
applications, 8
cloning vector, 130–132, 131*f*
flowchart, 132, 132*f*
homologous recombination, 17–18
melting temperature (Tm) and GC content, 3–4
modified primers, 16
multiplex PCR, 12
nested PCR, 8
PAGE, 8
primer degeneration, 13–16
primer length, 4–5
product length and placement, 6
real-time PCR, 10–12
recombineering, 16–17
RT-PCR, 9–10
sequence and secondary structures, 5–6
sequencing, 12–13
software tools, 18
target cloning, 8–9
thermodynamics, 2–3
Pellet bacteria, 24–25

Picking single colonies of cells
cloning rings, 223
duration, 221–222
flowchart, 223*f*
isolation, 221
selection media, 222
TrypLET Express, 222
Plasmid DNA isolation, bacteria
alkaline lysis method, 136
applications, 136
cell lysis, 139–141
centrifugation, 136
equipment, 136–137
flowchart, 139, 139*f*
harvesting, 139
materials, 137–138
restriction enzyme digestion, 136, 137*f*
Plasmid preparation kits
binding buffer through column, 27
clarified lysate to column, 26
DNA elution, 27
equipment, 24
flowchart, 24, 25*f*
lyse bacteria, 25–26
materials, 24
neutralization, 26
pellet bacteria, 24–25
washing column, 27
Plating electroporated cells
die-off, 220
duration, 220
flowchart, 221*f*
untransfected cells, 220
Polyacrylamide gel electrophoresis (PAGE), 8. *See also* DNA oligos purification, PAGE
Polyethylenimine (PEI). *See* Transient mammalian cell transfection, PEI
Polymerase chain reaction (PCR). *See also* PCR-based random mutagenesis; PCR primer design
aerosol barrier pipette, 296
analyze products, 297–298, 298*f*
DNA contamination control product, 296
duration, 295–296
equipment, 293
flowchart, 297*f*
materials, 293–294
protocol
caution, 294
duration, 294
flowchart, 295*f*
preparation, 294
quantitative (*see* Quantitative PCR (qPCR))
theory, 292
thermal cycler, 295
Preparative polyacrylamide gel electrophoresis
flowchart, 76, 77*f*
RNA purification, 75
Primer design, 282–283
Pythia, 6

Q

Quantitative PCR (qPCR)
amplification efficiency (AE), 283–284
control gene, 283
data analysis
absolute quantification, 287
normalization, 287
relative quantification, 287–288
equipment, 281
materials, 282
primer design, 282–283
probe choice
SYBR green I (SG), 285–286
TaqMan probes, 286–287
protocol
duration, 282
preparation, 282
theory
amplification cycles, 280–281, 280*f*
fluorescence, 280–281, 280*f*
product accumulation, 281
terminology, 281

R

Random primed labeling kit, 56–58, 58*f*
Relative quantification, 287–288, 288*f*
Resolution, labeled DNA products
denaturing polyacrylamide gels, 179, 180*f*
drying times, 182
duration, 179–182
electrophoresis, 179, 180*f*

flowchart, 183*f*
loading reactions, 182
sequencing plates, 182
Restriction digestion
ligated plasmid DNA, 136, 137*f*
PCR product and pENTR4 vector, 115
vector and insert
DNA gel purification kit, 91–92
DNA oligos, 91
flowchart, 93, 93*f*
gel-purified PCR product, 91
microcentrifuge tubes, 91
Restrictionless cloning
amplification primer, 126
equipment, 126
flowchart, preparation, 128*f*, 130
insert denaturation and annealing, 132–133
ligation and transformation, 133
materials, 126–127
PCR primer design and PCRs, 130–132
vector preparation, 128–129
Restriction mapping, 48
Reverse-transcription PCR (RT-PCR), 9–10
Roche 454 system, 205, 206
RT-PCR. *See* Reverse-transcription PCR (RT-PCR)

S

Sanger dideoxy sequencing, DNA
anneal primer, 176–177, 177*f*
equipment, 172–173
extension/termination, 178–179, 179*f*
labeled products resolution, urea-PAGE, 179–183
labeling, 177–178, 178*f*
materials, 173–175
protocol
caution, 176, 176*f*
duration, 175
flowchart, 176*f*
preparation, 175
theory, 172
Seed adherent cells, transfection, 232–233, 233*f*
Sequencing reaction and analysis, 116
Set up and run PCR, 243–245, 245*f*
Shear DNA to random fragments
caution, 192
duration, 191–192
flowchart, 192, 192*f*
sonication, 191
Site-directed mutagenesis
colony screening, 248
digestion, template DNA, 246
equipment, 242
isolate and sequence plasmid DNA, 247*f*
materials, 242–243
protocol
duration, 243
flowchart, 244*f*
preparation, 243
set up and run PCR, 243–245, 245*f*
theory, 242
transform *E. coli*, 246–247, 247*f*
Size selection, adaptor-ligated library
duration, 196–197
excised gel fragment, 197
flowchart, 198*f*
Invitrogen E-Gel system, 196
Small-scale transient transfection, 232
Southern blotting, DNA analysis
alkaline phosphatase-conjugated antibody, 61
alkaline treatment, 48
autoradiography, 61, 62*f*
cloning process, 48
colorimetric assay, 61
description, 48
detection techniques, 48
digoxigenin-11-dUTP, nonradioactive detection, 59, 60*f*
digoxigenin-labeled probe, 61
DNA denaturation and transfer, nylon membrane, 54–56, 57*f*
equipment, 49
flowchart, 53, 53*f*
gene rearrangements, 48
genomic DNA isolation, 53
labeled probe hybridization, 59–60
location, hybridized probe, 62, 63*f*
materials, 49–52
nonradioactive labeling, 48
random primed labeling kit, 56–58, 58*f*

Southern blotting, DNA analysis (*Continued*)
- restriction digestion and agarose gel electrophoresis, 54
- restriction mapping, 48

Stable cell lines
- duration, 118–122
- flowchart, 122, 123*f*
- initial transfection, 118
- liquid nitrogen storage, 122
- transfection efficiency, 120
- Western blot analysis, 122

Subclone
- mutagenized gene, 268, 269*f*
- and sequence PCR, 255–256, 257*f*

Suspension cells preparation, 236–237, 237*f*

SYBR green I (SG), 285–286

T

Taq DNA polymerase, 276

TaqMan probes, 286–287

Termination, DNA chains, 178–179, 179*f*

ThermoBLAST™, 6

Transfect cells, PEI 'Max', 234, 235*f*, 237–238, 238*f*

Transformation, *E.coli*. *See also* Chemically competent *E. coli* transformation
- Bunsen burner flame, 95
- flowchart, 96, 97*f*
- ligated plasmid, 94–95
- resistance gene, 94–95
- restrictionless cloning, 133, 134*f*
- transformed bacteria, 94–95

Transient mammalian cell transfection, PEI
- DNA and PEI 'Max', 234, 235*f*
- equipment, 229
- harvest cells and protein expression, 235–236, 236*f*, 238–239, 240*f*
- large-scale, suspension cells, 236
- materials, 229–230
- PEI 'Max', 237–238, 238*f*
- protocol
  - duration, 232, 232*f*
  - preparation, 231, 231*t*
- seed adherent cells, 232–233, 233*f*
- small-scale, 232
- suspension cells preparation, 236–237, 237*f*
- theory, 228

Troubleshooting PCR
- equipment, 272
- materials, 272
- protocol
  - DNA template, 276–277
  - dNTPs, 276
  - lack of PCR, agarose gel, 273*t*
  - long/short products, 274*t*
  - $MgCl_2$, 275–276
  - nonspecific band amplification, 277
  - primers, 276
  - Taq DNA polymerase, 276
  - weak amplification, 277
- theory, 271–272

Trypsinize and washing cells, 216–218, 218*f*

V

Visualization, PCR product, 308–309, 309*f*

W

Weak amplification, 277

Western blot analysis, 122

Western blotting, 31–32

Y

Yeast transformation
- electroporation, 31
- lithium acetate, 31
- log-phase growth yeast cultures, 31